Routledge Handbook on Cultural Heritage and Disaster Risk Management

This Handbook provides a comprehensive and interdisciplinary overview of the intersections between cultural heritage and disaster risks. It serves as a defining reference, presenting the key concepts and policy arena that disaster risk management and cultural heritage currently operate.

With 22 contributions from leading scholars and practitioners in the field, the chapters explore the various contexts for cultural heritage and disaster risk management, illustrated through case studies from around the world. The Handbook is organized into four parts: Section I includes Disaster Risk Management and Cultural Heritage, Section II helps to understand the context, Section III focuses on the challenges, and Section IV delves deep into the future prospects. This Handbook provides insights of a wide range of topics and themes, such as climate change, conflict, urbanisation, and the role of community, and examines the relationships with a range of sectors such as governance and policy, finance, infrastructure, shelter, and urban planning. It also presents critiques on issues that are often taken for granted, including technocratic approaches, nature/culture binary, the romanticisation of traditional knowledges, and the role of recovery and reconstruction. Insights into the future are also presented, and the Handbook concludes with a detailed agenda of proposed action to be taken in the field.

Offering critical reflections on the topic, this book caters to students, researchers, professionals, and policymakers in the fields of disaster studies, cultural studies, heritage studies, conservation, and geography.

Rohit Jigyasu is a conservation architect and risk management professional from India, currently working at ICCROM as Project Manager on Urban Heritage, Climate Change and Disaster Risk Management. Rohit served as UNESCO Chair holder professor at the Institute for Disaster Mitigation of Urban Cultural Heritage at Ritsumeikan University, Kyoto, Japan, where he was instrumental in developing and teaching International Training Course on Disaster Risk Management of Cultural Heritage. He was the elected President of ICOMOS-India from 2014 to 2018 and President of ICOMOS International Scientific Committee on Risk Preparedness (ICORP) from 2010 to 2019. Rohit has served as the Elected Member of the Executive Committee of ICOMOS since 2011 and was its Vice President from 2017 to 2020. Before joining ICCROM, Rohit was working with several national and international organizations such as UNESCO, UNISDR, Getty Conservation Institute and World Bank for consultancy, research, and training on Disaster Risk Management of Cultural Heritage.

Ksenia Chmutina is a Professor of Disaster Studies at the School of Architecture, Building and Civil Engineering at Loughborough University. Her research focuses on the processes of urban disaster risk creation and systemic implications of sustainability and resilience in the context of neoliberalism. Her research interests also include narratives and framings of disasters, intersectionality and vulnerability, and interlinkages between critical urban studies, cultural heritage, and disaster studies. A core part of her activities is science communication: she is a co-host of a popular podcast, 'Disasters: Deconstructed'. Ksenia uses her work to draw attention to the fact that disasters are not natural.

Routledge Handbook on Cultural Heritage and Disaster Risk Management

Edited by Rohit Jigyasu
and Ksenia Chmutina

LONDON AND NEW YORK

Designed cover image: © Getty Images

First published 2024
by Routledge
4 Park Square, Milton Park, Abingdon, Oxon OX14 4RN

and by Routledge
605 Third Avenue, New York, NY 10158

Routledge is an imprint of the Taylor & Francis Group, an informa business

© 2024 selection and editorial matter, Rohit Jigyasu and Ksenia Chmutina; individual chapters, the contributors

The right of Rohit Jigyasu and Ksenia Chmutina to be identified as the authors of the editorial material, and of the authors for their individual chapters, has been asserted in accordance with sections 77 and 78 of the Copyright, Designs and Patents Act 1988.

All rights reserved. No part of this book may be reprinted or reproduced or utilised in any form or by any electronic, mechanical, or other means, now known or hereafter invented, including photocopying and recording, or in any information storage or retrieval system, without permission in writing from the publishers.

Trademark notice: Product or corporate names may be trademarks or registered trademarks, and are used only for identification and explanation without intent to infringe.

British Library Cataloguing-in-Publication Data
A catalogue record for this book is available from the British Library

Library of Congress Cataloging-in-Publication Data
Names: Jigyasu, Rohit, editor. | Chmutina, Ksenia, editor.
Title: Routledge handbook on cultural heritage and disaster risk management / edited by Rohit Jigyasu and Ksenia Chmutina.
Description: Abingdon, Oxon ; New York, NY : Routledge, [2023] | Includes bibliographical references and index. | Identifiers: LCCN 2023034086 (print) | LCCN 2023034087 (ebook) | ISBN 9781032274805 (hbk) | ISBN 9781032274812 (pbk) | ISBN 9781003293019 (ebk)
Subjects: LCSH: Cultural property--Protection. | Hazard mitigation. | Emergency management.
Classification: LCC CC135 .R698 2023 (print) | LCC CC135 (ebook) | DDC 363.6/9--dc23/eng/20230816
LC record available at https://lccn.loc.gov/2023034086
LC ebook record available at https://lccn.loc.gov/2023034087

ISBN: 9781032274805 (hbk)
ISBN: 9781032274812 (pbk)
ISBN: 9781003293019 (ebk)

DOI: 10.4324/9781003293019

Typeset in Times New Roman
by KnowledgeWorks Global Ltd.

Contents

List of Figures — *viii*
List of Tables — *xi*
List of Boxes — *xii*
List of Abbreviations — *xiii*
List of Contributors — *xiv*
Preface — *xxi*

Introduction: Why Disaster Risk Management of Cultural Heritage — 1
KSENIA CHMUTINA AND ROHIT JIGYASU

SECTION I
Disaster Risk Management and Cultural Heritage — 7

1 **Disaster Risk Management Terms and Concepts** — 9
   LEE BOSHER

2 **Role of Intangible Attributes of Heritage in Disaster Risk Reduction** — 20
   SUKRIT SEN

3 **A New Approach to Cultural Heritage and Disaster Risk Reduction: A Review of International Policies** — 30
   GIOVANNI BOCCARDI

4 **Financing Disaster Risk Management for Cultural Heritage** — 41
   BARBARA MINGUEZ GARCIA

SECTION II
Understanding the Context — 61

5 **Heritage and Peacebuilding** — 63
   ELKE SELTER

6 Cultural Heritage, Climate Change and Disaster Risk Management 75
   WILL MEGARRY

7 Cultural Heritage and Urbanization in the Context of Disaster Risk
   Management in Istanbul 92
   EBRU A. GENCER

8 Vernacular Built Heritage and Disaster Resilience 100
   RAJENDRA AND RUPAL DESAI

9 Traditional Livelihoods for Climate Action and Disaster Risk Management 115
   SUKHREET BAJWA, TANAYA SARMAH, RANIT CHATTERJEE, AND RAJIB SHAW

SECTION III
Understanding the Challenges 127

10 All Fired up: The Inseparability of Nature and Culture
   in Disaster Risk Management 129
   STEVE BROWN

11 The Dangers of Romanticising Local Knowledge in the Context
   of Disaster Studies and Practice 147
   DEMET INTEPE, ROBERT ŠAKIĆ TROGRLIĆ, MARIA EVANGELINA FILIPPI,
   THIRZE HERMANS, HANNAH BAILON, AND ANUZSKA MATON

12 Challenges with Techno-Centric Approaches in the Implementation of
   Disaster Risk Management for Cultural Heritage 165
   DAVID A. TORRES AND GIUSEPPE FORINO

13 Development and Cultural Heritage in the Disaster Capitalism Era 178
   VICTOR MARCHEZINI, ANDREA LAMPIS, DANILO CELSO PEREIRA,
   AND ADRIANO MOTA FERREIRA

14 Cultural Heritage and Post-Disaster Recovery 199
   WESLEY CHEEK

15 The Politics of Post-Disaster Reconstruction of Heritage: Planning and
   Funding Mechanisms in Bhaktapur, Nepal, in the Aftermath of the 2015
   Gorkha Earthquake 211
   VANICKA ARORA

16 'Dark Heritage': Landscape, Hazard, and Heritage 225
   JAZMIN SCARLETT, MIRIAM ROTHENBERG, FELIX RIEDE, AND KAREN HOLMBERG

SECTION IV
Moving Forward 245

17 **Arts and Other Cultural Expressions as Tools for Disaster Risk Management** 247
CLAUDIA GONZÁLEZ-MUZZIO, CLAUDIA BEATRIZ CÁRDENAS BECERRA, AND BERNADETTE ESQUIVE

18 **'Planning for Disasters Facing Heritage at Risk: Ethics and Epistemes'** 263
FALLON SAMUELS AIDOO

19 **New Technologies and Disaster Risk Management for Cultural Properties** 279
HIROFUMI IKAWA

20 **Integrating DRM Considerations into Heritage Management Systems: Barriers and Opportunities** 289
LUISA DE MARCO

21 **Building Synergies for Cultural Heritage: Insights from Theory and Practice** 298
MONIA DEL PINTO AND CLINTON DEAN JACKSON

Conclusions: Challenges and Opportunities for Disaster Risk Management of Cultural Heritage 320
ROHIT JIGYASU AND KSENIA CHMUTINA

Index 323

List of Figures

1.1	DRM phases.	15
1.2	Qualitative risk analysis.	16
2.1	A brass moulder in his workshop using fire without any safety apparatus.	22
2.2	Stone carvers spilling out on vehicular road due to lack of space.	22
2.3	Children are working in brass moulding workshops.	23
2.4	Use of attic space in tea trading units.	24
2.5	Hanging of musician instruments from ceiling.	25
2.6	Milk trade happening from truck.	25
6.1	Infographics showing various direct climate impacts from well-known WH properties including: (A) Coastal erosion at Rapa Nui, (B) Efflorescence due to saline intrusion and rising aquifers in Bagerhat, Bangladesh, (C) Sea-level rise at Kilwa Kisiwani, Tanzania, and (D) Increased precipitation at the Old and New Towns of Edinburgh, Scotland.	79
6.2	The vulnerability cycle.	81
6.3	Images from Kilwa Kisiwani WH property including: (A) eroding shoreline showing archaeological deposits, (B) protective wall above the beach below the Malindi Mosque, (C) protective wall in from of the Gereza Fort, and (D) damage to the Gereza Fort from coastal erosion (foreground) and mangrove forests protecting the structure (background).	85
7.1	Topographical Istanbul during the Byzantine period.	93
8.1	Marathwada region on Deccan Plateau of Western India.	101
8.2	Tribal area of South Gujarat.	102
8.3	Stone walls not laced with timber.	103
8.4	Urban (valleys) 'Dhajji Divaari' – timber laced brick or stone masonry walls.	104
8.5	Adobe/mud block masonry with flat roof.	105
8.6	'Assam type' timber framing with cane infill mud plastered walls and pitched roof.	106
8.7	Severely damaged rubble masonry building in 2005 Kashmir Earthquake.	107
8.8	Vertical timber reinforcement anchored to timber band and the wall.	109
8.9	Severely damaged vernacular stone masonry house having RC columns in corners without beams.	110
8.10	Earthquake resistant house.	112
8.11	Retrofitting and restoration completed.	113
9.1	Complex interplay between intangible and tangible heritages.	117

List of Figures ix

10.1	Kiandra Courthouse, Kosciuszko National Park, following extensive restoration in 2010 (left) and following the damage sustained by bushfire (January 2020). The structure consists of the 1890s stone courthouse and associated police quarters and a 1950s chalet. (Source: Wikimedia Commons).	130
10.2	Diagram showing how pyrocumulonimbus clouds form above intensely hot bushfires. Pyrocumulonimbus clouds created by bushfires were reported during the Australian bushfires in 2019/2020. These events represented the entanglement of local natural and cultural histories, land management practices, and changing climate.	136
11.1	Dimensions of local knowledge for flooding in the lower Shire Valley in Malawi.	154
13.1	Economic growth drivers and threats/factors affecting the outstanding universal value of world heritage properties.	183
13.2	Number of world heritage properties inscribed each year by region.	185
13.3	Assets listed by the national heritage institute (Iphan), by Brazilian regions.	188
13.4	Culture heritage assets listed by the national heritage institute (Iphan) and the presence of civil defence units, disaster risk reduction (DRR), and climate change adaptation (CCA) plans, according to Brazilian regions.	193
13.5	Percentage of civil defence units, disaster risk reduction (DRR), and climate change adaptation (CCA) plans in cities with assets listed by the national heritage institute (Iphan), according to the Brazilian regions and their human development index (HDI).	194
14.1	Looking at the SunSun shopping area from Kaminoyama Hachiman Shrine. Minamisanriku, Japan. 9th March 2021.	200
15.1	(A) Signage prominently displaying the heritage agencies erected across reconstruction sites in Hanumandhoka (Kathmandu's Durbar Square). (B) Basantpur Durbar funded by Chinese government and right reconstruction of Agamchen funded by JICA.	215
15.2	Ongoing reconstruction of the Vatsala Durga Temple in Bhaktapur in 2019.	218
16.1	The two overlapping spheres of cultural heritage and geoheritage can be imagined to sit within a three-dimensional box (A). Geocultural heritage exists where the two spheres overlap on the x-axis; the y-axis grades from non-traumatic forms of heritage at the top to dark heritage at the bottom through various degrees of darkness; and the z-axis represents spatial scale, from the site level to the landscape level. Looking down the z-axis shows that all three forms of heritage represented here can also be dark heritage (B). Viewing the heritage box down the x-axis makes it clear that spatial scales and degree of darkness are both continuua and they allow for non-traumatic heritage sites, dark heritage landscapes, and everything in between (C). (Image by Miriam Rothenberg).	229
16.2	Modern ruins and volcanic deposits on the outskirts of Plymouth, 24 July 2018.	231
16.3a	Two plaques installed on the historic market house in Providence, RI, marking high water lines associated with the hurricane of 1938 and the great gale of 1815. Although the physical damage from both storms is no longer present in the landscape, the events have been commemorated in the built environment through the instalment of markers like these throughout Providence and Rhode Island.	233

List of Figures

16.3b	A so-called 'flood pillar' on the old harbour in Ribe, Denmark. Each band corresponds to a major flood that occurred in historical times. This is an example of a Northern European tradition commemorating traumatic coastal and River flood events in this way.	234
17.1	'Hypocentres of memory'.	249
17.2	Chillan after the earthquake, 1939.	250
17.3	'Bajo los escombros de Chillán (vals-canción)' (Under the debris of Chillán: waltz-song).	251
17.4	Mural 'Mexico to Chile' by Xavier Guerrero.	252
17.5	Spring festival celebrated in 2016.	252
17.6	'Volver a habitar' by DAN, licensed under CC BY-NC-SA 2.0.	254
17.7	Exhibition 'Seguimos inundadxs: archivos e imaginaciones del agua común'. (We are still flooded: Archives and imaginations on common water).	255
17.8	'Simple' and 'complex' figures representing the two snakes Kai-Kai and Treng-Treng.	257
18.1	Elevated EnergyStar® houses under construction in Pontchartrain Park five years after Hurricane Katrina as planned by Community Development Corporation (CDC) founders, descendants of the historic district's elderly homeowners, and prescribed by the CDC's funders, a collection of private, public, and philanthropic organizations.	269
19.1	Various methods for 3D measurement, from left: Laser scanning, photogrammetry using drone photos, Lidar app.	280
19.2	3D documentation using smartphone app.	282
19.3	Distribution of cultural heritage in Japan and type of damage.	285
19.4	Development of a monitoring and conservation system for cultural properties using machine learning.	286
21.1	Overview of the barriers recorded in the design stage.	304
21.2	Overview of the catalysts recorded in the design stage.	306
21.A1	Graph showing the distribution and type of barriers recorded with the stakeholder relation to the context.	318
21.A2	Graph showing the distribution and type of catalysts recorded with the stakeholder relation to the context.	319

List of Tables

1.1	Typology of hazards and threats	12
6.1	Glossary of key terms	76
6.2	CVI Africa summary table with key variables and outcomes	86
9.1	Sustainable livelihood: Influence of heritage and impact of climate change	122
13.1	Number of world heritage properties by region	185
13.2	Number of assets listed by Iphan, by type, until December 2021	187
13.3	Brazilian cities' profile in the sector of culture	189
13.4	Brazilian cities' profile for disaster risk management and climate change adaptation	191
21.1	Summary of relevant documents regulating the protection and safeguard of cultural heritage in relation to disaster risk	300
21.2	Start of fieldwork recordings per province	310

List of Boxes

1.1	Categorising the challenges experienced at cultural heritage sites	12
1.2	Time to ditch the 'disaster cycle'?	14
2.1	Case of creative adaptation of the paper mache masks of 'Chhau' dancers of Purulia West Bengal to fight Covid-19	25
4.1	Financing DRM for CH: The case of Japan	49
5.1	Rebuilding Timbuktu's heritage for peace	68
6.1	Values-based climate change risk assessment: Piloting the climate vulnerability index for cultural heritage in Africa	83
8.1	Way Forward	112
9.1	Case Study #1: Intertwined tangible and intangible heritage: Case of Bungamati in Nepal (Chatterjee, 2018)	118
9.2	Case Study #2: Community-led water harvesting system	119
9.3	Case Study #3: Environmental and cultural implications of climate change and the shifting of Apple Orchards in Kinnaur, Himachal Pradesh, India	120
9.4	Case Study #4: Sustainable livelihoods from natural heritage on islands	123
10.1	Australian Government Royal Commission into National Natural Disaster Arrangements Report ('Bushfires Royal Commission')	138
11.1	Local knowledge for flood risk management in Malawi	154
12.1	Approaches to DRM of CH in Mexico	168
13.1	Ouro Preto, a legacy of cultural heritage and disaster risk creation	192
14.1	The Former Disaster Management Center – Minamisanriku, Japan	200
14.2	Parades of New Orleans	201
15.1	Local planning and funding mechanisms	217
16.1	The dark heritage of Montserrat's Volcanic 'Exclusion Zone'	230
17.1	Research and education on disasters through art. DESARTES. CIGIDEN's Arts and Disaster Unit	248
17.2	Hazards and disasters in music, a tool for disaster risk reduction	258
18.1	Reducing the risk of washed-out Black heritage	268
19.1	Reconstruction of Mosul's ancient city and machine learning for monitoring	286
21.1	National Audit of Monuments & Memorials (NAMM) – A Prerequisite to Heritage Landscape Transformation Programme in South Africa	307

List of Abbreviations

ACA	Agency for Cultural Affairs, Japan
CH	Cultural Heritage
DRM	Disaster Risk Management
DRR	Disaster Risk Reduction
EIB	European Investment Bank
GFDRR	Global Facility for Disaster Reduction and Recovery
GIZ	German Corporation for International Cooperation Agency
HFA	Hyogo Framework for Action
ICCROM	International Centre for the Study of the Preservation and Restoration of Cultural Property
ICOMOS	International Council on Monuments and Sites
IPCC	International Panel on Climate Change
SDGs	Sustainable Development Goals
SFDRR	Sendai Framework for Disaster Risk Reduction
TCA	Techno-centric approaches
UIS	UNESCO Institute for Statistics
UNDRR	United Nations Office for Disaster Risk Reduction
UNEP	United Nations Environment Programme
UNESCO	United Nations Educational, Scientific and Cultural Organization
UNISDR	(Former) United Nations Office for Disaster Risk Reduction
UNU-EHS	United Nations University Institute for Environment and Human Security

List of Contributors

Fallon Samuels Aidoo, a preservation planner, interrogates dis/investment in real estate of significance to Black, Indigenous, and immigrant histories and futures. These interdisciplinary analyses – published recently in the *Journal of Environmental Studies & Science, Preservation & Social Inclusion*, and other edited volumes – build from consulting and community engagements. Currently an Assistant Professor of Real Estate and Historic Preservation at Tulane University, she previously taught urban planning and design at University of New Orleans, Northeastern University, Harvard, and MIT. Dr. Aidoo holds a PhD in Urban Planning (Harvard), MS in architectural history (MIT), and BS in structural engineering (Columbia).

Vanicka Arora is a Lecturer in Heritage at the University of Stirling, UK. She has recently completed her doctoral research at Institute for Culture and Society, Western Sydney University, where she studied trajectories of post-disaster reconstruction of built heritage in Bhaktapur, Nepal, following the 2015 Gorkha Earthquake. Vanicka is trained as an architect with a specialization in conservation of historic buildings and has over a decade of professional experience in India, where she worked in areas of planning and policy for disaster risk reduction of heritage sites, conservation, and adaptive re-use of built heritage, and urban regeneration.

Hannah Bailon holds a bachelor's degree in behavioural sciences from the University of the Philippines. She also completed her master's degree in societal resilience from the Vrije Universiteit Amsterdam, where she focused on climate analysis, modelling, and risk management. Currently, she works as a sustainability consultant in a top international consulting firm focusing on climate and reporting projects. Her research interest mainly lies in utilizing data science in climate risk and sustainability issues.

Sukhreet Bajwa is a development sector professional with over eight years of experience. She has worked with RIKA India Pvt Ltd and is currently working with EdelGive Foundation in climate action. She has a Masters in Disaster Management from Tata Institute of Social Sciences.

Claudia Beatriz Cárdenas Becerra is a Professor of French at the National University of Costa Rica and a Master of Science in Education (University of Panama). She has been a consultant and independent researcher for 25 years about disaster risk reduction in all the countries of Latin America and the Caribbean. She has specialized in gender equality, human rights, local development, etc. He currently works in Chile with the Risk and Disaster Management Corporation – GRID Chile and in the consulting company Ámbito Ltda. She is an active member of the Network of Social Studies in Disaster Risk Prevention-LA RED since 1998.

List of Contributors xv

Giovanni Boccardi is an Italian conservation architect and heritage consultant. Over 25 years, he worked with UNESCO, where he headed two regional desks at the World Heritage Centre and led the development of various policies on World Heritage and capacity building, the relation between World Heritage conservation and sustainable development, as well as on disaster risk reduction. Since 2014 until 2019, Giovanni was in charge of a new Unit within UNESCO's Culture Sector to coordinate and implement the Organization's actions to prepare for and respond to situations of emergency, including disasters and armed conflict.

Lee Bosher is a Professor of Risk at the University of Leicester's School of Business. He is an experienced researcher that has developed a portfolio of projects related to DRM and the inter-disciplinary integration of proactive risk management strategies into the decision-making processes of key stakeholders. Lee's recent research projects have been undertaken in the UK, India, Indonesia, Vietnam, China, and across Europe. Lee's books include *Disaster Risk Reduction for the Built Environment* (2017, with Prof Ksenia Chmutina), *Hazards and the Built Environment* (2008), and *Social and Institutional Elements of Disaster Vulnerability in India* (2007).

Steve Brown, PhD, is an Australian archaeologist and heritage specialist. He is a Senior Research Fellow at the University of Canberra and a Special Advisor with GML Heritage. Steve's work extends from the heritage of the everyday to World Heritage. He works with government and private industry and internationally with ICOMOS, IUCN, and ICCROM. Steve's research interests include cultural landscapes, nature cultures and the improved integration of natural and cultural heritage, protected area management, place attachment, and the heritage of landscapes with the imprint of Indigenous and settler-migrant communities. Steve lives on a 56-hectare property that he manages for conservation and love.

Wesley Cheek, PhD, is a Lecturer in Human Geography at Edge Hill University. He is a sociologist of disasters, focusing on community involvement in post-disaster reconstruction and critical urban theory. Wesley also holds a degree in historic preservation and conducts research focusing on the nexus of architectural history, cultural heritage, and disasters. His aim in his work is to reveal post-disaster reconstruction as a part of a larger process of the social production of risk and the reinforcement of existing inequalities. His book on community-based post-disaster reconstruction in Minamisanriku, Japan, will be published.

Ranit Chatterjee is an architect turned disaster management professional with a doctoral degree in environmental management from Kyoto University, Japan. He has been involved in various projects ranging from architectural planning and conservation to disaster risk reduction in Asia Pacific working with UN organizations, national and local governments, the private sector, and academia. Ranit is a visiting Associate Professor at Keio University in Japan. He co-founded RIKA India, RIKA Institute, and UINSPIRE Alliance. Ranit is an IRDR Young scientist fellow, a CEM member of the IUCN business and biodiversity group, and an Advisory member of UNDRR's Stakeholder Engagement Mechanism.

Ksenia Chmutina is a Professor of Disaster Studies at Loughborough University, UK. Her research focuses on the processes of urban disaster risk creation and systemic implications of sustainability and resilience in the context of neoliberalism. A core part of her activities is science communication. She is a co-host of the podcast 'Disasters: Deconstructed'.

Luisa De Marco, architect and PhD in Conservation of Monuments and Landscape, has been acting as ICOMOS World Heritage adviser since 2010 and has collaborated with national and international cultural heritage agencies as heritage and conservation expert and lecturer.

She worked for 20 years at the Italian Ministry of Cultural Properties, Activities and Tourism, holding responsibilities on disaster response, World Heritage Properties and cultural landscapes in the Liguria Region. In 2019 she completed a 26-month EU-funded Twinning Project on Cultural Heritage (Italy – Republic of Moldova) as resident twinning adviser. Luisa is the author of several publications concerning cultural heritage.

Monia Del Pinto is a Doctoral Prize Fellow in the School of Architecture, Building and Civil Engineering at Loughborough University, UK. Her research is at the intersection of architecture and planning, critical heritage studies, and disaster studies, and aims at advancing methodologies and tools for effective disaster risk reduction at the urban scale. Monia holds a PhD in Urban Planning and Disaster Risk Reduction from Loughborough University and an MSc in Architectural Engineering from the University of L'Aquila (Italy). She teaches architecture-related disciplines, has professional experience in post-earthquake reconstruction, and is active in Disaster Risk Management for Cultural Heritage.

Rupal Desai and **Rajendra Desai** are an architect-structural engineer couple. Both originally graduated in India and then pursued post graduate studies in the United States during 1970 to 1974. During the course of their early work in the rural India, they found the vernacular building systems having small carbon-footprint to be the most viable and sustainable for the people. Hence, they worked in different parts of the Subcontinent focusing on the vulnerability reduction of these building systems. These ideas were transferred to local communities through sensitization coupled with the education of the building artisans.

Adriano Mota Ferreira is a PhD candidate in the Graduate Program on Disasters at the Institute of Science and Technology at São Paulo State University (ICT/UNESP), São José dos Campos, Brazil. As an environmental engineer, his research is focused on forensic investigation approaches in disasters, especially with the use of action research, cartography, and participatory methodologies for disaster risk reduction.

Maria Evangelina Filippi, PhD, is a transdisciplinary researcher specializing in disaster risk reduction, climate change adaptation and resilience in cities, primarily in low-to-middle-income countries. Her work coalesces at the intersection of urban development planning, sustainability transitions, and socio-environmental justice, experimenting with participatory and action research approaches that recognize a diversity of knowledges and embrace an ethics of care in co-production processes. She has extensive experience collaborating and engaging with community-based and non-governmental organisations, city governments, and scientists for co-designing and implementing strategies, policies, and plans towards more sustainable, prosperous, equitable, and resilient urban futures for all. She is a member of the UK Alliance for Disaster Research (UKADR) Steering Committee and has actively contributed to high-level policy discussions and research that informed the UN Office for Disaster Risk Reduction (UNDRR) Global Platforms and Words into Action guides.

Giuseppe Forino is a Lecturer in Human Geography at Bangor University, Wales, UK. Giuseppe is an interdisciplinary human geographer working on disaster risk and climate change adaptation. He has published over 30 papers and academic journals. He is the co-editor of *Governance of Risk, Hazards and Disasters: Trends in Theory and Practice* (Routledge, 2018) and editor of *Disasters and Changes into Society and Politics. Contemporary Perspectives from Italy* (2023).

JC Gaillard is Ahorangi/Professor of Geography at Waipapa Taumata Rau/The University of Auckland and Research Fellow at the University of the Philippines Resilience Institute. He serves as editor of the journal *Disaster Prevention and Management* and sits on the Board of Directors of the NGO Center for Disaster Preparedness. JC's work centres on inclusion and power in disaster with a particular focus on ethnic and gender minorities, children, and people in detention. JC collaborates in participatory DRR trainings with local governments, NGOs, and other civil society organizations.

Barbara Minguez Garcia is a Disaster Risk Management and Cultural Heritage Specialist with over 13 years of international experience. She has worked at the UNESCO Emergency Preparedness and Response Unit, the World Bank and the Global Facility for Disaster Reduction and Recovery (GFDRR) in several projects and countries, including Central America region, Ecuador, Myanmar, Bhutan, Uzbekistan, Saudi Arabia, Iraq, and Japan. Earlier she worked at the Cultural Offices of the Embassy of Spain in Washington DC and New York, and collaborated with the US National Parks Service. Barbara is a member of the International Committee on Risk Preparedness (ICOMOS-ICORP), and mentor for ICCROM's First Aid and Resilience for Cultural Heritage in Times of Crisis and Net Zero: Heritage for Climate Action Programs.

Ebru Gencer is an Adjunct Associate Professor of Architecture, Planning and Preservation at Columbia University. She is also a Senior Urban Resilience Adviser at the World Bank, and an Associate Editor of the *Progress in Disaster Science Journal*. Previously, Dr. Gencer held positions at Columbia University's Earth Institute and the Euro-Mediterranean Climate Change Center in Venice, as well as at the Center for Urban Disaster Risk Reduction and Resilience (CUDRR+R), a non-profit research organization which she founded in New York. She is the author of several literature on the nexus between DRR, climate resilience, and sustainable urban development.

Thirze Hermans, PhD, is a research scholar and advisor at Wageningen Centre for Development Innovation and a research fellow at the Forest and Nature Conservation Policy group at Wageningen University focusing on transformative change, knowledge, and innovation processes in socio-ecological systems. Previously she was a research fellow at the University of Leeds where she still holds a visiting research fellowship on Agriculture and Climate Services.

Karen Holmberg is an archaeologist and volcanologist who specializes in volcanic contexts to examine the long-term experiences humans have had with environments that change unpredictably. She is interested in how the past can aid understanding of the environmental challenges and crises of the 21st century, particularly in the Global South. Her work has received awards and fellowships from Fulbright, Mellon, Wenner-Gren, Creating Earth Futures, Make Our Planet Great Again, and This Is Not a Drill award through the Tisch Future Imagination Fund. She teaches at the Gallatin School of New York University, where she also serves as the Scientific Director and co-founder of the Gallatin WetLab, an experimental initiative for public-facing teaching and a living art-science laboratory. Holmberg currently directs interdisciplinary field projects examining past environmental changes and future volcanic risks on coastlines in Patagonia (Chaitèn, Chile) and near Naples, Italy (Campi Flegrei); closer to home, she researches the radically transforming past and future coastlines of New York City.

Hirofumi Ikawa is a conservation architect, primarily concerned with the preservation of Japan's modern architectural heritage. His extensive professional experience includes leading numerous projects, most notably the conservation of Tokyo's 1917 Sewage Treatment Plant and the 1954 Memorial Cathedral for World Peace. Ikawa has also been instrumental in national research initiatives, using machine learning methods to assess architectural damage.

In his current role as Project Manager at ICCROM, he is tasked with developing educational modules for cultural heritage professionals.

Demet Intepe, PhD, is an interdisciplinary researcher and policy expert in climate justice and climate change adaptation. She received her Research MA from Leiden University in the Netherlands on migration and culture and her PhD from the University of Warwick in the United Kingdom on environmental justice and cultural works by marginaliszd communities. Her research interests include socio-environmental justice, cultural production, and local and Indigenous knowledge. She currently works at the nexus of climate change and development at Practical Action and actively participates in civil society working groups to inform the international negotiations under the UNFCCC around climate change adaptation and loss and damage.

Clinton Dean Jackson is the senior manager for heritage information, policy, and skills development at the South African Heritage Resources Agency, which is responsible for coordinating the management of cultural heritage resources in South Africa. Clinton holds a BA (Hons) degree in Archaeology from the University of South Africa and a certificate in Disaster Risk Management of Cultural Heritage from the UNESCO Chair Programme at the Ritsumeikan University in Japan. Clinton holds a keen interest in the intersection of technology and heritage management and how emergent technologies can aid in protecting heritage resources.

Rohit Jigyasu is a conservation architect and risk management professional from India, currently working at ICCROM as project manager on Urban Heritage, Climate Change, and Disaster Risk Management. He is also, at present, the Vice-President of ICOMOS International Scientific Committee of Risk Preparedness (ICORP). Rohit has been working with several national and international organizations for consultancy and research on Disaster Risk Management of Cultural Heritage.

Andrea Lampis is a FAPESP research fellow at the Institute of Energy and Environment (IEE/USP) of the University of São Paulo (2018–2023). His research focuses on the governance and justice implications of ongoing socio-technical transitions.

Victor Marchezini is a sociologist at the Brazilian Early Warning Center (Cemaden) and Professor at the Doctorate Program on Earth System Science, National Institute for Space Research, and at the Graduate Program on Disaster (ICT/Unesp). He was awarded a postdoctoral scholarship from Fapesp to serve as a visiting scholar at the Natural Hazards Center/University of Colorado-Boulder from July 2022 through June 2023. Over the past 20 years, Marchezini has focused on the sociology of disasters. He has experience in coordinating international networks of researchers, such as the free e-book project, *Reduction of Vulnerability to Disasters: From Knowledge to Action*.

Will Megarry, PhD, is a landscape archaeologist from Ireland and Senior Lecturer in Archaeology at the School of Natural and Built Environment at Queen's University Belfast. His research explores the intersections between culture, heritage, and climate change, and he has managed a wide range of research projects exploring climate vulnerability assessment, climate communication, and the integration of diverse knowledge systems in climate action. He remains an active field archaeologist with a long-running project exploring life on Neolithic Shetland. He is the focal point for climate change at the International Council on Monuments and Sites (ICOMOS).

María Bernadette Esquivel Morales is an architect. She received her master's degree in Conservation of Cultural Heritage for Development at the University of San Carlos de

Guatemala in 2018. She also completed the interdisciplinary master's degree in Environmental Management and Ecotourism, Social Impacts in Communities at the University of Costa Rica in 2008. Maria is an independent consultant and researcher with teaching and professional experience in restoration of buildings declared national heritage. She also works as the consultant in Cultural Heritage Management, especially on the formulation of Cultural Projects and Risk Management for Cultural Heritage. She has been the facilitator of awareness and training workshops in earth architecture and is a member of International groups PROTERRA, ICOMOS_ISCEAH, and CIAV, and the co-founder of the local association ADEPA Santo Domingo Cultural.

Anuszka Mosurska is a PhD researcher at the University of Leeds and co-chair of the Society for Applied Anthropology's Risk and Disaster Topical Interest Group. She previously studied at the Institute for Risk and Disaster Reduction at University College London where she looked at how different knowledges were used in disaster risk reduction. Currently, Anuszka is researching disaster and humanitarian governance in Indigenous contexts, as well as different conceptualizations of care in disaster risk reduction and humanitarian action. She has fieldwork experience in Northern Alaska where she helped with community-based participatory disaster management programs.

Claudia González-Muzzio is an architect from the Pontifical Catholic University of Chile, MSc in Environment, Science and Society at University College London, and has a PG Certificate in Sustainable Development from SOAS – University of London. Claudia has been the partner and director at Ámbito Consultores since 2002. Member of ICORP, ICOMOS Chile, and GRID Chile. Claudia works doing consultancy and research in conservation, heritage management, and disaster risk management for cultural heritage as well as on DRM and land planning in general. Claudia has contributed to prepare management plans for protected areas in Chile and coordinated studies for the nomination of Qhapaq NÞan and the Chinchorro culture to the World Heritage List.

Danilo Celso Pereira is a geographer and PhD candidate at the University of São Paulo (USP), Brazil. He develops research on public policies for the preservation of cultural and natural heritage.

Felix Riede is a German-born and British educated with a PhD in Archaeology from Cambridge University. His work focuses on climate change and disasters in the past and on the Anthropocene, all from a distinctly archaeological perspective. Volcanic eruptions and their interactions with human societies have been a long-standing preoccupation, on which he has published widely. Since 2009, Felix has worked at Aarhus University in Denmark where he now is Professor.

Miriam Rothenberg is a junior research fellow at the University of Oxford. Miriam's research investigates the impacts of the 1995–present Soufrière Hills eruptions on the Caribbean island of Montserrat and the ways that a detailed understanding of post-eruption landscapes and materialities can inform archaeological interpretations both locally and by comparison (e.g., the ancient Mediterranean). With graduate degrees in both archaeology and geology, Miriam's research interests include contemporary archaeology, social volcanology, landscape archaeology, ruins and ruination, and memory studies.

Tanaya Sarmah is an architect – urban planner – Disaster Risk Management professional by training. She has a PhD from the Indian Institute of Technology Kharagpur. She was associated with RIKA India in the capacity of a Project Officer. Currently, she is a research fellow in Disaster Risk Modelling at Cranfield University, United Kingdom. She is working on 'Management of Disaster Risk and Societal Resilience', which aims to study the impacts of droughts on the water-energy-food nexus in Africa, with a special emphasis on the agrarian communities and energy infrastructure systems.

Jazmin Scarlett, PhD, was an interdisciplinary volcanologist. Her research focused on the historical and social context of volcanic eruptions in history in the Caribbean and Europe, as well as geoheritage, cultural heritage, and science communication. She now works in the civil service.

Elke Selter, PhD, works as the coordinator for 'heritage in crisis' with the Belgian federal government's Royal Institute for Cultural Heritage. She is also a researcher with the British Institute of International and Comparative Law. Earlier, Elke worked for over 15 years with various agencies of the United Nations in conflict and disaster-affected places around the world. Elke holds a PhD in International Politics from SOAS, UK, with research focusing on the interests of international organizations, the Security Council, and the International Criminal Court in post-war heritage and the intent to link heritage with R2P-based interventions.

Sukrit Sen from Kolkata, India, is a heritage manager by profession and a musician by passion. He is trained in Tabla, an Indian Percussion Instrument, and has been associated with Indian Classical Music for over two decades. Given his background in music and architecture, Sukrit takes a keen interest in the linkages between tangible and intangible heritage, exploring them to engage with communities and have a more holistic outlook towards heritage conservation strategies. This approach informs his recent engagement with Disaster Management, observing the role of traditional knowledge and other intangible aspects in risk reduction practices. Sukrit is currently associated with the Living Waters Museum and ICCROM. He has contributed to several national and international projects and conferences and holds a few academic publications to his credit.

Rajib Shaw is a Professor at the Graduate School of Media and Governance, Keio University, and is a Director & Co-founder at RIKA India. He has a rich academic background and experience of over 25 years in the field of Environment Sciences (Disaster Management & Climate Change). He is the Coordinating Lead Author, Asia Chapter, IPCC 6th Assessment Report, a recipient of the United Nations Sasakawa Award for DRR in 2022, and Norio Okada from the Integrated Disaster Risk Management (IDRIM), 2022 and Pravasi Bharatiya Samman Award (PBSA) 2021 from the President of India.

David Torres is MSc in Risk, Disaster and Resilience from University College London and a BA in Cultural Heritage Conservation from Mexico's National School of Conservation. Since 2010, he has been part of the National Institute of Anthropology and History, where he was part of the emergency response for the protection of cultural heritage after the 2017 earthquakes. David has also been a lecturer in DRM for cultural heritage in several Universities, such as the National School of Anthropology, and has published in national and international journals and magazines.

Robert Šakić Trogrlić is a research scholar at the International Institute for Applied Systems Analysis (IIASA) in Austria. Educated in engineering and social sciences, he takes an interdisciplinary lens to explore the assessment and management of multi-hazards and multi-risks, local knowledges, and their role in community-based disaster risk reduction. He received his PhD from Heriot-Watt University in the United Kingdom. He has extensive practical and research experience working in risk reduction and climate resilience in the Global South.

Preface

In Heritage We See Disaster

JC Gaillard

Waipapa Taumata Rau, Aotearoa

There are two ways of seeing heritage. One is static and an outcome. The other is fluid and a process. The former has been the dominant, if not the only perspective that has informed disaster studies and disaster risk reduction; that is, heritage has been considered as a frozen legacy of the past; a physical expression of culture that needs to be preserved as much as many other dimensions of the social fabric. As such, heritage is often essentialized in monuments to preserve and restore folklore and traditions that need to survive the impact of disaster.

A brief etymological foray reveals that heritage, in both its Latin and subsequent French origins, is, in fact, about passing, sharing, receiving, and succeeding through time. The predominance of both these verbs and time in all etymological definitions underlines the importance of looking at heritage as a process in the *longue durée*. A process that shapes our identity. We, as living beings, are indeed inherently a product of the past but one that is always changing. It is about our being a biological legacy of our parents and ancestors, yet continuously aging. It is also about our understanding of the world, our culture and social ties, our economic endeavours, and political life that all reflect the multiple influences of the past and present together.

As a result, our heritage is us and our societies as a whole; a marker of our identity that is in perpetual movement, fluid, and dynamic. Monuments, traditions, and folklore that we aim to preserve in disaster are, therefore, the tip of the iceberg – an essentialized legacy that hides the critical importance of apprehending heritage as a process to ultimately understand who we are and how we interpret the world.

It is in this perspective that heritage is necessarily crucial to disaster studies and disaster risk management. It is a lens that allows us to unpack what disasters are across cultures and societies. It provides an entry point to reinterpret disaster from many different perspectives grounded in their local context. Heritage should thus no more be a sectoral dimension of disaster studies that we research as we study, let's say, early warning or gender, for the sake of examples that, in fact, cannot be essentialized either. Heritage should rather be the overarching frame through which we look at how people experience hardship and harm. A frame to revisit what a disaster is, whether or not the term makes sense beyond its Western cradle.

Approaching heritage as a process and a dynamic lens to understanding disaster prevents us from falling in the nativist trap that an essentialized perspective only associated with monuments, traditions, and folklore may lead to. Our contemporary identities and understanding of the world are indeed inherently hybrid or creole. They reflect the multiple influences of passing, sharing, and receiving across time but also across geographical locations and cultures. As such, looking at heritage as a dynamic process allows us to (re-)connect cultural differences, the local and global, the present and past.

Heritage is, therefore, an epistemology – An epistemology of the *Tout Monde*, in Glissant's sense, or of the *Pluriverse*, in Mignolo's and Escobar's; one that allows to look at and through the hybridity and *créolité* of the contemporary world, including of how people understand and experience what we call disaster; ne that is inherently relational and fluid rather than isolating and rigid; ne that allows to capture the multiple and complex perspectives on hardship, suffering, and harm; one that acknowledges the diversity and unequal relations of power across cultures and societies reflected in precolonial, colonial, and contemporary interactions and legacies.

In fact, the very concept of disaster and the (neo)colonial origin of its usage in many societies around the world nowadays illustrates the critical need for such epistemology. Because the concept has been imposed upon societies that do not necessarily have a word in their own language to capture what the concept means in the West, it is necessarily being reinterpreted through and confronted to local understandings of the world and the multiple past and contemporary influences that shape them. Seeing disaster in heritage therefore allows to reject common sense and recognize hybridity and *créolité* in our interpretation of hardship, suffering, and harm.

Recentring disaster studies around heritage and fostering its epistemological potential is a prerequisite to reframing disaster risk management. Disaster risk management is about people; hence, it is all about the heritage that shapes their identity. Heritage should thus be more than a sector, a box to tick in a plan or in a report. Because heritage is people, heritage must shape disaster risk reduction, from providing the lens to identify risk to defining actions to enhance people's wellbeing within their unique yet hybrid or creole cultural context. Only if framed through heritage can disaster risk reduction be relevant and meaningful to people.

Seeing and addressing disaster in and through heritage is a postcolonial agenda. It acknowledges that there cannot be any universal understanding of a concept that reflects one epistemic tradition, that is, that of the West and its project of modernity carved in Europe's 18th century. It recognizes that hardship, suffering, and harm are experienced differently across cultures and societies and, hence, that enhancing people's wellbeing requires grounded perspectives. Seeing and addressing disaster in and through heritage ultimately opens up new grounds for a more meaningful future. Heritage thus becomes a gateway to the future rather than a legacy of the past.

This postcolonial agenda that draws upon heritage as an epistemology forms the ethos of the Disaster Studies Manifesto and its companion Accord which many of us came up with a few years ago. It is within this perspective that this book by Rohit Jigyasu and Ksenia Chmutina, two of the most critical scholars of heritage in disaster, is so important. It constitutes a launch pad for our agenda. Indeed, the diversity of articles compiled hereafter testifies of the holistic dimension of heritage, of its central role in shaping contemporary identities, of its potential to understanding disaster and framing disaster risk reduction through a new lens. A lens for us all to see disaster in heritage.

Introduction

Why Disaster Risk Management of Cultural Heritage

Ksenia Chmutina and Rohit Jigyasu

The scope of cultural heritage has gradually expanded over time. From merely grand monuments and archaeological sites, cultural heritage is increasingly associated with the way of life and people's interrelationships with natural contexts; it includes vernacular buildings, historic settlements, cultural landscapes, industrial sites as well as those that are associated with painful past such as wars or accidents. Museums, libraries, and archives (and the collections they hold) are also important components of heritage, as are intangibles such as knowledge, beliefs, and value systems as well as rituals and practices, that have strong impact on people's way of life.

However, this vast range of heritage is also increasingly exposed to various natural and human induced hazards such as earthquakes, floods, fires, hurricanes, landslides, terrorism, and armed conflicts. In recent years, disasters have caused extensive damage to cultural heritage. These include floods in Pakistan and Nigeria in 2022 that damaged many heritage buildings and sites including those on the world heritage list, floods in Belgium and Germany in July 2021 that caused significant damage to many historic settlements, fires in the World Heritage Sites of Shuri Castle in Japan in 31st October 2019 and Notre Dame Cathedral in Paris on 15th April 2019 and in the National Museum of Brazil on 3rd September 2018 that significantly damaged the historic built fabric as well as collections of great heritage value. Damage to important heritage sites such as Old San Juan in Puerto Rico and Old Havana and its fortifications in Cuba due to Hurricane Irma that ravaged Caribbean region in September 2017, huge loss of heritage sites due to devastating earthquakes in Central Mexico in 2017, central Italy and Myanmar in 2016, and Nepal in 2015 further brought forward the need to undertake immediate measures to mitigate such a massive loss to heritage due to disasters. Moreover, recent conflicts have also caused enormous damage to cultural heritage in places such as Syria, Iraq, and Ukraine. Covid-19's impacts on cultural heritage – and in particular, museums and arts – have also been devastating.

Besides loss to the material fabric, disasters also affect intangible heritage including traditional knowledge, practices, skills, and crafts that ensure continuity of living cultural heritage as well as means for its maintenance and conservation. However, to date the focus has largely been on tangible heritage without taking into account a broader context that includes intangible dimension and the social, economic, geographical, and institutional dimensions of the region in which heritage is situated (Jigyasu, 2016; Jigyasu and Sen, 2022).

Heritage not only gives identity to community but also makes direct and significant contribution to sustainable development across its economic, social, and environmental dimensions (Boccardi and Duvelle, 2013; UNESCO, 2017): culture is directly addressed in Sustainable Development Goal (SDG) 11 – Make cities and human settlements inclusive, safe, resilient and sustainable, and SDG 13 – Take urgent action to combat climate change and its impacts.

It has also been recognised as a key dimension of and an asset for disaster risk management (UN, 2015; Hellgate et al., 2016). Various characteristics of cultural heritage, especially

DOI: 10.4324/9781003293019-1

traditional knowledges gained through trials and errors over time, have demonstrated its enormous potential to enhance capacities of communities (Hermans et al., 2022). Lessons learnt from the past and indigenous understandings of local hazards, climate, natural resources, and geography are embedded in cultural heritage and can be used in restoration work as well as in new construction (Ravankhah et al., 2017, Okubo, 2018).

Yet, the increasing pressures on cultural heritage, especially in urban areas, due to the geo-physical environment, carrying capacity, and socio-economic developments are likely to reach a critical point in the near future and require urgent action. We thus need a more holistic diagnosis and treatment as well as a better understanding of the historical, social, political, and technological contexts of heritage to promote improved practices for the guardians of cultural assets in historic urban areas.

Current Status of DRM of CH

Disaster Risk Management for Cultural Heritage is a relatively new area of disaster risk scholarship and practice. The global attention to challenges faced by cultural heritage in the context of disaster risk first came to light when floods in Florence in 1966 led to the destruction of millions of masterpieces of art and rare books. International Centre for the Study of the Preservation and Restoration of Cultural Property (ICCROM) launched a massive operation to salvage heritage (Hamburg, 2018). Another landmark event was the 1995 Kobe earthquake, when wooden heritage was lost due to post-earthquake fires (Toki et al., 2004).

Yet, until recently, the fields of cultural heritage and disaster risk management had virtually no connection, and there was no comprehensive effort to reduce disaster risks to cultural heritage until the World Conference on Disaster Risk Reduction in 2005 to commemorate the 10th anniversary of Kobe earthquake. During this conference, a thematic meeting on heritage and disasters was organised jointly by UNESCO, ICCROM, ICOMOS (International Council on Monuments and Sites) and Ritsumeikan University, Kyoto. This was probably the first time that the issue of cultural heritage was discussed in a conference on disaster risk management. There has been a massive progress since then: the Sendai Framework on Disaster Risk Reduction (SFDRR) adopted in 2015, for the first time explicitly included reference to cultural heritage (Target C-6), providing a boost to the ongoing efforts in this area.

Several countries, such as Myanmar and Bhutan (supported by international organisations such as World Bank and UNESCO) have taken initiatives towards preparing disaster risk management plans tailored to the characteristics of their cultural heritage sites and addressing the constraints and opportunities offered by them. Many other initiatives lead by international organisations such as #unite4heritage by UNESCO, #culturecannnotwait by ICCROM, and 'ICORP on the Road' by ICOMOS, have brought out success and failures in disaster risk management of cultural heritage.

The challenge, however, is to implement these and other global initiatives; this requires considerable building of capacities at international, national, and local/community levels and the setting up of the necessary institutional mechanisms, complemented by data collection and monitoring, as well as the recognition of a political nature of cultural heritage. In spite of noteworthy advances in this area in terms of research, training and advocacy, development of international policies, tools, and knowledge resources, many challenges remain.

Main Challenges of Disaster Risk Management of Cultural Heritage

The challenges are manyfold and require collective action as well as appreciation of both disaster risk management and cultural heritage from an interdisciplinary perspective, through intergovernmental coordination and support of various national and international actors.

One of the key challenges is the lack of sensitivity and appropriateness to local contexts. Whilst the challenges posed by disasters for research as well as practice are shared by many cultural heritage sites around the world, there are no 'silver bullets' for good solutions (Bosher et al., 2019). It is thus important to understand what strategies are being adopted to mitigate disaster risks – and which of these strategies can be adapted into different contexts.

The way the risks are assessed is yet another challenge. Mitigating risks to cultural heritage necessitate comprehensive risk assessment that takes into consideration multiple hazards and multiple – social and physical – vulnerabilities are often intertwined. Although it is important to consider physical aspects of vulnerability, the social, political, economic, institutional, and attitudinal dimensions of vulnerability should not be overlooked, as these often point towards underlying root causes linked to issues related to inherent power relations within community(ies), transparency and accountability in governance, etc. that need to be addressed for long term vulnerability and risk reduction – or else we run the risk of recreating vulnerabilities in spite of investing time and resources in mitigation measures.

This, however, needs to be done with care as the focus on vulnerabilities or the lack of capacity tends to emphasise what people cannot do as opposed to 'what they can do – and this patronises the work of many organisations and community efforts that are not taken into account. Such focus on vulnerabilities also tends to ignore indigenous and traditional knowledges (Jigyasu and Sen, 2022).

Vulnerability and capacity assessments (VCA) have become popular in disaster risk management of cultural heritage; if carried out appropriately, taking local context into account, they can play an essential role in enhancing local people's confidence making their opinions heard and taken into account (Chmutina et al., 2021). Nevertheless, there is still seldom a focus on assessing local capacity – and when it takes place, there is often an over-emphasis on communities' resilience that hides suffering. After all, VCA is merely a management tool that is based on the pre-defined concepts (whether they make sense locally or not) and quantitative and/or demographic indicators that help ticking the box. People and their heritage cannot always be understood through standard criteria and methodologies designed by outsiders. An interpretation of an outsider would always reflect their own knowledge, assumptions, and values, thus 'creating false "stories" that fit her or his expectations' (Bhatt, 1998, p. 71).

We also need to remember that capacity and vulnerability are not on the same spectrum (Gaillard et al., 2019): building capacity does not reduce vulnerability, as vulnerability is ingrained in political and social systems. Realistically, disaster risk management and cultural heritage practitioners and researchers can rarely address the societal root causes of vulnerability. This, however, does not mean that vulnerabilities should be neglected when developing capacity; both should be addressed in tandem.

Capacities differ: often those who are seen as the most marginalised and vulnerable can make up for their lack of access to economic and political resources by relying upon strong social and human resources – and these are often founded in heritage. Local capacity, however, should not be romanticised: whilst traditional knowledge can help dealing with frequent flooding, it is not enough to prevent a long-term flooding. Moreover, in some cases, traditional knowledge and practices can enhance vulnerability: deeply rooted discriminatory socio-cultural values and traditions exclude some community groups from participation.

We also need to focus on mainstreaming cultural heritage into broader disaster risk management processes – and we hope that this Handbook will contribute to this. Cultural heritage cannot be seen in an exclusive manner devoid of its larger social and economic context. After all, heritage is an inherent part of human identity that is closely intertwined in day-to-day lives of people. Cultural heritage is not static in time; rather is continuously evolving in response to changes including those induced by disasters.

4 *Introduction*

We mostly still think and work in silos: cultural heritage is considered as a separate sector as are disaster risk management, urban planning, climate change, sustainable development, and many other areas concerned with socio-political issues. But such a compartmentalised approach increases duplication of efforts and decreases efficiency. Many of the solutions that are proposed in disaster risk management for new urban developments can be appropriate for climate change adaptation measures for cultural heritage; many tools can be shared and applied holistically, thus mainstreaming cultural heritage in sustainable development efforts (Chmutina et al., 2017; Chmutina et al., 2019).

The political dimension of heritage is also critical: heritage is a construct, and its definition and recognition are linked to the individuals or communities who define and value it. Yet, this value is often decided by the stakeholders with more power. That is why, whilst the role of the community is an important part of any disaster risk management efforts, very often, the idea of communities in many of disaster risk management approaches is rather simplified. We need to ask ourselves more often: who's heritage are we protecting? Marginalisation is an often-unspoken issue in cultural heritage, and by claiming to engage with the 'community' we are hiding a lot of root causes of vulnerability. Communities are never homogenous: the information asymmetries and unequal distributions do not only exist between a community and other political actors but also within the community (Pelling, 1998). Therefore, the idea of heritage is closely linked to the inherent power relations in the community. It would be pertinent to ask whether vulnerability and risk reduction of one aspect of heritage is not at the expense of heritage belonging to the weak and marginalised. The issues of power, donors (who bring what for what?), and leadership also play a role in deciding whose heritage is protected by whom and how.

Inclusion is extremely hard to ensure, particularly in those societies without a history of public involvement in decision-making or in deeply divided communities. Political neglect and social discrimination are often evident when capacities of certain marginalised groups are ignored. Therefore, a practical shift towards mainstreaming disaster risk management of cultural heritage requires trade-offs at multiple levels, between and among various sectors and between and among various sections of the community and their representatives. This would also necessitate efforts towards addressing root causes of vulnerability of those communities who are true bearers of heritage. This also means that professionals from different disciplines need to recognise each other's issues and vocabularies – as well as the issues, knowledges, and vocabularies of the local peoples. Although we attempt to provide universally accepted definitions of terms related to disaster risk management, these are not very useful (Chmutina et al., 2020; Gaillard, 2022). The malleability can, however, be made useful if we can connect people to work in a cross-disciplinary and collaborative manner.

About This Book

Indeed, disaster risk management of cultural heritage is still facing many challenges. But there are also many good practices that could serve as case studies of what needs to be appreciated and adapted to the local contexts. We hope that this Handbook provides a comprehensive interdisciplinary overview as the intersection of cultural heritage and disaster risks has gained increasing traction over the years, as researchers and practitioners have been elaborating on the scope and providing critical reflections on various aspects of the subject area.

This Handbook is a team effort: the chapters included here offer the expertise of academics and practitioners from around the world. The chapters provide an insight into the complexities of disaster risk management of cultural heritage – but the Handbook does not claim to be exhaustive. Instead, we hope that it would invite the curious reader to further study and explore the exciting world of heritage and disasters.

The Handbook is divided into four sections. Section I introduces the key concepts and policy arenas within which DRM of CH currently operates. Section II explores various contexts for cultural heritage and disaster risk management such as climate change, conflict, urbanisation as well as the role of community. Section III offers a critique on some of the issues that are taken for granted, including technocratic approaches, nature/culture binary, romanticising of traditional knowledges as well as the role of recovery and reconstruction. Finally, Section IV gives an insight into the future – where do we need to go from here? Case studies from around the world are included to illustrate various issues and contexts in all chapters. Each chapter also contains a brief summary of key points to guide the reader. The Handbook concludes with a detailed agenda for action for the next decade for multiple audiences, both in terms of research and practice.

References

Bhatt, M.R. 1998. Can vulnerability be understood? In Twigg, J. and Bhatt, M.R. (Eds.), *Understanding Vulnerability: South Asian Perspectives*, Intermediate Technology Publications, London, pp. 68–77.

Boccardi, G. and Duvelle, C. 2013. Cultural Heritage and Sustainable Development: A Rationale for Engagement, available at: http://www.unesco.org/new/fileadmin/MULTIMEDIA/HQ/CLT/images/HeritageENG.pdf

Bosher, K., Kim, D., Okubo, T., Chmutina, K. and Jigyasu, R. 2019. Dealing with multiple hazards and threats on cultural heritage sites: An assessment of 80 case studies. *Disaster Prevention and Management*, 29(1), pp. 109–128.

Chmutina, K., Jigyasu, R. and Bosher, L. 2017. Integrating DRR including CCA into the delivery and management of the built environment. In Kelman, I., Mercer, J. and Gaillard, J.C. (Eds.), *Routledge Disaster Risk Reduction and Climate Change Adaptation Handbook*. Routledge, Abingdon, UK,, Ch. 25.

Chmutina, K., Jigyasu, R. and Okubo, T. 2019. Securing future of heritage by reducing risks and building resilience. *Disaster Prevention and Management*, 29(1), pp. 1–9.

Chmutina, K., Sadler, N., von Meding, J. and Abukhalaf, A.H.I. 2020. Lost (and found?) In translation: Key terminology In disaster studies. *Disaster Prevention and Management*, 30(2), pp. 149–162.

Chmutina, K., Tandon, A., Kalkhitashvili, M., Tevzadze, M. and Kobulia, I. 2021. Understanding the interconnection of cultural heritage, vulnerabilities, and capacities participation. *International Journal of Disaster Risk Reduction*, 53. DOI: https://doi.org/10.1016/j.ijdrr.2020.102005

Gaillard, J.C. 2022. *The Invention of Disaster*. Routledge, London.

Gaillard, J.C., Cadag, J.R. and Rampengan, M.M.F. 2019. People's capacities in facing hazards and disasters: An overview. *Natural Hazards*, 95, pp. 863–76.

Hermans, T.D.G., Šakić Trogrlić, R., van den Homberg, M.J.C., Bailon, H., Sarku, R. and Mosurska, A. 2022. Exploring the integration of local and scientific knowledge in early warning systems for disaster risk reduction: A review. *Natural Hazards*, 114, pp. 1125–1152.

Jigyasu, R. 2016. Reducing disaster risks to urban cultural heritage: Global challenges and opportunities. *Journal of Heritage Management*, 1(1), pp. 59–67.

Jigyasu, R. and Sen, S. 2022. *Words in Action: Using Traditional and Indigenous Knowledges for Disaster Risk Reduction*. UNDRR. Available at: https://www.undrr.org/publication/words-action-guidelines-using-traditional-and-indigenous-knowledges-disaster-risk

Okubo, T. 2018. Traditional knowledge of disaster resilient designs in world cultural heritage, Japan for cultural conservation and disaster mitigation. *Abitare La Terra*, No. 46 & 47, pp. 32–33.

Pelling, M. 1998. Urban flooding in Guyana. *Journal of International Development*, 10, pp. 469–486.

Ravankhah, M., Chmutina, K., Schmidt, M. and Bosher, L. 2017. Integration of cultural heritage into disaster risk management: Challenge or opportunity for increased disaster resilience. In Albert, M.-T.,

Bandarin, F. and Pereira Roders, A. (Eds.), *Going Beyond Perceptions of Sustainability*, Heritage Studies No. 2. Springer, Switzerland.

Toki, K., Okubo, T. and Izuno, K. 2004. Protection of cultural heritages from post-earthquake fire. In: Proceedings of the 13th World Conference on Earthquake Engineering, Vancouver, B.C., Canada, August 1–6, Paper No. 2781.

UN. 2015. Sendai Framework for Disaster Risk Reduction, UN, Geneva.

UNESCO. 2017. World heritage convention and sustainable development, WHC/17/41.COM/5C, available at: https://whc.unesco.org/archive/2017/whc17-41com-5C-en.pdf

Hamburg, D.A. 2018. The florence floods of today. In Conway, P. and Conway, M.O. (Eds.), *Flood in Florence, 1966: A Fifty-year Retrospective*, Michigan Publishing, University of Michigan Library, Ann Arbor, MI.

Section I
Disaster Risk Management and Cultural Heritage

1 Disaster Risk Management Terms and Concepts

Lee Bosher

Disasters are often perceived to be caused by natural hazards but typically there are key underlying socio-economic and development factors that turn hazards and human decision-making into a disaster (O'Keefe et al. 1976). Therefore, the likelihood, magnitude, and impact of a disaster often depend on the decisions made when maintaining cultural heritage, and when undertaking development activities in the vicinity of cultural heritage sites. These choices relate to the ways that natural systems are treated, the location and types of developments and methods used in maintenance and development. Each decision and action can make our cultural heritage more or less prone to disasters.

This chapter explores the key disaster risk management (DRM) concepts and terms and provides a critical insight into the (traditional) phases of disaster and risk management and how these can be of relevance to cultural heritage.

DRM Terms and Concepts

The likelihood of disasters occurring can be reduced by decreasing the exposure to hazards, lessening vulnerability of people and sites/locations, sensible management of land and the environment (i.e. buffer zones and catchments), and by improving preparedness and early warning for adverse events.

According to the UNDRR (2022),[1] **disaster risk management** (DRM) is the application of disaster risk reduction policies and strategies to prevent new disaster risk, reduce existing disaster risk, and manage *residual risk*, contributing to the strengthening of resilience and reduction of disaster losses. Thus, **disaster risk reduction** is the concept and practice of reducing disaster risks through systematic efforts to analyse and reduce the impacts of disasters and is a central component of the broader concept of DRM.

In order to fully appreciate the concept of DRM, the following terms and concepts should first be understood before then being conceptualised in relation to the DRM phases:

Risk

Risk is a probability of an event and its negative consequences.

UNDRR (2022)

The word 'risk' has two distinctive connotations: in general usage, the emphasis is usually placed on the concept of chance or possibility (for instance, 'the risk of an accident'); whereas in technical settings, the emphasis is usually placed on the consequences, in terms of 'potential losses' for some particular cause, place, and period (Bosher & Chmutina 2017). People do not necessarily share the same perceptions of the significance and underlying causes of different risks, especially when it comes to the often contested values that can be attributed by different groups to cultural heritage.

Disaster risk is defined as the potential loss of life, injury, or destroyed or damaged assets which could occur to a system, society, or a community in a specific period of time, determined probabilistically as a function of hazard, exposure, vulnerability, and capacity. This definition reflects the concept of disasters as the outcome of continuously present conditions of risk. Disaster risk comprises different types of potential losses which are often difficult to quantify. Nevertheless, knowing the prevailing hazards and the patterns of population and socio-economic development, disaster risks can be assessed, mapped, and therefore reduced in broad terms at least.

Some risks may be considered ***acceptable***: these are potential losses that a society or community considers tolerable in the context of existing social, economic, political, cultural, technical, and environmental conditions. There are also risks that can be impossible to manage – these are called '***residual risks***'.

Vulnerability

> Vulnerability relates to the conditions determined by physical, social, economic and environmental factors or processes which increase the susceptibility of an individual, a community, assets or systems to the impacts of hazards.
>
> UNDRR (2022)

Vulnerability has many aspects including various physical, social, economic, and environmental factors. Examples may include poor construction or maintenance of buildings, inadequate protection of assets, lack of public information and awareness, limited official recognition of risks and preparedness measures, and disregard for wise environmental management. It can vary significantly for cultural heritage over time, and is specific to each location even if other factors (e.g. economic development) are similar. Vulnerability is most often associated with poverty, but it can also arise when people are isolated, insecure, and defenceless in the face of risk, shock, or stress. People can differ in their exposure to risk as a result of their social group, gender, ethnicity, religion, age, and health, along with a myriad of other factors (see Adger 2000; Aldrich 2012; Berkes et al. 2000; Bosher et al. 2007; Cutter et al. 2003; Enarson & Chakrabarti 2009; Fordham 2008; Kelman 2020; Wisner et al. 2004).

Mitigation

> Mitigation is the lessening or minimising of the adverse impacts of a hazardous event.
>
> UNDRR (2022)

Mitigation implies that whilst adverse impacts of hazards often cannot be prevented fully, their severity can be substantially reduced by various strategies and actions. Mitigation measures encompass structural (e.g. engineering techniques, hazard-resistant construction, or maintenance) and non-structural approaches (e.g. improved environmental policies, early warning procedures/systems, and public awareness). It should be noted that in the context of climate change policy, 'mitigation' is defined differently as the term is used for the reduction of greenhouse gas emissions that are the source of climate change.

Preparedness

Preparedness is the knowledge and capacities developed by governments, response and recovery organizations, communities and individuals to effectively anticipate, respond to, and recover from the impacts of likely, imminent or current disasters.

UNDRR (2022)

Preparedness takes place within the context of DRM and aims to build the capacities needed to efficiently manage all types of emergencies and achieve orderly transitions from response through to sustained recovery. Preparedness should be based on a comprehensive analysis of disaster risks and establishment of effective early warning systems. It includes such activities as contingency planning, stockpiling of equipment and supplies, the development of arrangements for coordination, evacuation, and public information plus associated training and field exercises. For cultural heritage assets, this work includes providing visitors (possibly people that don't speak/read the local language) with understandable signposting and support if an evacuation is required. In order to be effective, preparedness actions must be supported by formal institutional, legal, and budgetary capacities and implemented thoroughly at the national, regional, and local levels (Jeggle & Boggero 2018).

Recovery

Recovery is restoring or improving of livelihoods and health, as well as economic, physical, social, cultural and environmental assets, systems and activities, of a disaster-affected community or society, aligning with the principles of sustainable development to avoid or reduce future disaster risk.

UNDRR (2022)

Recovery includes rehabilitation and reconstruction activities that can begin soon after the emergency phase. Recovery should be based on pre-existing strategies and policies that facilitate clear institutional responsibilities for recovery action and enable participation of key stakeholders.[2] Recovery programmes, coupled with the heightened public awareness and engagement after a disaster, offer a valuable opportunity to develop and implement disaster risk reduction measures. If key local stakeholders are involved in the decision-making, this can be an important opportunity to incorporate traditional and/or locally acceptable risk reduction measures that will not compromise cultural heritage values.

Hazards and Threats

There are many ways with which different disciplines define hazards and threats, but for the purposes of this book, the definitions that will be used to distinguish between the two main causes of disasters are the simple descriptors of:

Hazard: is primarily a 'natural' source of potential danger
Threat: is primarily a 'human induced' source of potential danger

(Bosher & Chmutina 2017)

Disasters are typically classified into natural and human induced (sometimes also called 'manmade'). 'Natural disasters' is a common term used, particularly by the media, as it relates to

12 Lee Bosher

Table 1.1 Typology of hazards and threats

Natural hazards		Human induced threats	
Geophysical hazards	Earthquakes	*Malicious*	War
	Volcanic eruptions		Terrorism
	Tsunamis (*inc. Seiches*)		Arson
	Landslides		Civil unrest
	Subsidence		Vandalism
Hydro-meteorological hazards	Floods	*Non-malicious*	Ineffective planning
	Coastal erosion		Poor quality construction
	Hurricanes/cyclones/typhoons		Rapid urbanisation
	Tornadoes		Pollution
	Extreme temperatures		Epidemics
	Storm surges		Industrial 'accidents'
	Drought		Corruption
	Fires		

Source: After Bosher and Chmutina (2017).

disasters that appear to have been caused by hazards of natural origin such as extreme weather, geophysical phenomena, or epidemics. However, it is important to recognise that these so called 'natural disasters' are rarely very natural because there tend to be many important human induced factors that have converted the natural hazard into a disaster (i.e. low-quality buildings, poor locational planning, etc.); see O'Keefe et al. (1976) and Chmutina and Von Meding (2019) for further discussions on this matter.

Natural hazards[3] are typically split into two categories, namely; (1) geo-hazards and (2) hydro-meteorological hazards. Table 1.1 provides a list of the most common geophysical-hazards and hydro-meteorological hazards that occur globally.

The magnitude of natural hazards tends to be determined by key factors such as meteorology (which is influenced by the changing seasons), topography, hydrology, geology, biodiversity (of flora and fauna), and tidal variations (caused by lunar and meteorological influences, coastal topography and influenced by the type and locality of coastal developments). These processes are typically benign and provide the basis for people to exist in harmony with their natural environment. However, infrequently (and some would suggest more frequently due to climate change and poorly regulated urban development) natural hazards impact upon our cultural heritage causing damage, deaths, disruption, and irreplaceable losses to our precious heritage (see Box 1.1).

Box 1.1   Categorising the challenges experienced at cultural heritage sites

A study by Bosher et al. (2019) adopted 6 analytical categories (including social, cultural, environmental as well as technical factors) when reviewing the underlying hazards and threats experienced at 80 cultural heritage sites (across 45 countries). The study found that in all the cases, multiple, often deep rooted, challenges were experienced, and some challenges were significantly more prominent than others (see table IV, ibid). The most prevalent challenges were technical (90 per cent), infrastructure (74 per cent), and governance (71 per cent). The study also found that whilst the underlying causes of disasters at those sites were indeed complex (and typically context specific), there were nonetheless many similarities, to the problems encountered and approaches to risk reduction, regardless of the location.

Phases of Disaster Risk Management

Disaster risk management aims to avoid, lessen, or transfer the adverse effects of hazards through activities and measures for prevention, mitigation, and preparedness. According to Alexander (2002) in 'Principles of Emergency Planning and Management', disaster management was traditionally divided into pre- and post-disaster components containing four phases: Mitigation, Preparedness, Response/Relief, and Recovery.

Pre-Disaster Phases

Mitigative (Preventative) Adaptations – Structural and non-structural measures undertaken to limit the adverse impact of hazards/threats. Ideally, this includes the identification of potentially damaging physical events, phenomenon, or human activity (Bosher & Chmutina 2017). For cultural heritage, this can be an overlooked phase because there may be a perception that if something has been around for hundreds or thousands of years, then it should be safe. However, the impacts of climate change (through changing weather/rainfall patterns) and urbanisation (through development encroachment and pollution) can have a deleterious impact on cultural heritage sites and thus should be considered when regularly undertaking risk assessments.

Preparedness – The knowledge and capacities developed by governments, response and recovery organizations, communities, and individuals to effectively anticipate, respond to, and recover from the impacts of likely, imminent, or current disasters (UNDRR 2022).

Post-Disaster Phases

Response/Relief – Actions taken directly before, during, or immediately after a disaster in order to save lives, reduce health impacts, ensure public safety, and meet the basic subsistence needs of the people affected (UNDRR 2022). In the post-disaster response phase, a key challenge is how to salvage heritage properties, which are at risk of demolition and to assess their damage (Bosher et al. 2019). Recent years have witnessed the development and evolution of the Post Disaster Needs Assessment (PDNA) framework. The PDNA has been set up by key international agencies to harmonize post-disaster assessment methods to better support governments and affected populations with a coordinated approach (Jeggle & Boggero 2018). More on the PDNA will be provided in Chapter 4.

Recovery (Sometimes Called 'Recovery and Rehabilitation') – Decisions and actions typically taken after a disaster with a view to restoring or improving the pre-disaster living conditions of the stricken community (Bosher & Chmutina 2017). A key challenge during the recovery phase is how to repair and retrofit affected properties, sites, and assets and undertake reconstruction that respects tangible as well as intangible heritage values (Ravankhah et al. 2017). Chapters 15 and 16 provide a more in-depth perspective of recovery and reconstruction.

Often, these phases have been conceptualised in the form of a 'disaster cycle'. The phases described by Alexander in 2002 (and defined above) have rightly endured (even if the terminology used has not always been consistent) as they align well with key operational aspects of DRM. Clearly, it has also been noted that these phases are not self-contained phases but, in reality, are substantially overlapped and interconnected (Aguirre & Lane 2019; Contreras 2016; Neal 1997).

Moving away from the 'Disaster Cycle'

Bosher et al. (2021, also refer to Box 1.2) explain that the persistence of the 'disaster cycle' in much of the extant literature has contributed towards a conceptualisation of disasters that

> **Box 1.2 Time to ditch the 'disaster cycle'?**
>
> Bosher et al. (2021) suggest that there is a need to move away from the two dimensional and overly simplistic 'disaster cycle' but that this is done in a way where the key phases (that they acknowledge invariably overlap with each other) are not only maintained but better represented. A key suggestion is that the management of disasters needs to be conceptualised in a way that takes on board some key principles which can be difficult to represent in the two-dimensional disaster cycle, namely:
>
> a *Move away from a Closed Loop* – There needs to be scope to include the idea that moving through the phases can include new inputs and activities that may lead to no further disasters. Thus, the loop is not closed, it is open and moves towards the future.
> b *Avoid Making the 'Disaster Event' a Key Component of the Concept* – The key weakness of the 'disaster cycle' as it is often represented is that the cycle starts/ends in a disaster. This implies that a disaster is required to instigate risk reduction activities (which is not the case) or all the efforts to undertake mitigation and preparedness activities will inevitably end up in another disaster (which should also not be the case).
> c *Factor in Temporal and Resource Considerations* – There is a need to consider how much time (relatively) is being devoted to DRM activities and there is scope to integrate hazard mitigation and risk reduction activities into post-disaster rehabilitation and reconstruction activities, as after all the boundaries between the phases are fuzzy (Contreras 2016).
> d *Be Flexible Enough to Factor in Underlying Root Causes of Vulnerability and Drivers of Disaster Risk* – The disaster cycle, as typically portrayed, does not provide sufficient scope for important socio-economic and political aspects to be factored into it, thus providing an overly simplistic representation of how disasters are not only managed but how they can be created.
> e *Acknowledge the Role of Complex Systems* – Systems thinking challenges linear cause and effect relationships and aims to better understand different systems' states and system levels. Such an understanding will facilitate adaptive and transformative behaviour which is a hallmark for effective disaster risk management – the ability to change.

undervalues the beneficial impacts of pre-disaster risk reduction activities that are often associated with mitigation (i.e. structural and non-structural risk reduction measures) and preparedness (i.e. emergency planning and capacity building) activities. The phases, therefore, tend to frame a disaster as a one-off event and a technocratic problem that can be 'managed' rather than a process that is neither linear nor cyclical but instead multi-dimensional and evolving (also see Chapter 13).

Current thinking represents four phases of Disaster Risk Management (DRM) in a more interlinked manner where disaster impacts are not shown as it is felt that a disaster impact should not really be required to instigate thinking and actions for DRM (Figure 1.1):

- *Hazard Identification* – Identification of potentially damaging physical events, phenomenon or human activity. As part of this, it is important to ascertain what the underlying vulnerability factors are that may lead to problems for cultural heritage.

Figure 1.1 DRM phases.
Source: Bosher and Chmutina (2017).

- **Mitigative Adaptations ('Hazard Mitigation')** – Structural and non-structural measures undertaken to limit the adverse impact of hazards/threats.
- **Preparedness Planning** – Activities and measures taken in advance to ensure effective response to disasters if/when they occur.
- **Recovery and Rehabilitation** – Decisions, plans, and actions taken after a disaster with a view to restoring or improving the pre-disaster conditions of the affected sites, people, or livelihoods.

Risk Management Elements

Typically, risk management consists of five elements, performed more or less, in the following order (adapted by Bosher and Chmutina (2017), from ISO 31000 'Risk management – Principles and guidelines' [British Standards Institution 2009]):

1 *Identify, Characterise, and Assess Hazards/Threats* – The aim of this stage is to begin recognising the threats and hazards to which a location is exposed to. There is a wide range of multi-hazard/threat identification approaches (many of which can be hazard/location/resource specific) including:

- Hazard mapping
- Assessment of recent events
- Assessment of historical events
- Local knowledge
- Scientific monitoring/forecasting (i.e. using sensors, computational modelling)

2 *Assess the Vulnerability of Critical Aspects of Your Cultural Heritage to Specific Hazards/Threats* – This is the process of assessing the susceptibility of the intrinsic properties. For instance, these properties could be the structure, materials, construction, lives, and livelihoods that might be impacted by the hazards/threats, and thus that can lead to a disastrous situation. Bear in mind that these disastrous situations can be relatively sudden onset (such as a lightning strike or tornado) or something that has formed over an extended period of time (such as encroachment of development to a heritage site that increases overcrowding, fire risk, or flood risk).

3 ***Determine the Risk (i.e. the Expected Consequences of Specific Hazards/Threats on Specific Assets)*** – This stage fundamentally aims to answer three questions (Bosher & Chmutina 2017):

   a What can happen? ('What can go wrong?')
   b How likely is it?
   c What are the consequences? ('How bad could it be?')

   In addition to the above standard three questions, it is encouraged that a fourth question should be added to the list:

   - How much uncertainty is present in the analysis? (In other words, 'How reliable are the answers to questions 3a–3c?')

   Quantitative risk analysis can use probabilistic estimates for many undesired events and the risk is then determined as the mathematical expectation of the consequences of the undesired events. The aim of a probabilistic logic (or probability logic) is to combine the capacity of probability theory to handle uncertainty with the capacity of deductive logic to exploit structure. The expected result is a richer and more expressive formalism with a large range of possible application areas.

   In contrast, qualitative risk analysis focuses on the likelihood of a hazard(s) occurring and the consequences if something occurs, and by adopting a simple 'Low' to 'High' range can provide a straightforward visual and easy to understand approximation of the risk scale (Figure 1.2).

4 ***Identify Ways to Reduce Those Risks*** – The aim of this stage is to then identify courses of action that can address and treat the hazards/threats and risks associated with them. These actions can include structural and non-structural measures. For heritage sites, **structural measures** could include seismic isolation technologies (as used for The Heisei Chishinkan Wing at the Kyoto National Museum in Japan) or coastal defences (as used to protect the Shore Temple from the Bay of Bengal in Mahabalipuram, India). **Non-structural measures** may include using real-time monitoring on rivers to support early flood warning which is then also aligned with well-established and practised evacuation procedures (e.g. for people and portable heritage).

Likelihood	Consequences				
	Insignificant	Minor	Moderate	Major	Severe
Almost certain	M	H	H	E	E
Likely	M	M	H	H	E
Possible	L	M	M	H	E
Unlikely	L	M	M	M	H
Rare	L	L	M	M	H

Figure 1.2 Qualitative risk analysis.
Source: Bosher and Chmutina (2017).

> **Structural Measures** – Any physical construction to reduce or avoid possible impacts of hazards, or application of engineering techniques to achieve hazard-resistance and resilience in structures or systems;
>
> **Non-Structural Measures** – Any measure not involving physical construction that uses knowledge, practice, or agreement to reduce risks and impacts, in particular through policies and laws, public awareness raising, training, and education.

5 *Prioritise Risk Reduction Measures* – Once the potential courses of action have been identified, it is important to prioritise the most suitable options. Thus, the objective of this stage is to assist in identifying the most appropriate intervention(s) for a given project. The prioritisation will depend on a number of factors that are likely to be highly specific to each project; these include (but are not limited to):

- Costs versus benefits of identified interventions
- Technical feasibility
- Social feasibility and acceptability
- Alignment with protection of key heritage values/attributes
- Proportionality of identified interventions
- Complementarity with measures introduced to mitigate other hazards
- Corporate social responsibility
- Business continuity
- Legal and statutory requirements

Summary

It is widely appreciated that cultural heritage is not only an important component of a country's history and identity, but it can also provide drivers of economic sustainability and can be central to supporting societal resilience. However, an increasing number of disastrous events, as highlighted in the introduction to this book, have emphasised the vulnerability of cultural heritage worldwide. Thus, it is increasingly being recognised how important it is to undertake proactive measures using a phased approach that can reduce risks to cultural heritage from disasters through DRM measures (Jigyasu et al. 2013). There have been some recent initiatives (such as an increased appreciation of using traditional knowledge where relevant and the development of the PDNA framework) that have helped heritage practitioners to incorporate DRM into cultural heritage practice. However, there is still more work to be done on better understanding how DRM can be incorporated into the rich variety of cultural heritage assets and attributes, especially in a way that recognised the positive role that cultural heritage can play in DRM.

Key Points

- Disaster risk management is a multi-disciplinary concept that requires the inputs of many stakeholders, over and beyond the traditional 'disaster management' and cultural heritage disciplines.
- There is now more recognition that we (collectively) need to change the way we (re)develop our cities, infrastructure, and buildings, and to be more conscious of how we can best integrate DRM into cultural heritage.

- There are well established frameworks for (disaster) risk management that can be adopted or adapted to the needs of those responsible for cultural heritage sites/assets.
- Finally, we also need to remember that it is a case of not just considering what DRM can do for cultural heritage but also considering what cultural heritage can do for DRM!

Notes

1 Most of the definitions are taken from the UNDRR (2022) https://www.undrr.org/terminology unless otherwise indicated. This is not to suggest that the UNDRR terms are perfect, but for the context of this edited book, the use of such terms in a regular way across chapters should hopefully keep readers (especially those from different disciplines) on a suitably consistent path.
2 A 'Stakeholder' is anybody who can affect or is affected by an organisation, strategy, or project. Stakeholders can be internal or external and they can be formally or informally involved or affected.
3 An in-depth analysis of each of these hazards/threats is beyond the scope of this chapter. For more detailed information about different types of hazards and some key threats you may wish to refer to the book 'Environmental Hazards' by Keith Smith (2013).

References

Adger, W.N. (2000), Social and ecological resilience: Are they related? *Progress in Human Geography*, Vol. 24 No. 3, pp. 347–364.
Aguirre, B.E. and Lane, D. (2019), Fraud in disaster: Rethinking the phases. *International Journal of Disaster Risk Reduction*, Vol. 39, p. 101232.
Aldrich, D.P. (Ed.). (2012), *Building Resilience. Social Capital in Post-Disaster Recovery.* The University of Chicago Press: Chicago.
Alexander, D.E. (Ed.). (2002), *Principles of Emergency Planning and Management.* Oxford University: Oxford.
Berkes, F., Folke, C. and Colding, J. (Eds.), (2000), *Linking Social and Ecological Systems: Management Practices and Social Mechanisms for Building Resilience.* Cambridge University Press: Cambridge.
Bosher, L.S. and Chmutina, K. (Eds.). (2017), *Disaster Risk Reduction for the Built Environment.* Wiley: London.
Bosher, L.S., Chmutina, K. and van Niekerk, D. (2021), Stop going around in circles: Towards a reconceptualisation of disaster risk management phases. *Disaster Prevention and Management*, Vol. 30 No. 4/5, pp. 525–537.
Bosher, L.S., Kim, D., Okubo, T., Chmutina, K. and Jigyasu, R. (2019), Dealing with multiple hazards and threats on cultural heritage sites: An assessment of 80 case studies. *Disaster Prevention and Management*, Vol. 29 No. 1, pp. 109–128. Available online https://www.emerald.com/insight/content/doi/10.1108/DPM-08-2018-0245/full/html (Accessed 10th June 2022).
Bosher, L.S., Penning-Rowsell, E. and Tapsell, S. (2007), Resource accessibility and vulnerability in Andhra Pradesh: Caste and non-caste influences. *Development and Change*, Vol. 38 No. 4, pp. 615–640.
British Standards Institution (Eds.). (2009), *Risk Management: Principles and Guidelines.* British Standards Institution Group: London.
Chmutina, K. and Von Meding, J. (2019), A dilemma of language: 'natural disasters' in academic literature. *International Journal of Disaster Risk Science*, Vol. 10 No. 3, pp. 283–292.
Contreras, D. (2016), Fuzzy boundaries between post-disaster phases: The case of L'Aquila, Italy. *International Journal of Disaster Risk Science*, Vol. 7 No. 3, pp. 277–292.
Cutter, S.L., Boruff, B.J. and Shirley, W.L. (2003), Social vulnerability to environmental hazards. *Social Science Quarterly*, Vol. 84 No. 2, pp. 242–261.
Enarson, E. and Chakrabarti, P.D. (Eds.), *Women, Gender and Disaster: Global Issues and Initiatives.* SAGE Publications India: New Delhi.

Fordham, M. (2008), The intersection of gender and social class in disaster: Balancing resilience and vulnerability. In Philips, B.D. and Morrow, B.H. (Eds.), *Women and Disasters: From Theory to Practice*. Xlibris: Philadelphia, pp. 75–98.

Jeggle, T. and Boggero, M. (Eds.). (2018), *Post-Disaster Needs Assessment*. GFDRR, World Bank: Washington DC.

Jigyasu, R., Murthy, M., Boccardi, G., Marrion, C., Douglas, D., King, J., O'Brien, G., Dolcemascolo, G., Kim, Y., Albrito, P. and Osihn, M. (2013), 'Heritage and resilience: Issues and opportunities for reducing disaster risks', *4th Session of Global Platform for Disaster Risk Reduction*, Geneva, May.

Kelman, I. (2020), *Disaster by Choice: How Our Actions Turn Natural Hazards into Catastrophes*. Oxford University Press: Oxford.

Neal, D.M. (1997), Reconsidering the phases of disaster. *International Journal of Mass Emergencies and Disasters*, Vol. 15 No. 2, pp. 239–264.

O'Keefe, P., Westgate, K. and Wisner, B. (1976), Taking the naturalness out of natural disasters. *Nature*, Vol. 260 No. 5552, pp. 566–567.

Ravankhah, M., Chmutina, K., Schmidt, M. and Bosher, L.S. (2017), Integration of cultural heritage into disaster risk management: Challenge or opportunity for increased disaster resilience. In Albert, M.-T., Bandarin, F. and Pereira Roders, A. (Eds.), *Going Beyond – Perceptions of Sustainability in Heritage Studies No. 2, Volume 5 of the Heritage Studies Series*. Springer: Cham.

UNDRR (2022), *Terminology on Disaster Risk Reduction*. UNISDR: Geneva. Available at: https://www.undrr.org/terminology (Accessed 12th February 2022).

Wisner, B., Blaikie, P., Cannon, T. and Davis, I. (Eds.). (2004), *At Risk: Natural Hazards, People's Vulnerability and Disasters*. Routledge: London.

Suggested Reading

Chmutina, K., Jigyasu, R. and Okubo, T. (Guest Eds.). (2020), Special issue on 'Securing future of heritage by reducing risks and building resilience'. *Disaster Prevention and Management*, Vol. 29 No. 1. https://www.emerald.com/insight/publication/issn/0965-3562/vol/29/iss/1 (Accessed 11th March 2022).

Dean, M. and Boccardi, G. (2015), Sendai implications for culture and heritage in crisis response. *Crisis Response*, Vol. 10 No. 4, p. 54.

Ishida, Y., Kim, D., Konegawa, T. and Fukagawa, R. (2020), Consideration of issues and solutions related to the coexistence of cultural properties and the real life of the neighborhood. *Disaster Prevention and Management*, Vol. 29 No. 1, pp. 86–108.

Jigyasu, R. (2016), Reducing disaster risks to urban cultural heritage: Global challenges and opportunities. *Journal of Heritage Management*, Vol. 1 No. 1, pp. 59–67.

Smith, K. (2013), *Environmental Hazards: Assessing Risk and Reducing Disaster*, 6th Edition. Routledge: London.

UNESCO (2010), *Managing Disaster Risks for World Heritage*. UNESCO: Paris.

2 Role of Intangible Attributes of Heritage in Disaster Risk Reduction

Sukrit Sen

Laura Jane Smith defines heritage as *'a social process concerned with the creation and maintenance of certain social and cultural values'* (Smith, 2006). Heritage is broadly classified into tangible and intangible, where tangible heritage represents the physical and built aspects of cultural heritage such as archaeological sites, historic monuments, etc., whereas the intangibles include the immaterial expressions of culture such as practices, knowledge, skills – as well as the instruments, objects, artefacts, and cultural spaces that have developed as a response to the historical and social evolution of the communities and groups involved. Although there have been many discourses that talk about the creative and destructive tensions due to the lack of sync between the tangible and intangible attributes in heritage management practices over the years, it has been observed that most decision-making bodies have upheld intangible heritage at the forefront in developing suggestions for heritage management in current times. Thus, it can be safely inferred that intangible heritage does have the potential to provide for a larger framework within which tangible heritage could take its shape and significance and draw parallels between cultural values and cultural variables as explained by Arjun Appadurai where he states that, *'intangible heritage because of its very nature as a map through which humanity interprets, selects, reproduces and disseminates cultural heritage was an important partner of tangible heritage. More important it is a tool through which the tangible heritage could be defined and expressed [thus]transforming inert landscapes of objects and monuments turning them into living archives of cultural values*' (Museum international, 2004).

Growing concern has been expressed in the last few years over the threat to cultural diversity and the loss of tangible and intangible cultural assets posed by human and climate-induced hazards. Intangible cultural legacy serves as a crucial foundation for the identity, well-being, and sustainable development of communities as well as for mutual understanding both within and between them, as the international community has come to recognise that conflicts and natural catastrophes may, therefore, have a significant impact on fundamental parts of people's life. Alongside human-induced hazards, since disasters brought on by climate change are occurring more frequently and with greater frequency, protecting the intangible aspects of heritage is proving to be particularly challenging. Loss of oral traditions, languages, traditions, and beliefs is a common consequence as a result. As we see in the case of Aranmula Kannadi in Kerala, India, an 18th-century craft form that was severely affected during the 2018 floods leading to a huge loss of livelihoods for the involved community due to complete shutdown of the practice for over two months and due to the massive impact on their instruments and raw materials which is crucial for the continuity of the practice (Kutoor, 2018; Raghunath, 2019). Taking the case of Sub-Saharan Africa, studies in the last ten years show how extreme weather events caused by climate change are now also having a noticeable impact on the distribution of medicinal plant species not only leading to the unavailability of the main ingredient but also the traditional healers. Furthermore, the considerable extinction of these important medicinal species has serious ramifications leading to the disruption of economic and socio-cultural harmony within many local communities, which heavily rely on traditional medicines to meet their basic medical needs (Maroyi, 2013).

While hazards are a huge threat to cultural heritage, particularly intangible, various vulnerabilities play a major role in exacerbating the impacts of disasters. The same was seen in the case of Chitpur, a region in the historic urban fringe of Kolkata, India during a study conducted during my master's thesis that looked at the *role of intangible cultural heritage in disaster management practices*. Chitpur is a neighbourhood in the northern side of Kolkata that has been relevant historically even before the arrival of the British, predating the city itself. It is one of the busiest areas in the city and is one of the biggest economic hubs of Calcutta with a good mix of old and new, royal and grounded, residential and commercial buildings giving a very heterogeneous texture to the area.

My study focuses on the intangible attributes of the site that is housed within these beautiful structures. They are exhibited through a number of socio-cultural practices, urban crafts, and historic trades that have continued over generations, as shown in the images (Figures 2.1–2.6). It is the diversity of the various communities and their continuing practices that make Chitpur a creative industry today and plays a major role in not only adding up to the intangible cultural heritage values on site but also shaping up the aesthetic texture of the neighbourhood, as they have contributed much towards the existence of these buildings in current times. The social, economic, and environmental values of these historic practices have undergone transformation over time but they have been able to continue for so many years because of their sustainable and creative approaches and have been able to generate livelihoods by bringing in social cohesion within the communities.

Kolkata, apart from a few cyclones, hasn't faced many catastrophic events in the last few decades that would affect the studied neighbourhood or its livelihoods significantly. During the study, it was found that although fire and water logging due to rainfall were among the two main hazards that may impact the practices and trades mentioned above, the transformation that the historic urban setup has gone through in the previous years has given rise to several underlying vulnerabilities that pose threat to the continuity of the historic practices and trades.

The vulnerabilities established were due to:

a **Lack of Maintenance of Buildings and Public Services** – Almost every building in the historic neighbourhood of Chitpur houses an outlet for either a historic trade or an urban craft, as mentioned above. Due to the transition in time, most of these buildings have worn off structurally due to age and a lack of maintenance. Due to several residents moving out of the area, most of the residential spaces also get repurposed for commercial purposes and storage encouraging incongruent interventions that not only lead to the addition of dead weight to these buildings making them structurally weak but also loss of visual character. On speaking with my respondents it was also inferred that most of the spaces used for these trades and crafts are still on rent which is very menial and has been so for ages due to rent control laws, hence neither do the traders or craftspersons nor the owners want to invest on retrofitting spaces that need immediate attention. A few buildings have also been identified as 'dangerous' by the Kolkata Municipal Corporation but they continue to be used as workshops, storage units, and in some cases, outlets for the practices despite cases of structural collapse on similar buildings in the vicinity posing a huge threat to human lives and the practices in general.

Figure 2.1 A brass moulder in his workshop using fire without any safety apparatus.
Source: Author's image.

In many cases due to the over usage of buildings, the space available for the craftsmen and traders is often very restricted. As a result of which the workshops have inadequate ventilation and lighting, resulting in the artisans to spill out on the roadside, putting their lives at risk. Additionally, in some stretches, fences have been put on footpaths to restrict pedestrians onto the vehicular road, but due to its design, pedestrians prefer walking on the road due to the lack of accessibility through those fences making it risky for both the vehicles as well as the artisans who spill out on the road to work. In terms of services and infrastructure, the wiring and electrical systems both at a building and public level are very old and arranged in a very haphazard manner. The electrical poles that support these wirings on the road are old and the substations lack proper concealing. All these electrical units lack maintenance and are often exposed to environmental weathering and human-induced ill practices like urination and garbage dumping around it. As a result of which it leaves behind a possibility of pounding, short-circuits, and the breakouts of fire.

Figure 2.2 Stone carvers spilling out on vehicular road due to lack of space.
Source: Author's image.

Figure 2.3 Children are working in brass moulding workshops.
Source: Author's image.

b **Lack of Financial Returns** – Although the crafts and trades have managed to perish even today, they often fail to keep up with the pace of a very rapidly growing industry of modern-day alternatives and services which are more cost-effective and readily available. As a result, traditional practices such as the ones studied tend to get outdated with time. Hence, crafts like sweet meat mould making, musical instrument making, and aluminium moulding suffer largely due to a loss in demand. Moreover, all the practices were stationed there for the steady market that Chitpur itself provided, for it housed a few of the wealthiest families in the city. Due to the rapid emigration of these residents abroad and to newer parts of the city, the in-house market which formed the backbone of their demand has been lost. Trades like that of tea, brass, and knives suffer majorly due to the same further resulting in a lack of interest in current generations to continue these practices in the future.

   Such aspects lead to other indirect vulnerabilities such as the use of cheap labour due to a lack of resources often involving children where they are exposed to tasks not fit for their age. For example, the brass makers of Chitpur were seen indulging in child labour where young boys were made to work with acid to treat brass in return for a menial daily wage.

c **Lack of Coordination between Decision-Making Bodies** – From an institutional point of view, the biggest challenge for such areas is the lack of coordination between the different governing bodies and the ill/lack of implementation of guidelines and regulations in such historic urban neighbourhoods. As a result of which, chaos is a daily episode in such areas risking the lives of the craftsmen and traders. In terms of the practices, no formal institution recognizes these trades or crafts as an entity that contribute to the development or building character and maintaining social cohesion of the city. Hence, they all fall under the informal sectors which provide these traders or practitioners with practically no regulations, laws, or capacity building that they could fall back on to get any sort of protection that encourages them to take these practices forward. Additionally, this also excludes them from any requirements from following bylaws in the outlets that regulate against the threats of using fire, acids, or other substances that may be hazardous for human use.

d **The Carelessness of People** – The practices discussed above mostly don't provide much financially due to a diminishing market as a result of which there is a lack of optimism within the bearers of such cultural practices as well as the governing bodies when it comes to investing in reducing disaster risks. It is for the same that everyone in the practice is negligent and careless. Despite understanding that it's a matter of great concern, they are very confident about their lifestyles considering they haven't faced any sort of causalities yet. However, respondents have reported that due to a lack of financial growth, many traders in the area often indulge in unethical practices like planning a fire in his/her establishment to claim insurance and make up for the losses incurred in business which is not only fatal but may also end up harming others in the vicinity too.

Creative Adaption to Reduce Risk – The study concludes that there is a direct connection between the tangible and intangible attributes of Chitpur as a result of which all the identified practices are exposed to different types of risks including discontinuity for some of them due to the underlying vulnerabilities which may be catalysed in case of hazards. However, despite current restraints, these practices have been able to adapt creatively to different issues to remain relevant.

Water logging is one of the most prevalent risks in Chitpur and tends to affect several practices in different ways to which practitioners too have responded in respective ways to minimize the impacts of the same and optimise the situation in different ways.

During the excessive accumulation of water on the streets, it tends to enter the shops. While it does affect their business, most of them are used to the situation now. However, to reduce the logged water from entering the shops, almost every outlet has found ways to reduce it by either increasing the height of the plinth level or by building a higher threshold that acts like an embankment. Storage units, especially for the tea, which are very sensitive to moisture have been set up in attic spaces taking advantage of the high ceiling levels of the old houses to prevent the incoming water from destroying them. Leather too is sensitive to moisture, and hence in the case of the musical instrument makers where the rising dampness is a result of the clogged water, shelves have been designed in a particular manner so that the tablas, an Indian percussion instrument, are not in physical contact with the wall. The structure of the tablas is often made in a way that they can be hung from the ceiling. This way, they not only make an efficient use of the lack of space for storage but also restrict contact with the damped walls of the buildings.

Milk being an essential commodity, the milk traders have their ways of coping with excessive rainfall and water logging, so while all other practices shut down, they are the only ones to still function using smart and innovative ways: The lack of customers being the biggest challenge, the price of milk is reduced (from Rs. 150/lt to Rs. 60/lt) to increase demand.

Figure 2.4 Use of attic space in tea trading units.
Source: Author's image.

Figure 2.5 Hanging of musician instruments from ceiling.
Source: Author's image.

The milkmen often use plastic containers instead of aluminium ones which are lighter and easier to tread over water. To avoid contamination, the traders often conduct business from trucks which are generally elevated to a level and avoid contact with the dirty clogged water when transferred from one vessel to another.

Similar vulnerabilities are seen impacting cultural heritage both tangible and intangible in every part of the globe and similar creativity and adaptability have also been observed in many other parts of the world to keep practices pertinent in times of crisis (Box 2.1).

Figure 2.6 Milk trade happening from truck.
Source: Author's image.

Case of creative adaptation of the paper mache masks of 'Chhau' dancers of Purulia West Bengal to fight Covid-19

Chhau is a semi-classical Indian traditional dance form with both martial and folk art origins. It is prevalent in the states of West Bengal and Jharkhand in the eastern part of India and also happens to be on UNESCO's Representative List of Intangible Cultural Heritage of Humanity. Apart from the rituals during 'Chaitra Parab' or spring harvest festival, around 40,000 people from around the areas of East India are involved with this dance form and they perform in and around their village at various times of the year. Since this art form is predominantly practised and performed by the men in the community, masks play a very important role especially when they play special or female characters.

The masks are made by the artists of the 'Sutradhara' community who use 8–10 layers of soft paper in organic glue to make them light weight and easy to wear and dance. Depending on the shape and size of the mask, they can take up to a month to be made from scratch. Around 11,000 tribals are actively involved in the mask making process. Spring being the main time when this dance is performed, the mask and the dressmakers are totally dependent on the dancers who purchase them during these times. Being small cottage industries, these communities are often dependent on bank loans for the production of these items and preparations for the same start from the beginning of the year.

However, with the outbreak of the global pandemic just in the beginning of spring 2020, not only were all the performances cancelled but around 20,000 people were left unemployed. At a point when the artisans were confused with what is to be done with all the unsold masks and huge debts, capacity building organisations in the field of crafts, such as 'Kalamandir', came forward with an innovative idea of making paper mache masks, that cover just the mouth and nose. This adaptation not only helped create an opportunity for selling these masks again and helping the artisans overcome the situation but also released the pressure of the huge demand for surgical face masks and have them available for the front line helpers who needed it more than normal citizens.

To increase the credibility of these masks, the artisans also consulted with medical teams and introduced a few changes by adding holes in the nostrils and use three layers of cotton cloth instead of paper to act as a filter and make it more durable. The artists also innovated the item by making smaller ones and designing them as per mythological heroes for the children. The idea went viral on social media and pulled a lot of attention and investment in the same CITATION Mit20\l 16393 (Mitra, 2020).

Contribution of Intangible Aspects of Heritage in Building Resiliency – From the above discussions, it is evident that intangible aspects of heritage particularly play a versatile role in every stage of the disaster cycle. However, the focus on these aspects in general disaster management practices is insufficient. For the same reason, traditional skills, crafts, cultural practices, vernacular construction systems, etc., as well as historic locations, cultural landscapes, and living heritage components, are disappearing. This situation is harmful not just for the preservation of cultural assets but also because traditional knowledge, skills, and cultural practices have frequently helped people cope with disasters and recover from them, and their absence has frequently increased further *vulnerabilities* as we see in the case of traditional buildings that performed well in comparison to newer modern day buildings in the case of the 2001 and 2003 earthquake in Bhuj, India and Boumerdes, Algeria respectively (Jigyasu, 2002; EERI, 2003). The expansion of the definition of cultural heritage stresses that these intangible aspects despite not falling under neat categories must be recognised while taking protective measures (Haider, 2008; Jigyasu, 2005). Another very significant role that intangible attributes of heritage play is to contribute to one's social and psychological comfort. A visible presence of heritage within communities adds a sense of connection to both the past and present and reassures people that they are a part of something. People develop emotional connections around social interactions influenced by culture and its physical manifestation (Murakami, 2011). As seen in the case of the nuclear power plant accident in Futaba, Japan, in 2011, where the victims were relocated to Tsukuba City and the revival of their festivals and traditional dance form 'Nagareyama' helped them cope with the psychological distress that they were going through (Takakura, 2018). Similarly, the potential for traditional management systems to ensure community cooperation during disaster recovery is enormous. Following the 2015 Nepal earthquake, this was amply illustrated when networks of traditional 'Guthi' (community trusts) supported local communities as they moved from the response to the rehabilitation phase (Jigyasu, 2019).

From all the examples, we can conclude that intangible cultural heritage often acts as a binder for the community and its built heritage and as seen in the above case of Kolkata, risks on either of these attributes are directly proportional to the underlying vulnerabilities in either of the attributes. However, when narrowing down to disaster management practices, it has been observed that despite social aspects having made a mark in the conventional DRM discourses, there is a lack of implementation of the same in the field. While much of the literature and comments from heritage experts stress on the connections that exist between the tangible and intangible cultures, they also note that international and national protection efforts in disaster rehabilitation and reconstruction are still largely focused on preserving built heritage, for example, monuments.

Chitpur, like many other historic neighbourhoods around the world, has developed over the years in such a way that the buildings, the practices within, and the people associated with it are what give the space its current character. These spaces become a node of identity for almost all the citizens of the city, mainly for their historic relevance which is why all these values put together help them develop a psycho-social connection with the site. As seen in the case of Chitpur and many other similar cases that the practices prevalent in such historic neighbourhoods are vulnerable due to various factors and are prone to negative impacts by various hazards which may in a way lead to the degeneration or loss of the practices. Such historic practices being manifested within buildings, the built aspects contribute to these impacts directly and indirectly in the form of physical vulnerabilities. At times it has been noticed that the practice itself contributes indirectly by becoming a catalysing agent to the risk. The practices and their proponents, in most cases, develop a number of adaptations after years of exposure to certain hazards and vulnerabilities, but they might not be enough to cope in case of a catastrophic event and considering the economic condition of these practices, they may not be able to cope up at all.

Hence current disaster risk management procedures that take into account the intangible aspects of heritage should not only be a crucial component on which decisions shall depend but also be a vital chapter in the broader heritage management and contemporary developmental practices. If a disaster management plan is to be developed for a particular site, every acting factor, direct or indirect should be considered alongside relevant indicators for a holistic approach to reducing the different vulnerabilities. When it comes to recommendations it can also be stated that the proposals if implemented well can also act as steps that ensure the safeguarding of the intangible cultural heritage in the long run.

It is very important to identify the current and future needs for sustenance for such practices on site further shedding light on its economic relevance, as financial security plays a key role in developing community resilience. As exemplified by the Covid-19 outbreak, where the intangible sector suffered the most especially on the economic front. Hence apart from the different mitigation proposals made at structural, planning, and policy levels; empowering the artisans, craftspersons, and traders on financial grounds through training programmes would help reduce vulnerabilities and provide capacities to manage disasters. Thus, the economic sustainability of the intangible aspects of heritage needs to be brought to the centre as a right, essential for human development. It is for the same, the first step towards developing any plan is the identification of all the practices and their associated patrons and provide them with livelihoods, respect, dignity, and recognition. We cannot look at intangible aspects of heritage and its bearers in isolation, solely based on what sells, but try to identify the true essence of it and its relationship with the people around it and the environment. Culture has always been and will always be about its custodians from the beginning to the end. A people-based approach is therefore a definite need of the hour to break the wrong perception that dealing with heritage is an elitist affair. Looking from a broader point of view, social, cultural, and environmental factors constitute the pillars of sustainability and therefore looking at the culmination of all three aspects when developing proposals for the mitigation of risks to these practices is necessary. This process also demands the participation of a robust multi-disciplinary team and collaboration between governmental and non-governmental institutions. Intangible aspects of cultural heritage should therefore be considered a crucial resource in establishing linkages between disaster management proposals and sustainable development goals.

Key Points

- Disaster risk reduction should be looked beyond buildings with a multi-disciplinary team.
- Intangible heritage is equally affected by hazards and hence decision-making bodies should initiate efforts to reduce its vulnerabilities.
- The role of intangible cultural heritage is vital in disaster risk management, particularly in building community resilience, and hence must be included in broader heritage management plans.
- If opportunities are provided to safeguard intangible heritage practices holistically, it may help in retrieving financial stability, especially in post-disaster situations.
- People associated with intangible practices should not only be engaged in raising awareness but their capabilities must be mobilized in recovery processes as well.

References

Museum international. (2004). Intangible Heritage. (pp. 4–197). Blackwell Publishing. Retrieved from https://unesdoc.unesco.org/ark:/48223/pf0000135852
EERI. (2003). *The Boumerdes, Algeria, Earthquake of May 21, 2003*. Retrieved from https://www.eeri.org/site/images/lfe/pdf/algeria_20030521.pdf
Haider, H. (2008). *Helpdesk Research Report: Intangible Heritage and Post-Disaster Protection*. Governance and Social Development Resource Centre.
Jigyasu, R. (2002). *Using Traditional Knowledge to Reduce Disaster Vulnerability of Earthquake Prone Rural Settlements in India and Nepal*. Norwegian University of Science and Technology, Trondheim.
Jigyasu, Rohit (2005) Towards developing methodology for integrated risk management of cultural heritage sites and their settings. In: 15th ICOMOS General Assembly and International Symposium: 'Monuments and sites in their setting – conserving cultural heritage in changing townscapes and landscapes', 17–21 Oct 2005, Xi'an, China.
Jigyasu, R. (2019). *Does Cultural Heritage Make More Resilient Cities?* URBANET. Retrieved from https://www.urbanet.info/does-cultural-heritage-make-more-resilient-cities/
Kutoor, R. (2018). *Flood Steals Sparkle from Aranmula's Dream Mirror*. Kerala: *The Hindu*. Retrieved from https://www.thehindu.com/news/national/kerala/flood-steals-sparkle-from-aranmulas-dream-mirror/article24810888.ece
Maroyi, A. (2013). Traditional use of medicinal plants in south-central Zimbabwe: review and perspectives. *Journal of Ethnobiology and Ethnomedicine*. Retrieved from https://ethnobiomed.biomedcentral.com/articles/10.1186/1746-4269-9-31

Mitra, B. (2020). *Tribal Artisans Modify Chhau Masks to Fight Coronavirus. Times of India.* Retrieved from https://timesofindia.indiatimes.com/city/kolkata/tribal-artisans-modify-chhau-masks-to-fight-coronavirus/articleshow/75329926.cms

Murakami, Y. (2011). Disaster Risk Management of Cultural Heritage Based on the Experience of the Great Hanshin Earthquake. *Cultural Heritage Protection Cooperation Office, Asia-Pacific Cultural Centre for UNESCO (ACCU Nara).*

Raghunath, A. (2019). *A Strong Revival. Deccan Herald.* Retrieved from https://www.deccanherald.com/sunday-herald/sunday-herald-articulations/a-strong-revival-724653.html

Smith, L.J. (2006). *The Uses of Heritage.* London: Routledge.

Takakura, Hiroki. (2019). The Role of Intangible Cultural Heritage in the Disaster Recovery in Fukushima; https://www.researchgate.net/publication/332142991_The_role_of_intangible_cultural_heritage_in_the_disaster_recovery_in_Fukushima

3 A New Approach to Cultural Heritage and Disaster Risk Reduction

A Review of International Policies

Giovanni Boccardi

Introduction

This book is based on the assumption that to fully understand cultural heritage and its role in disaster risk reduction (DRR), it is necessary to adopt an interdisciplinary and holistic approach that would integrate the principles and practices of both areas. This is because heritage – as demonstrated in Chapter 10 – is not just a liability in the face of disasters, which needs to be protected in itself, but also a potential resource to strengthen resilience and reduce disaster risks for communities in general. In other words, in addition to devising the most appropriate strategies to protect heritage from risks associated to potential hazards, it is necessary to also understand, and harness, the multiple ways in which a well-preserved heritage, both tangible and intangible, can contribute to reducing the vulnerability of assets and people within our historic environments.

The twofold nature of this relationship, unfortunately, is not always understood within the two sectors (UNESCO et al., 2010, page 2). On the one hand, disaster risks tend to be neglected by heritage organizations, who give priority to other issues that are perceived – often wrongly – as more visible and urgent. On the other hand, many disaster risk policies and plans tend to overlook heritage in their strategies and operations, considering it as a secondary concern. There seems to be a lack of awareness of the mutual relevance of heritage and DRR, which is reflected in a certain disconnect between the two communities, especially at national and local levels.

Indeed, one of the main challenges facing heritage practitioners wishing to strengthen disaster risk management at their properties, even beyond the availability of resources and skills, is to garner the support of policy- and decision-makers, from both sectors, and build the institutional coordination mechanisms that are required to make any strategy and operational plan truly effective.

What can be done, thus, to help 'break' the indifference and have the link between heritage and DRR finally recognized as a priority?

An important reference and resource, in this respect, can be found within a large set of recent – and less recent – international policies, spanning from human rights and sustainable development to DRR and cultural heritage itself. These policies are not necessarily known by the average heritage or DRR practitioners, who often are only familiar with the legal and administrative processes related to their sector of activity. When reading them attentively, however, we realise that – taken together – they contain, at a fundamental level, the key concepts and ideas that are necessary to effectively advocate for the integration of cultural heritage in DRR.

The chapter aims therefore at reviewing such policies and exploring the different ways in which these may help lay a solid foundation in support of a new approach to cultural heritage and DRR. The reasoning behind this is that since States have agreed to and ratified such policies within their respective intergovernmental fora, and more importantly, they have committed to applying them, then it is important to draw attention to them, harness their potential and use them, to the greatest possible extent, to strengthen the arguments in support of our cause.

Human Rights, Sustainable Development and Climate Change Policies

At a fundamental level, the relevance of cultural heritage for DRR may be implicitly inferred by various international policies concerning human rights and sustainable development, such as *the Universal Declaration of Human Rights (1948)*,[1] the *International Covenant on Economic, Social and Cultural Rights (1966)*,[2] the *United Nations Declaration on the Rights of Indigenous Peoples (2007)*[3] and, more recently, the *2030 Agenda for Sustainable Development (2015)*,[4] among others.

The intrinsic importance of cultural heritage, and hence the need to safeguard it, including from disasters, is affirmed first of all within Human Rights normative texts and policies. Special provisions concern, of course, cultural rights, which are considered as integral to fundamental Human Rights in general.

Article 27/1 of the *Universal Declaration of Human Rights* reads as follows:

Everyone has the right freely to participate in the cultural life of the community, to enjoy the arts and to share in scientific advancement and its benefits.

Article 15 of the *International Covenant* reiterates the right of everyone to take part in cultural life and, in its second paragraph, clarifies that:

The steps to be taken by the States Parties to the present Covenant to achieve the full realization of this right shall include those necessary for the conservation, the development and the diffusion of science and culture.

Although a single definition of cultural rights is not provided, according to the Office of the High Commissioner for Human Rights (OHCHR), these are deemed to 'protect the rights for each person, individually and in community with others, as well as groups of people, to develop and express their humanity, their world view and the meanings they give to their existence and their development through, inter alia, values, beliefs, convictions, languages, knowledge and the arts, institutions and ways of life. They are also considered as protecting access to cultural heritage and resources that allow such identification and development processes to take place'.[5] Protecting cultural rights, therefore, implies preserving cultural heritage and making it accessible to people, in particular when they are facing an exceptional and traumatic situation as it may be the case in the aftermath of a disaster.

The *United Nations Declaration on the Rights of Indigenous Peoples*, in particular, emphasises the importance for Indigenous Peoples of maintaining their traditional ways of life, and, notably in its Art. 31, their right to 'maintain, control, protect and develop their *cultural heritage, traditional knowledge and traditional cultural expressions* ...' (emphasis added), as well as to protect the associated intellectual property of such knowledge.

In their reports since 2009, the last two UN Special Rapporteurs in the field of cultural rights (Ms Farida Shaheed and Dr. Karima Bennoune) have further explored the role of cultural heritage as a critical component for the realization of Human Rights,[6] notably in situations of emergency. Starting from 2016, the reports have focused on the issue of the intentional destruction of cultural heritage in connection with the conflicts affecting the Middle East and Sub-Saharan

Africa. In 2020 and 2021, respectively, the Special Rapporteur examined the questions of climate change and of the Covid-19 pandemics and their effects on the ability of people to access and enjoy their cultural heritage. The work of the UN Special Rapporteurs has made a fundamental contribution to clarifying the connection between cultural heritage and Human Rights, particularly when communities are under stress, giving substance to the principles expressed by the Universal Declaration of Human Rights and the International Covenant on Economic, Social and Cultural Rights and providing, at the same time, essential policy recommendations to Member States.

Turning our attention from Human Rights to Sustainable development, the *2030 Agenda for Sustainable Development,* adopted in 2015 by the UN General Assembly, acknowledges the important role of culture across many of the Sustainable Development Goals (SDGs), with heritage directly addressed in Goal 11. Target 4, under this SDG, refers explicitly to the need to '*strengthen efforts to protect and safeguard cultural and natural heritage*', notably because of its importance in fostering inclusive, safe, resilient, and sustainable cities and human settlements.

Beyond this recognition of the intrinsic value of cultural heritage, these policies also acknowledge the instrumental role that heritage can play as an enabler and a driver of human rights and sustainable development in general, which of course, are also relevant for DRR.

All Human Rights and Sustainable Development policy instruments identify full inclusion, respect, and equity as critical elements for the quality of life and wellbeing of individuals and groups. Because cultural heritage is, by definition, something very close to people's identities, at the heart of their concerns and deeply imbued with emotions, protecting and integrating it in sustainable development efforts is in itself a way of fostering ownership and participation, acknowledging rights and ultimately empowering local communities and indigenous groups in relation to essential decisions concerning critical aspects of their lives. In this regard, cultural heritage plays a key enabling role for the promotion of Human Rights and Sustainable Development. This, of course, is especially true when communities face difficult times in their life, for example, during emergencies associated with disasters or conflicts.

But cultural heritage can also make a *direct* contribution to various other SDGs, some of which are particularly relevant for DRR. Despite the fact that heritage is not specifically mentioned by other SDGs and the related targets (something that many have regretted), its implicit role has been reaffirmed by several resolutions by the United Nations General Assembly and elicited in a number of publications.

Since 2010, indeed, the UN General Assembly has adopted a series of important Resolutions emphasizing the importance of Culture for Sustainable Development and encouraging all Member States to develop and apply appropriate national policies and operational strategies to harness such potential.[7] Many of these Resolutions draw attention to the significant role of culture, and heritage, to foster resilience and to the need to protect heritage from disasters. The most recent of such Resolutions, UN GA Res. 76/214, adopted on 17 December 2021,[8] reaffirms:

> Paragraph 5 (c). That culture contributes to environmental sustainability, since the protection of cultural and biological diversity and natural heritage is important for sustainable development, and that support for traditional systems of environmental protection, and resource management, can contribute to the increased sustainability of fragile ecosystems and the conservation, preservation and sustainable use of biodiversity and to avoiding land degradation and addressing climate change.

The Resolution also calls on Member States:

> Paragraph 17 (i). To accelerate efforts to protect cultural and natural heritage from extreme weather events, sea level rise, desertification and other threats exacerbated by climate change, which jeopardize its integrity and preservation for present and future generations.

In its 2019 publication 'Culture/2030 Indicators' (UNESCO, 2019), moreover, UNESCO claimed that culture (and heritage) makes, in fact, a meaningful contribution to most of the 17 SDGs agreed under the Agenda 2030, and identifies a series of possible indicators to measure this. The indicators are arranged by UNESCO around four thematic dimensions, the first of which is 'environment and resilience'. Under this dimension, a series of indicators are put forward based on the assumption that natural and cultural heritage, both tangible and intangible, can help mitigating disaster risks, including those associated with climate change, support resilience and enhance the adaptation capacities of communities.

More recently, the International Committee on Monuments and Sites (ICOMOS) published a 'Policy Guidance for Heritage and Development Actors' (ICOMOS, 2021) to demonstrate how cultural heritage may contribute to all the UN SDGs and to assist heritage and development practitioners in harnessing this potential. The document contains 17 sections (1 for each SDG). For six of these (notably 1, 4, 11, 13, 15, and 16), a clear link between heritage and disasters is identified, and case studies are presented to illustrate this relationship.

Another important policy area having a strong connection with heritage and DRR is climate change. Under the *United Nations Framework Convention on Climate Change,* a fundamental breakthrough was the adoption, in 2015, of the so-called *Paris Agreement.*[9] Climate change, of course, is considered one of the key driving factors in a growing number of disasters worldwide, notably those associated with extreme meteorological events. Through its long-term effects and their consequences on physical and socio-economic aspects of our environment, moreover, climate change is responsible for the increased vulnerability of many structures and sites in the face of hazards. The *Paris Agreement* contains an important reference to traditional knowledge, a key component of intangible cultural heritage, as follows:

> Art. 7, paragraph 5 – Parties acknowledge that adaptation action should follow a country-driven, gender-responsive, participatory and fully transparent approach, taking into consideration vulnerable groups, communities and ecosystems, and should be based on and guided by the best available science and, as appropriate, traditional knowledge, knowledge of indigenous peoples and local knowledge systems, with a view to integrating adaptation into relevant socioeconomic and environmental policies and actions, where appropriate.

Although they do not focus specifically on heritage and disaster risks, therefore, international policies in the areas of Human Rights, Sustainable Development and Climate Change make a convincing case for protecting cultural heritage, including from disasters, and acknowledge its inherent value as a resource for strengthening the resilience of our societies.

The Sendai Framework for Disaster Risk Reduction and Other DRR Policies and Tools

An obvious place to look when trying to find references to cultural heritage in relation to disaster risks is, of course, the *Sendai Framework for Disaster Risk Reduction* (SFDRR),[10] adopted by the UN General Assembly in 2015. As the most important international policy document in this area, the Sendai Framework is meant to orient and monitor efforts by Member States to reduce disaster risks until 2030.

A certain recognition of the importance of culture and heritage in DRR was contained already within the predecessor of the Sendai Framework, i.e. the *Hyogo Framework for Action (HFA)*,[11] adopted in 2005, which called for the protection of 'culturally important lands and structures' (Paragraph 4, (ii), (f)).

The *Sendai Framework*, however, represents a step forward compared to the HFA since it integrates consideration for culture (and heritage), side by side with other major societal concerns (e.g. economic, social, health, and environmental), throughout the entire policy document. References to cultural assets and measures, moreover, are more systematically and organically integrated within the *Sendai Framework*, from its guiding principles to the four main 'priorities for action', with cultural heritage specifically mentioned under main priorities 1, on 'Understanding disaster risks', and 3, which is about 'Investing in disaster risk reduction for resilience'.

In relation to Priority 1, Paragraph 24 (d) of the SFDRR, which concerns national and local level policies, calls on Member States:

> To systematically evaluate, record, share and publicly account for disaster losses and understand the economic, social, health, education, environmental and cultural heritage impacts, as appropriate, in the context of event-specific hazard-exposure and vulnerability information.

Under priority 3, Paragraph 30 (d) of the SFDRR calls on Member States:

> To protect or support the protection of cultural and collecting institutions and other sites of historical, cultural heritage and religious interest.

Traditional knowledge, a critical component of intangible cultural heritage, is referred to explicitly in two provisions of the SFDRR. Paragraph 24 (i) urges Member States:

> To ensure the use of traditional, indigenous and local knowledge and practices, as appropriate, to complement scientific knowledge in disaster risk assessment and the development and implementation of policies, strategies, plans, and programs of specific sectors, with a cross-sectoral approach, which should be tailored to localities and to the context.

Paragraph 36 (a) (v), on the other hand, notes how:

> Indigenous peoples, through their experience and traditional knowledge, provide an important contribution to the development and implementation of plans and mechanisms, including for early warning.

In addition to integrating a concern for culture as a sector in its own right, moreover, the SFDRR acknowledges, first of all in its guiding principles (paragraph 19 (d)), the need to apply a culturally sensitive approach to DRR across all sectors, as the HFA had already done, drawing on existing knowledge.

These advances, albeit regarded as still inadequate by some, were the product of a renewed understanding of the multiple ways in which 'culture', in the wider anthropological sense of the term, may affect how societies perceive and prepare for disaster risks, and respond to actual disasters. An important contribution, in this regard, was made by the 2014 *World*

Disaster Report (IFRC, 2014), issued by the *International Federation of the Red Cross Societies* (IFRC). This publication explained very well the critical importance of putting culture at the centre of DRR planning and implementation through a compelling analysis of numerous case studies.

A concrete – and practical – result of the new understanding of the importance of culture in DRR was the addition, in 2017, of a specific chapter on culture (including cultural heritage) as part of the interagency *Post-Disaster Needs Assessment* (PDNA), a government-led process supported by the United Nations system in cooperation with the World Bank and the European Union (GFDRR et al., 2017). The PDNA is not a policy in the strict sense of the term, but it is nevertheless very important as it introduces a new methodological approach to assessing the effects and impacts of a disaster on the culture sector and planning for recovery, which reflects a holistic understanding of the complex interrelation between heritage and disaster risks.

The PDNA methodology, indeed, implies an assessment of the effects of a disaster – for each sector – against four different dimensions: physical assets; the provision of goods and services; governance; and new risks and vulnerabilities. It also involves an assessment of the broader human development and macroeconomic impact of the disaster on the society, in relation to the culture sector. Similarly, as it concerns post-disaster recovery planning, the PDNA is not limited to the rehabilitation of the material attributes of affected cultural heritage but envisages a comprehensive strategy to restore all the connections between heritage and society across the social, economic, and environmental dimensions, incorporating moreover the concept of 'building-back better'.

This represents a significant innovation compared to traditional approaches to post-disaster assessment of cultural heritage, still prevalent within the heritage sector, which tend to focus exclusively on physical damage to cultural heritage properties and, occasionally, on the related cost for restoration.

At the time of writing, only a dozen PDNA Culture exercises had been undertaken, all within developing countries, under the coordination of UNESCO and in close collaboration with the concerned national authorities and other Intergovernmental Agencies. This is due to the fact that, so far, PDNAs can only take place at the request of the Governments of countries affected by major disasters in the context of international humanitarian and aid programmes. The methodology of the PDNA, however, has relevance in its own right and could be usefully applied to any post-disaster assessment where cultural heritage is at stake, including outside a formal PDNA process.

Cultural Heritage Policies

The last set of international policies that will be examined in this chapter are those from within the culture sector itself, particularly those developed by UNESCO.

A concern for disasters caused by natural and human-caused hazards is expressed in general terms within many of UNESCO's cultural Conventions, including the 1954 *Hague Convention for the Protection of Cultural Property in the Event of Armed Conflict*,[12] the 1970 *Convention on the Means of Prohibiting and Preventing the Illicit Import, Export and Transfer of Ownership of Cultural Property*,[13] the 1972 *Convention Concerning the Protection of the World Natural and Cultural Heritage*,[14] and the 2003 *Convention on the Safeguarding of Intangible Cultural Heritage*.[15] All these international agreements have developed over the years specific provisions and recommendations addressing, in particular, the issue of disasters.

Among these normative instruments, the above-mentioned 1972 *World Heritage Convention* is undoubtedly the one which has produced the largest number of technical guidance and policy documents. A *Strategy for Reducing Disaster Risks at World Heritage Properties*[16] was adopted in 2007 by the World Heritage Committee (the governing body of the 1972 Convention), which laid out the policy implications of the 2005 *Hyogo Framework for Action* for the tangible heritage sector. A resource manual on *Managing Disaster Risks for World Heritage* was later published in 2010 (UNESCO et al., 2010), and a number of resources were developed to assist Member States in implementing the 2007 Strategy.[17] The resource manual is currently under review to reflect advances in policies since 2010. International organizations concerned with cultural heritage, such as the *International Centre for the Study of the Preservation and Restoration of Cultural Property* (ICCROM), ICOMOS or the *International Council of Museums* (ICOM), have also addressed the question of disaster risks and produced several guidelines and capacity building tools.

As far as intangible cultural heritage is concerned, the 2003 *Convention for the Safeguarding of the Intangible Cultural Heritage* sets out, in Chapter VI of its *Operational Directives*,[18] the principles and measures to be implemented by State Parties at the national level, both to harness the contribution of intangible cultural heritage (ICH) for DRR and community-based resilience to natural disasters, as well as to mitigate the impact of natural disasters on ICH practices and their transmission.

Despite the wide ratification of all these Conventions, the silo approach inherent in their scope (each dealing with a specific typology of cultural heritage or with a specific kind of hazard) and implementation mechanisms had shown serious limitations at the operational level particularly in the context of major disasters. Especially following the devastating impact of the December 2004 South Asian tsunami and other major catastrophic events, indeed, it became apparent that the sectorial approach of the individual Conventions was no longer adequate to address the challenges posed by disasters to culture and heritage. If culture and heritage were to be truly considered essential components of human rights and resilience and key instruments for recovery and sustainable development, then there had to be a way to integrate them, regardless of typological distinctions, into international and national DRR and humanitarian policy and operational frameworks.

Within UNESCO, therefore, a reflection was launched aimed at developing a more integrated approach to dealing with cultural emergencies, including at the policy level.

Following the social unrest, political changes, and major upheavals that resulted from the so-called 'Arab Spring' in the Middle East, the attention was initially placed on emergencies associated with conflicts. Confronted by the major devastations, intentional destructions, and looting affecting many heritage properties and Museums in countries such as Syria, Iraq, Mali, Libya, and Yemen, the Member States of UNESCO adopted, in November 2015, a '*Strategy for Reinforcing UNESCO's Action for the Protection of Culture and the Promotion of Cultural Pluralism in the Event of Armed Conflict*'.[19] This document is important because, for the first time, it considers culture as a whole, including heritage and creativity, in relation to a specific challenge, that of armed conflicts. It is also innovative in introducing, as one of its two stated objectives, the incorporation of 'the protection of culture into humanitarian action, security strategies and peacebuilding processes by engaging with relevant stakeholders outside the culture domain'.

Two years later, in 2017, the same Member States adopted an *Addendum*[20] to the above-mentioned Strategy, of which it became an integral component, focusing, however, on disasters caused by natural and human-induced hazards. In its introduction, the *Addendum* notes the '*growing appreciation of the dual role of culture in disasters – on the one hand as key consideration in risk prevention, and on the other hand as a contributing factor in enhancing resilience*'. In line with the main Strategy of which it had become an integral component, the *Addendum* identifies two main objectives:

1. Strengthen the ability of Member States to prevent, mitigate and recover the loss of cultural heritage and diversity as a result of disasters caused by natural and human-induced hazards. This will be done primarily through capacity building initiatives, as well as support to preparedness, response, and recovery;
2. Incorporate consideration for culture into the DRR sector and humanitarian action related to disasters by engaging with the relevant stakeholders outside the cultural domain. This will involve the development of partnerships and tools, and engaging with UN-wide processes to encourage a culturally sensitive approach to DRR, which would draw on culture to strengthen resilience in the face of disasters.

The document also defines specific provisions to assist Member States in implementing the cultural dimension of the Sendai Framework, across its four priority areas of action, with multiple references to heritage. An Action Plan[21] was developed by UNESCO to implement the Strategy and its Addendum, with an initial life span of six years, which laid the ground for numerous programmes and projects carried out by UNESCO and its partners over the past years.

The integration of an 'addendum' on disasters to a strategy initially dealing with armed conflict should not surprise us. First of all, recent research (Peters and Budimir, 2016) has shown how conflicts and disasters have mutually reinforcing effects since they aggravate vulnerabilities that play a critical role in both occurrences. Secondly, within the culture sector, policies and guidelines related to disaster risks are often similar to those recommended for emergencies associated with armed conflict or social unrest. Classic measures to protect heritage, such as continuous maintenance, the drawing of inventories, the construction of safe storages, and training on first aid procedures, have relevance for both conflict and disaster scenarios. The resilience that comes from a well-looked-after cultural heritage, moreover, will support communities in all kinds of traumatic situations and help them bounce back and recover.

In fact, it has been argued (Hortolf, 2018) that a key component of resilience is precisely the ability of humans to reassess their priorities and values in the face of irreversible losses and 'culturally' adapt to a new, unpredictable scenario, for example through the recovery and reconstruction of their heritage following a major traumatic event.

This concept, which goes counter the traditional view that any heritage loss is by definition 'irreplaceable', has been incorporated within another recent international policy document, the '*Warsaw Recommendation on Recovery and Reconstruction of Cultural Heritage*',[22] adopted initially at an International Conference organized by UNESCO in Warsaw (2018) and then welcomed by the World Heritage Committee at its 42nd Session in Manama (Bahrain, 2018). This lays out a set of principles to guide post-disaster and post-conflict recovery and reconstruction of cultural heritage. The *Warsaw Recommendation* is important and innovative in that it marked a shift from what was perceived as an excessive focus on maintaining the authenticity of the material attributes of cultural heritage properties affected by a disaster, which in most cases precluded to communities affected by a disaster or conflict the possibility of reconstructing their lost cultural heritage assets. The Recommendation embraces instead a broader approach taking into account the role of heritage as a key developmental, security, and humanitarian issue, and recognizing:

> Paragraph 6 – ... the legitimate aspiration of concerned communities to overcome the trauma of conflicts, war and disasters by reconstructing as soon as possible their cities and villages – and particularly their affected cultural heritage – as a means to reaffirm their identity, restore their dignity and lay the conditions for a sustainable social and economic recovery.

More significantly, the Warsaw Recommendation acknowledges that:

> Paragraph 9 – ... each generation has the right to contribute to human legacy and to the wellbeing of present and future generations, including through adaptation to natural and historic processes of change and transformation.

Another important development, following the adoption of the above-mentioned Strategy and Addendum, is the attempt – at the international level – to integrate consideration for cultural heritage already within the early humanitarian phase of post-disaster interventions, including immediately after an event, during search and rescue operations. With this objective in mind, UNESCO and ICCROM have been cooperating over the past years with the Secretariat of the International Search and Rescue Advisory Group (INSARAG), part of the United Nations Office for the Coordination of Humanitarian Affairs (OCHA), in order to include a section on cultural heritage within the INSARAG Guidelines (currently under finalization).

ICCROM, moreover, has included disaster risk management of cultural heritage in situations of conflicts, disasters, and complex emergencies as one of the objectives of its first strategic direction for 2018–2025 titled 'Focusing on World Concerns for Cultural Heritage'.[23] Several activities include annual international training course on disaster risk management of cultural heritage, carried out in collaboration with Ritsumeikan University, Kyoto, since 2006 as well as regional training courses in Southeast Asia, Latin America, and Africa, publication of guidelines and methodological kists. Also as part of ICCROM's Flagship Programme called 'First Aid and Resilience for Cultural Heritage in Times of Crisis (FAR)',[24] heritage practitioners and first-aid respondents from the humanitarian and security sectors have been brought together for integrating heritage both as a concern and as a resource in the crucial days following a traumatic event.

Beyond policies developed within the perimeter of its cultural sector, finally, UNESCO has promoted the establishment of a number of other cross-cutting policies on key development issues relevant to DRR, drawing the necessary implications across its wider mandate (Education, Sciences, Social Sciences, Culture, and Communication). These include, to cite only the most recent, a *UNESCO Strategy for Action on Climate Change*[25] (2017) and a UNESCO *Policy on Engaging with Indigenous Peoples*[26] (2018), the latter containing a dedicated section on DRR. All of these comprehensive UNESCO policy documents refer to culture, heritage, and traditional knowledge (sometimes called 'local and indigenous knowledge') extensively.

Conclusions

In conclusion, a review of the nexus between cultural heritage and DRR within international policies on a wide spectrum of issues shows a growing awareness of the importance of heritage as a precious resource to be protected, both in its own right and for its possible contribution to resilience and recovery. Taken all together, in fact, these policy documents provide a strong foundation for leveraging heritage as a resource for DRR.

In many ways, one might argue that the importance of culture in development (and thus also in DRR as one of its most critical aspects) has never been more discussed and acknowledged, compared to previous decades, within development circles and government policies. Cultural heritage, as the essence of the cultural identities and world views of many peoples around the world, in all their diversity, has definitely been part of this discussion.

The implementation at national and local levels of these policies, however, appears to be still very limited. In part this is due to a lack of understanding of the issues. Another major obstacle is the inherent difficulty of bringing together organizations and people from different sectors to work within a common legal, administrative, financial, and technical framework. Addressing this challenge requires the raising of awareness and the building of capacities among policy- and decision-makers, practitioners, and civil society.

What heritage and DRR policymakers and managers need, in this context, is clear practical guidance and training opportunities – supported by compelling examples and tools – that would help them moving from words to action in implementing the Sendai Framework. Advocacy is also very important, if not preconditional, to create awareness and consensus around the need to do more towards the integration of cultural heritage within DRR policies and operations, and vice versa.

The strongest case for leveraging culture and heritage in DRR can be made when considering together all the policies and recommendations, combining sector-specific and cross-cutting themes, as a sort of large 'corpus' of normative texts. Too often, indeed, those working in a certain sector are only familiar with, and feel themselves committed to, the policies and related goals established for their specific area of work. Nowadays, however, the inextricable connections and interactions among many different issues and processes force us to take into account a much broader spectrum of considerations. At the end of the day, it is the same Member States that adopt all these policies on different, but closely interrelated issues, and it should be, therefore, their responsibility to implement them all in a coordinated fashion.

Key Points

- Heritage, tangible and intangible, is both a valuable asset that deserves to be protected from the impact of disasters, and a precious resource that can considerably strengthen the resilience of communities in disaster prevention, response, and recovery.
- Cultural heritage and disaster risk reduction, as sector of activities, are therefore highly relevant to each other's mandate but are unfortunately not enough integrated at institutional and implementation levels.
- International policies related to sustainable development, humanitarian aid, disaster risk reduction, and heritage protection recognize this critical connection and offer valuable guidance to address this challenge.
- By familiarizing with and implementing such international policies, thus, decision-makers and practitioners may seize opportunities for synergies among sectoral institutions and significantly enhance the effectiveness of their work.

Notes

1 Accessible from https://www.un.org/en/udhrbook/pdf/udhr_booklet_en_web.pdf
2 Accessible from https://www.ohchr.org/en/professionalinterest/pages/cescr.aspx

3 Accessible from https://www.un.org/development/desa/indigenouspeoples/wp-content/uploads/sites/19/2018/11/UNDRIP_E_web.pdf
4 Accessible at https://sustainabledevelopment.un.org/post2015/transformingourworld
5 From the website of OHCHR, accessible from https://www.ohchr.org/en/issues/culturalrights/pages/internationalstandards.aspx (last accessed on 13 November 2020).
6 These reports are accessible from https://www.ohchr.org/EN/Issues/CulturalRights/Pages/Annual-Reports.aspx
7 UNESCO has compiled all these important resolutions in a single web page, accessible from https://en.unesco.org/culture-development/public-policies (last accessed on 27 January 2022).
8 Accessible from https://undocs.org/en/A/RES/76/214
9 Accessible from https://unfccc.int/process-and-meetings/the-paris-agreement/the-paris-agreement
10 Accessible from https://www.preventionweb.net/files/43291_sendaiframeworkfordrren.pdf
11 Accessible from https://www.preventionweb.net/sendai-framework/hyogo/
12 Accessible from http://www.unesco.org/new/en/culture/themes/armed-conflict-and-heritage/convention-and-protocols/1954-hague-convention/
13 Accessible from https://en.unesco.org/fighttrafficking/1970
14 Accessible from https://whc.unesco.org/archive/convention-en.pdf
15 Accessible from https://ich.unesco.org/en/convention
16 Accessible from https://whc.unesco.org/archive/2007/whc07-31com-72e.pdf
17 See the relevant page from the World Heritage Centre's website: https://whc.unesco.org/en/disaster-risk-reduction/
18 Accessible from http://www.unesco.org/culture/ich/en/Directives/6.GA/170
19 Accessible from https://en.unesco.org/system/files/unesco_clt_strategy_en.pdf
20 Accessible from https://unesdoc.unesco.org/ark:/48223/pf0000259805
21 Accessible from https://unesdoc.unesco.org/ark:/48223/pf0000247706
22 Accessible from https://whc.unesco.org/en/news/1826
23 https://www.iccrom.org/iccrom-strategic-directions-and-objectives
24 See: https://www.iccrom.org/section/disaster-resilient-heritage/first-aid-and-resilience-cultural-heritage-times-crisis-far
25 Accessible from https://unesdoc.unesco.org/ark:/48223/pf0000259255
26 Accessible from https://unesdoc.unesco.org/ark:/48223/pf0000262748

References

GFDRR, UNDP, EU, World Bank, 2017. *Post-Disaster Needs Assessments Guidelines: Volume B – Culture*. Published online and accessible from: https://recovery.preventionweb.net/publication/post-disaster-needs-assessments-guidelines-volume-b-culture (last accessed on 09 September 2022).

Hortolf C., 2018. *Embracing change: How cultural resilience is increased through cultural heritage*. World Archaeology, 50:4. Taylor and Francis, Abingdon, Oxon.

ICOMOS, 2021. *Heritage and the sustainable development goals: Policy guidance for heritage and development actors*. Paris. Accessible online from: https://openarchive.icomos.org/id/eprint/2453/ (last accessed on 09 September 2022).

IFRC, 2014. *World disasters report: Focus on culture and risks*. Geneva. Accessible online from: https://www.ifrc.org/document/world-disasters-report-2014 (last accessed on 09 September 2022).

Peters K., and Budimir M., 2016, 'When disasters and conflict collide: Facts and figures', Overseas Development Institute, London. The publication is accessible online from: https://odi.org/en/publications/when-disasters-and-conflicts-collide-facts-and-figures/

UNESCO, 2019. *Culture 2030/Indicators*. Paris, Accessible online from: https://unesdoc.unesco.org/ark:/48223/pf0000371562 (last accessed on 09 September 2022).

UNESCO, ICOMOS, ICCROM and IUCN, 2010. *Managing disaster risks for world heritage*. World Heritage Resource Manual Series, Paris. Accessible online from: https://whc.unesco.org/en/managing-disaster-risks/

4 Financing Disaster Risk Management for Cultural Heritage

Barbara Minguez Garcia

Introduction – the Double Challenge of Finding Finance for DRM-CH

The development of **disaster risk management (DRM)** as a discipline and a sector is quite recent. Awareness on the concept of 'natural' disasters – today surpassed with the international agreement to stop considering disasters as natural events (Chmutina and Von Meding, 2019; Mena, 2020; Mizutori, 2020; UNDRR, 2021) – started developing during the 70s–80s, and the period 1990–1999 was called *The International Decade for Natural Disaster Reduction* (UNDRR, 2022). Since then, it has kept evolving from an initial focus on just crisis response and relief, to a whole process studying the fundamentals of disaster risk, as interaction of natural hazards, exposure, vulnerability, and capacities, and with a focus on disaster risk reduction (DRR), improving preparedness and ensuring faster and more efficient emergency response. Likewise, recovery processes have evolved with new concepts such as *building back better* (Hallegatte et al., 2018; UNDRR, 2022a; UNISDR, 2017) to increase resilience of people and assets.

Institutionally, the initial 1994 United Nations (UN) World Conference on Disaster Risk Reduction (WCDRR) was followed by the 2005 conference in Kobe – which adopted the Hyogo Framework for Action 2005–2015 – and 2015 one in Sendai, which established the current Sendai Framework for Disaster Risk Reduction 2015–2030 (UNDRR, 2015). In its 3rd priority of action, the Sendai Framework highlights the need of investing in DRR for resilience, defending that 'public and private investment in disaster risk prevention and reduction through structural and non-structural measures are essential to enhance the economic, social, health and cultural resilience of persons, communities, countries and their assets, as well as the environment' (UNDRR, 2015, p. 18).

However, investing in DRM remains a challenge in several countries, although the understanding and agreements on the effects of climate change are contributing to the validation and reinforcement of the DRM sector.

The Disaster Risk Finance sub-sector is indeed growing fast to address the fiscal impacts and economic losses caused by natural hazards, while supporting countries to increase their financial resilience to disasters (Financial Protection Forum, 2015). Programs such as the World Bank Disaster Risk Financing and Insurance (DRFI) (World Bank, 2022) help governments to develop and implement comprehensive financial protection strategies.

Indeed, according to the Organisation for Economic Co-operation and Development (OECD), 'a wide range of approaches to the financial management of disaster risks have been implemented across economies, reflecting differing levels of disaster risk and economic development. However, a number of common challenges were identified across economies which suggests the need for further investment in developing comprehensive approaches to disaster risk financing' (OECD, 2015, p. 12).

DOI: 10.4324/9781003293019-6

In 2019, the Deutsche Gesellschaft für Internationale Zusammenarbeit (GIZ)[1] launched a toolkit targeting policymakers responsible for DRM to guide them through practical steps on how to identify and choose the best disaster risk finance instrument for different circumstances (Meenan et al., 2019). Several models and schemes for disaster risk financing and insurance are also presented in Clarke et al. (2015), and a recent publication by the UN University Institute for Environment and Human Security (UNU-EHS, 2021) provides a brief overview of climate and disaster risk financing instruments currently available.

The protection and conservation of **cultural heritage (CH)** face their own challenges in terms of financing worldwide. Culture is rarely a priority in national budgets, maybe because its implications are not usually understood in its whole potential. However, the United Nations recognised in 2015 the role of culture as an enabler and driver for sustainable development, mentioning culture in the preamble of the 2030 Agenda and in several Sustainable Development Goals (SDGs) targets. In this regard, the UNESCO Institute for Statistics (UIS) developed an indicator for the 'total per capita expenditure on the preservation, protection, and conservation of all cultural and natural heritage, by source of funding (public, private), type of heritage (cultural, natural) and level of government (national, regional and local/municipal)' (UIS, 2021, p. 4).

CH in all its forms (see Chapter 2) – including tangible, movable and immovable, and intangible[2] – is key to societies, although it still seems difficult to justify in budgets. Similar to DRM, the economics of heritage is also quite recent, appearing in the framework of UNESCO discussions on the links between cultural policy and economic development in the '60s and '70s, and consolidating in the '90s through the application of the theory and practices of economics to the analysis of heritage decisions (Throsby, 2012). Likewise, the establishment of the World Heritage Convention (WHC, 2022) served to make countries realise the need of allocating resources to develop and implement measures for heritage protection.

The financing of CH conservation and management, particularly in reference to iconic CH assets and World Heritage sites, tends to be a stand-alone exercise linked to governmental agencies at various levels, which usually set priorities and budgets and may be joined by philanthropic sources, such as donors and sponsors (Burnham, 2019). In this context, building over the base of a new generation of entrepreneurs and philanthropists, the **Cultural Heritage Finance Alliance** (CHiFA, no date) was created, aiming to build a global marketplace to open opportunities for capital investment in historic places. It offers different financial products and loans, seeking to empower local partners, establish frameworks for CH management sustainability, and connect physical conservation with social and cultural revitalisation (Burnham, 2019).

Still, showing the economic benefits of investing in culture is not easy, particularly when competing with other sectors. However, a World Bank publication (Licciardi and Amirtahmasebi, 2012) showed how investments in heritage have a positive return and contribute to urban liveability, attracting talent and providing an enabling environment for job creation. Indeed, multilateral organisations have also supported CH projects, such as the World Bank in Lebanon (World Bank, 2019) and Jordan (World Bank, 2020); and the European Investment Bank (EIB Institute, 2022) financing CH through investment loans, framework loans and funds, also funding ancillary services to culture and CH activities, such as infrastructural access and mobility to sites (European Commission, 2021).

The role – and power – of culture and CH in managing disasters and creating resilient communities is an even more recent acknowledgement, which still seeks to be broadened and accepted in order to make **disaster risk management for cultural heritage (DRM-CH)** become a well-recognised sector to be included into national and local budgets worldwide, as well as fully supported internationally.

The **Sendai Framework** includes some important references to culture and heritage; this is salient for the DRM-CH scope. Actually, in the context of the WCDRR held in Sendai on 14–18 March 2015, a special session on Resilient CH was organised by UNESCO, the Japanese Agency for Cultural Affairs (ACA), the Japanese National Institutes for Cultural Heritage (NICH), the International Centre for the Study of the Preservation and Restoration of Cultural Property (ICCROM) and the International Council on Monuments and Sites (ICOMOS) (UNESCO, 2015). The participants highlighted the potential of CH to contribute to the priorities for DRR, from understanding disaster risks and strengthening governance to investing in resilience and preparing for response, recovery, and reconstruction. Among the recommendations, the group emphasised the importance of fostering partnerships between the CH sector and the wide range of DRR stakeholders, including local governments, humanitarian organisations, and private sector.

Finding support to finance the development of DRM plans and strategies specifically targeting CH is probably a challenge as bigger as the ones posed by natural hazards or the pass of time over heritage itself. However, there are some mechanisms in place, and the constant advocacy of the increasing number of DRM-CH professionals, together with the inventiveness and creativity inherent to the culture sector, are fostering solutions to finance DRM for CH.

International Support to Finance DRM for CH

While DRM for CH is still finding –or fighting– its own place in national budgets, the support of international organisations is fundamental to mobilise resources, even if little, to help reducing risks in cultural contexts. Yet, the organisations which mandate are specifically reducing disaster risk, e.g. UNDRR and GFDRR, are not always engaged with the culture sector, nor consider CH as an independent thematic area, such as transport or education. In some cases, it is considered as a component under other umbrella section, such as general infrastructure or social sector. This lack of consideration and awareness of the potential of culture as a key sector for creating resilient communities is also reflected in the lack of financial support and resources destinated to it.

The culture sector often needs to justify its relevance and profitability, particularly when seeking support from international development banks and agencies. Nevertheless, some organisations specifically target culture in the framework of DRM, particularly to provide support in emergency situations.

Some of the main international programs and initiatives that help financing the integration of DRM and CH, are mentioned here, organised following the DRM phases as before, during, and after a disaster – although risk identification and reduction should be implicit in each DRM-CH action.

Risk Understanding and Risk Reduction

The biggest challenge in pre-disaster situations is making people outside the DRM sphere understand what risk is and why it is so important to invest in reducing it. In the CH context, understanding the risk can become even more difficult since the exposure often refers to heritage assets that have been there during centuries or even millennia. On the other hand, from the DRM perspective, the main challenge may be establishing the economic value of CH in order to accurately design investments on DRM. Understanding disaster risk for CH sites and assets is actually the very first step to reduce risk. Cross-sector collaborations, therefore, are essential to share knowledge and create awareness.

Various programs based on multi-sectoral cooperation demonstrate the relevance and success of this kind of initiatives, as the following examples show. The **UNESCO Chair for Cultural Heritage and Risk Management** is held by the Institute of Disaster Mitigation for Urban Cultural Heritage at Ritsumeikan University in Kyoto, Japan (R-DMUCH, 2022). Its annual International Training Course on DRM of CH, despite not directly providing investments in countries, provides participants with financial support to travel to Japan, by covering travel and accommodation expenses, to learn from Japanese and international experts and start preparing their DRM-CH projects, which will be implemented once back in their countries. This support facilitates the development of activities connecting DRM and CH at local or national levels since all participants are coming from either discipline, and help establishing synergies and collaborations between practitioners of both fields. As part of this program, an online **Training Guide** (R-DMUCH, no date; see also: Jigyasu and Arora, 2013) was developed to provide additional support. It is organised through a sequence of modules following the DRM phases, and subdivided into sections containing interactive resources and information, to learn and, at the same time, help developing a course on DRM of CH.

During the period 2017–2020, the GFDRR **Resilient Cultural Heritage Program**, part of the Japan-World Bank Program on Mainstreaming DRM in Developing Countries (World Bank, no date), supported World Bank teams and client countries to develop projects connecting DRM and CH, following the Japanese example. This program mobilised US$700,000 and was divided into three components: knowledge exchange, operational support, and knowledge development. In April 2017, nine teams with members from DRM, urban development, culture, and tourism disciplines, travelled to Japan to participate in a Technical Deep Dive on Resilient CH and Tourism (GFDRR, 2017). The outcomes from this training were applied to different World Bank projects and activities. Following that experience, the teams were able to request additional support to develop specific technical assistances related to their projects. It was the case of Bhutan (Minguez Garcia et al., 2018), Myanmar (Bagan DRMP, 2018), Uzbekistan (GFDRR, 2020), and Central America and Dominican Republic (Lejtreger and Minguez Garcia, 2020), among others. Finally, the program included the development of several materials, from which the paramount output was a publication reflecting the Japanese experience on DRM for CH (Newman et al., 2020). Likewise, through the **Understanding Risk** (UR, 2022) community, some support was provided to the integration of CH within the DRM and, particularly, the Risk Identification discipline.[3]

In order to develop disaster risk assessments for CH, some methodologies and tools have been and are currently being developed, to help a wide range of stakeholders from different backgrounds and with different levels of capacity and support to conduct those evaluations and be able to have solid results frameworks, to prepare risk reduction and emergency preparedness measures and strategies. Among others, in 2010, the UNESCO World Heritage Centre together with the three advisory bodies, ICCROM, ICOMOS, and IUCN, launched the manual *Managing Disaster Risks for World Heritage* (UNESCO, 2010). It provides step-by-step guidelines to understand DRM and identify and reduce risks, as well as prepare to response.

The UNDRR program **Making Cities Resilient** (MCR 2030, no date) developed **Ten Essentials** (ibid), directly connected with the Sendai Framework priorities, presented as critical and independent steps to be undertaken in order to create and maintain resilience in urban contexts. An Addendum to this Scorecard methodology on CH has been just published (UNDRR-MCR2030, 2022),[4] including inputs from a practical exercise carried out in some historic cities in Peru (UNESCO, 2021, pp. 16–17).

Likewise, the **Capacity for Disaster Reduction Initiative** (CADRI, no date) is an UN-led global partnership that provides capacity building through a multidisciplinary pool of experts in several socio-economic sectors, including culture, to help reduce disaster and climate risks in countries. Governments can send their requests through the UN System in their country to access the services through the UN Resident or Humanitarian Coordinators. Currently, a tool is being finalised to facilitate countries' capacity diagnosis on DRR, including within the culture sector.

Similarly, as part of the World Bank **City Strength Diagnostic** program (World Bank, 2017), a sectoral module on CH was developed in 2018 (World Bank, 2018). It provides specific topics, including finance, and related guiding questions to help authorities and other stakeholders to assess the level of integration between CH and DRM. For example, in terms of financing, some of the key questions are: *What are the funding sources for CH? How are they administered? Do they include prevention measures from disasters and/or rehabilitation/ recovery plans? Is there contingency financing available for urgent stabilization measures as well as for restoring CH assets after disasters? Are the cultural assets protected by some kind of insurance?*

It highlights some aspects such as the need of including specific investment in risk prevention/mitigation measures as part of the usually sustained and predictable funding for CH maintenance and conservation, at the same time that fostering coordination among agencies/institutions, both public and private, to ensure funding. It also emphasised how a contingency fund can enable fast actions over CH to avoid irreversible losses, as well as the importance of including heritage specialists in the rescue and restoration of the affected cultural assets, to avoid misunderstandings and potential additional damages. Proper assessments in the first place, may help avoiding unnecessary expenses due to mistakes.

Emergency Preparedness and Response

Due to the several disasters and crisis directly impacting and damaging culture and CH, there is a gradual increase in recognising and prioritising the need of investing in emergency preparedness to allow for a quick and effective response. Several international mechanisms can be activated in emergency situations.

The UNESCO **Heritage Emergency Fund** (HEF, 2021), established in 2015, is a multi-donor and non-earmarked funding mechanism, which was designed to enable UNESCO to respond in a fast and effective way to crises – either caused by armed conflicts or disasters caused by natural and human-induced hazards – worldwide. It finances preparedness and response activities in the domains of the UNESCO Culture Conventions, i.e. immovable cultural and natural heritage, movable CH, cultural repositories, underwater CH, intangible CH, and the diversity of cultural goods, services, and expressions. To benefit from the HEF support, the request can be made through the UNESCO Field Office in the country, or the Secretariat of one of the UNESCO Culture Conventions, always in consultation with the national authorities. It can be requested at any time of the year, up to the maximum amount of US$100,000, and the approval relies on the Assistant Director-General for Culture (or their representative) as the authority delegated by the Director-General. The main beneficiaries of HEF are: 'i) Least Developed Countries as defined by the UN Economic and Social Council's Committee for Development Policy; ii) Lower-middle-income Countries as defined by the World Bank; iii) Level 3 emergencies as designated by the UN Emergency Relief Coordinator (ERC) in consultation of the Principals of the Inter-Agency Standing Committee (IASC); iv) Countries where a mission coordinated by the UN Department of Peace-Keeping Operations (UNDPKO) is based' (HEF, 2021, pp. 1–2).

Regarding World Heritage Sites, the **World Heritage Convention** also provides support through its **International Assistance** (WHC, 2022), in particular the Emergency Assistance program, to be requested by the States Parties National Commissions for UNESCO, Permanent

Delegations to UNESCO, or an appropriate governmental Department or Ministry. This can cover properties included on the World Heritage list as well as the list of World Heritage in Danger, which have suffered or are in imminent danger to suffer damage. The financial assistance can be used to respond to a disaster, to prevent or mitigate potential impacts, and also to assess whether a site is exposed to an imminent threat.[5]

The **World Monument Fund**'s **Crisis Response Program** (WMF, 2022) supports the implementation of medium-large scale recovery projects in CH sites physically affected by the impacts of disasters and armed conflicts, as well as other disruptive events. This program aims to strengthen the capacity of local communities from CH sites affected by emergencies, by providing resources for early recovery actions, build social cohesion and community resilience, and create economic opportunities. WMF also advocates to take action drawing on CH to address some of today's most challenging issues – including climate change, underrepresentation, imbalanced tourism, and post-crisis recovery – through the partnerships with local communities, funders, and governments.

The **ICCROM First Aid and Resilience for Cultural Heritage in Times of Crisis** (FAR) (ICCROM, 2022) program, in addition to deliver practical trainings, create capacity, and strengthen the network of first aiders to CH, provides financial support to participants to help developing their projects – some examples from the 2019 FAR are reflected in *A Story of Change* (Hashem and Ambani, 2021). Furthermore, the FAR Network counts over 800 cultural first aiders from 87 countries, who can be reached and deployed in case of emergency, supported by FAR and other ICCROM programs. In the FAR framework, ICCROM has developed and tested methodologies and tools (ICCROM, 2018, 2020) and offers a thorough repository of resources, as well as opportunities to develop courses and trainings, and work on emergency response situation in collaboration with a long list of partners.

To recall the importance of seeking fundraising and sponsorship, this program counts also with the support of other institutions, such as the Swedish Postcode Foundation, the Prince Claus Fund and the Smithsonian Institution. These last two have also additional mechanisms to support CH during a crisis.

The **Prince Claus Fund** established in 2003 the **Cultural Emergency Response** (CER, 2022)[6] to quickly act over CH either threatened, damaged, or destroyed in crisis situations by providing quick financial support to local actors to stabilise the situation, prevent further damage, and implement basic repairs. CER relies on its international professional network to identify and implement actions on cultural emergencies in direct cooperation with local partners in the affected communities.

The **Smithsonian Cultural Rescue Initiative** (SCRI, no date) works to protect CH impacted by crises, through cultural rescue work, both in the United States of America (USA) and internationally, including Haiti, Syria, Iraq, Egypt, Mali, and Nepal. In addition to providing disaster training for heritage colleagues, first responders, and military personnel worldwide and launching rescue campaigns when severe crisis strikes, the initiative also supports and fosters research.

Resilient Recovery

In the aftermath of a disaster, establishing the economic value of damages and losses[7] in cultural assets is complex due to the number of associated values beyond the economic one, resulting in one of the main challenges that difficult the integration of CH into recovery plans. Some efforts have been made to help solving this complication, such as the development of an indicator to estimate economic losses from damaged CH (Romão and Paupério, 2019), but still need to be widespread in order to systematise the inclusion of CH as a key sector in resilient recovery strategies.

An important step was the integration of Culture as one of the social sectors in the **Post-Disaster Needs Assessment** (PDNA) (GFDRR et al., 2017) methodology – which was developed in 2005 building on the previous Damage and Loss Assessment (DaLA) (GFDRR, no date). The Culture PDNA Guidelines highlights the contribution of this sector to the effectiveness and sustainability of the recovery program and provides guidance to develop the assessment. The estimation of the economic valuation of damage to public cultural assets mainly refers to Vecvagars' (2006) approach based on a variation of the *benefit transfer method* – which calculates the value of a good or service based on the replacement cost – that would calculate the value *based on the creation of a new, possibly different and/or enhanced cultural asset*, instead of a replica or reconstruction of the original. However, the PDNA guidelines states that, when possible, the repair and reconstruction of damaged CH should be carried out having the cost calculated based on actual market prices of labour, materials, and management. Regarding the estimation of the economic value of losses, some references would be the reduction of revenues due to temporary closure or non-availability of CH and events; losses related to the disruption in the production of cultural and creative industries; the cost of measures to ensure continuity and access to culture; the cost of temporary emergency measures to prevent further damage; and others associated with mitigating new risks and increased vulnerabilities as resulted from the disaster. This contributes to determining the country's macro-economic impact, which for culture would reflect the decrease in percentage of the cultural sector's contribution to national GDP and potential losses of cultural good exports. This is an exercise to be developed at a national level.

The estimation of the overall cost – then financial need – for resilient recovery strategies should integrate *building back better* (BBB)[8] considerations. The PDNA methodology recommends that the cost for the recovery plan corresponds to the estimation as a result of the damage and loss assessment, and the BBB costs should be proportionate to the costs of recovery and reconstruction needs, realistic compared to the financial envelope pledged by the government and international development partners, as well as consider the country capacity and feasibility of actions and achievements.

While the development of a PDNA is led and owned by the government of the affected country, and assisted by a multi-disciplinary, multi-agency team (including World Bank, GFDRR, UN Agencies, European Commission, and other relevant stakeholders), the application of the guidelines might be helpful for other stakeholders in post-disaster situations. An additional knowledge resource is the **Guide to Developing Disaster Recovery Frameworks** (GFDRR, 2015) which dedicates its Module 4 to Financing for Recovery.

In terms of integrating culture and CH into post-crisis recovery actions, the **Culture in City Reconstruction and Recovery** (CURE) Framework, jointly developed by UNESCO and World Bank (2018), aims to provide guidance on operational recovery through four phases: (i) Damage and Needs Assessment and Scoping; (ii) Policy and Strategy; (iii) Financing; and (iv) Implementation. The overall objective is to provide policymakers and practitioners with a systematic, integrated approach to design and implement a participatory city reconstruction and recovery strategy with culture at its core.

Regarding the financing aspects of post-crisis urban recovery, the CURE framework recommends trying to deploy a **combination of public and private funds**. Ideally, it would start with an upfront investment by the public sector for rehabilitation, seeking to attract private sector investment. Despite not always generating market-viable returns directly, investments in CH and urban resilience can indirectly have a positive impact on a city's economic growth and can boost private sector confidence in the reconstruction process. In other cases, governments or international donors can develop risk transfer or credit enhancement, as well as guaranteed mechanisms to enable investment security, or use concessional finance by shifting the investment risk-return profile and reducing risk with flexible capital and favourable terms. It is also important to look into the long-term recovery, since stable and robust investment plans can eventually emerge several years after the crisis, becoming attractive for private sector financing.

Regarding the identification of funding resources, in the case of major events it is usually a national-level agency who oversees the disbursement of available funds for local recovery, although the local government should have reliable public finance management systems to be able to use the funds efficiently and timely. While financing post-crisis recovery processes can be similar to capital investment plans, the overall process should be quicker and more flexible since the conditions may change rapidly and waiting for central government budget decisions may create problematic delays.

Some additional ideas extracted from the CURE include the use of:

- incentives or regulations at city level to create attractive real estate markets and encourage post-crisis redevelopment where the private market is not yet strong enough to invest;
- fiscal tools to facilitate the exchange of funds between the government and the community to promote reconstruction;
- transfer of development rights between an owner (whose rights are limited due to historical significance of the structure) and a developer (in another part of the city): the authorised, but unbuilt floor area of the historic buildings can be transferred to certain districts with proper higher density, so the conservation cost is borne by more parties and not just the owner;
- grants to historic buildings' owners for specific purposes such as rehabilitation, and to non-profit organisations working on historic preservation and conservation; and similar to grants, local government could provide low-cost loans for conservation of historic buildings;
- tax-based incentives (such as the alleviation of an owner's property tax or income tax burden) to encourage development in post-crisis situations where the land market is not strong; they can be given to private developers or individual owners to stimulate real-estate markets (UNESCO and World Bank, 2018, p. 47).

To close this section, it is worth recalling that the support from the international community might make the difference in many cases, particularly in countries struggling with other many challenges and even crises, in addition to disaster risks. Situations of fragility, conflict, and violence (FCV) usually put additional stress in countries trying to develop DRM initiatives, although on the other hand, the disaster-conflict nexus brings opportunities for peacebuilding and social cohesion around the fight to avoid disasters (GFDRR, 2022; ISS, no date).

Some ideas from the *disaster risk financing for humanitarian action* (Anticipation Hub, no date) might be applied to CH contexts.

Good Practices from Regional, National, and Local Levels

Some examples from around the world illustrate good practices on funding DRM for CH, also including some innovative ideas and mechanisms that can indirectly contribute to finding additional financing, e.g., by raising and communicating specific issues and cases to create awareness and attract donors and supporters. Box 4.1 illustrates the case of Japan.

The **European Union** (EU), also through the European Commission (EC), is actively supporting and funding DRM for CH through several programs and initiatives (EC, no date). 2018 was declared the *European Year of Cultural Heritage* (EC, 2018), and the EU Civil Protection

Box 4.1   Financing DRM for CH: The case of Japan

Japan is probably one of the countries that invests the most in DRM-CH. Sitting in the Pacific Ring of Fire, the archipelago counts over 100 active volcanoes and is also prone to earthquake, tsunami, typhoon, floods, landslides, and fire, keeping at high disaster risk its rich heritage from millennia of history.

Aware of this situation, also learned from major disaster experiences, the country has set a strong system to protect its CH,[10] including financing support. Japan counts indeed with three types of budgets regarding DRM for CH in its system: (1) the ordinary budget for CH from the Agency of Cultural Affairs (ACA, no date); (2) a special budget for disaster recovery managed by the Reconstruction Agency (2020); and (3) a revised or supplementary budget (Newman *et al.*, 2020, pp. 27–29).

1 The **ACA budget** is divided into five components, whence the one dedicated to conservation (41%) is divided into three, from which the one for *transmission and utilisation of cultural properties by appropriate maintenance* (61%) is mostly allocated to DRM-related activities. Precisely, among the seven subcategories, five have specific activities connected to DRM: (a) conservation and repair for heritage structures, including activities to enhance fire and crime prevention measures, and earthquake-proof countermeasures; (b) similar for arts and crafts, including also the installation and maintenance of disaster and crime prevention facilities; (c) reinforcement of groups of traditional buildings, including conservation and repair activities for making buildings resilient against earthquakes, as well as installation of disaster prevention facilities; (d) management of registered cultural properties, supporting the maintenance of nationally registered cultural properties, such as through inspections of fire alarms and fire extinguishing facilities that are required by law; and (e) conservation, maintenance, and utilisation of historic sites, providing support to owners and managers, including repair work to buildings and the installation of disaster prevention facilities within important cultural landscapes.

2 The **special disaster recovery budget** was established in March 2011 as a response mechanism after the Great East Japan Earthquake (GEJE) and Tsunami. It is managed by the Reconstruction Agency, and through it, ACA specifically uses it for the recovery of cultural properties affected by the GEJE.

3 Finally, the **revised or supplementary budget** is set after the regular budget is approved to include adjustments in order to address unexpected or inevitable changes, which makes it very useful and frequently used in the face of a disaster situation. While the ordinary budget is mainly used for DRM activities that are planned in advance, this option provides a faster and more flexible use of money in case of crisis, enabling ACA to rapidly respond to cultural properties both damaged and at risk.

Additionally, subnational governments[11] may request **subsidies** from ACA's ordinary budget, that may vary according to different types of cultural properties, and can include covering costs related to DRM for CH. For example, for works developed by the subnational government on important cultural properties, the subsidy can cover up to 50 percent of the costs for regular repairs, maintenance, and DRM measures, particularly related to disaster preparedness; if these are carried out by not-for-profit organisations or individuals, the subsidy may raise up to 85 percent. In situations of disaster recovery, the subsidy may include an additional 20 percent, not exceeding the maximum of 85 percent of the total cost, except in the case of severe disasters, where the subsidy can cover up to 90 percent.

In sum, **DRM is considered as an integral part of CH protection and management**. For additional information about ACA's Protection of Cultural Properties, see *Awareness Raising on Protection of Cultural Properties* (ACA, no date).

Among the lessons learned from the Japanese experience, other practices may help financing, as well as making investments more efficient, in terms of developing DRM for CH. For example, investing in interagency cooperation highly improves performance throughout the DRM process: this would include investing in communication channels and collaboration actions connecting actors at different levels before a disaster occurs, as the Japanese example of interaction ACA-Prefecture-Municipality. This can be reinforced through explicit budgets and incentive mechanisms, as seen in Japan, to build on the ability of all stakeholders to proactively protect CH while reducing costs at the same time that potential losses from disasters.

In this regard, another crucial recommendation is investing in the engagement of local communities to get them involved in initiatives and activities related to all DRM phases for CH: risk identification, risk reduction, emergency preparedness and response, and resilient recovery, to improve the protection and resilience of both CH and people. This improves the performance of DRM functions while building social capital. Japan has also developed low-cost options for capacity building, drills, low-tech solutions, and engagement tools such as the Disaster Imagination Game (DIG),[12] as well as measures and equipment that can be used and maintained by the local population and neighbours, such as *gravity pressure water supply facilities*, using natural water and gravity for water pressure, incorporating hydrants that can be freely used by residents.[13]

The aim behind these initiatives is to ensure that the individuals from local communities are motivated, prepared, and ready to act in case of emergency to protect themselves and their CH assets, and also help visitors. This indirect way of investment in DRM for CH has proven very effective in Japan.

[*Main source: Resilient-Cultural-Heritage-Learning-from-the-Japanese-Experience.pdf*]

Mechanism (EUCPM, 2022) includes CH under its mandate. Additionally, the EU recently published a guide to help identify funding opportunities for CH (EC, 2021a). Some key European projects are mentioned as follows and listed in the references section.

The program **Protecting Cultural Heritage from the Consequences of Disasters (PROCULTHER)** was co-funded by the Directorate-General for European Civil Protection and Humanitarian Aid Operations (DG-ECHO), for the period 2019–2021, as a consortium of partners around four priorities: creating a pool of experts, strengthening synergies with other projects, raising awareness and developing trainings, and promoting new initiatives. Similarly, **ARCH** is a European-funded research project launched in 2019, which aims to better protect CH areas from hazards and risks, with a focus on climate change, and by developing DRM frameworks in historic areas.

Under Horizon 2020, the EU's research and innovation funding programme 2014–2020, with nearly €80 billion budget, two projects directly targeted DRM for CH: the **STORM** project provides novel predictive models and tools to European CH stakeholders facing climate change and natural hazards, for effective risk identification associated to climatic conditions for CH. The project also created a cooperation platform to foster collaborations on collecting data, enhancing knowledge, and improving methodologies. The **HYPERION** project focuses on delivering better preparedness, as well as faster efficient emergency responses and sustainable recovery processes for historic areas, using existing tools, services, and new technologies, through an integrated resilience and multi-hazard assessment platform.

Financing impact on regional development of cultural heritage valorisation (FINCH) is a project focused on the protection and preservation of CH and its environment, aiming to foster its potential impact on growth, job creation, and long-term social and economic benefits.

Moving to the North American continent, as part of the above-mentioned SCRI, the **Heritage Emergency National Task Force** (HENTF) is a partnership of 60 USA organisations and federal agencies, co-sponsored by the Federal Emergency Management Agency (FEMA) and the Smithsonian Institution. It is a good example of public-private partnerships to ensure that museums, libraries, archives, and historic sites have the tools and resources to prepare for, respond to, and recover from disasters.

An innovative project brought together local communities and researchers from Thailand, Nepal, and the United Kingdom (UK) to boost disaster resilience through indigenous knowledge-sharing (UNDRR, 2022b). The **Political Capabilities for Equitable Resilience** (POLCAPS) (SEI, 2022) project is funded by UK Research and Innovation, a non-departmental public body sponsored by the UK government, through the Global Challenges Research Fund Collective Programme (UKRI, 2022). It has created linkages between the three countries, for example, by using the historical knowledge from a small farming community in Thailand – that turned an initial threat of floods and seawater intrusion into resources instead – by sharing this experience to provide other communities with different perspectives to face their own challenges in their local context. Also aiming to foster resilience through traditional knowledge and practices, the Global Heritage Fund through its **Environmental Resilience** program (GHF, 2022), has supported some projects to document traditional practices on social resilience and climate adaptation.

Similarly, the initiative **ICORP on the Road** (ICORP, 2021) travels around the world meeting the protagonists of inspiring stories about post-crisis response and recovery of CH, and producing documentaries, campfire talks, and exhibitions enhancing the efforts and social support to protect CH. This initiative aims to showcase the experiences of those professionals and local communities to raise awareness on the role of CH as key contributor to ensure more inclusive, safe, resilient, and sustainable communities, connecting to the Sustainable Development Goals (SDGs).[9]

Conclusion – Brainstorming on Opportunities and Alternatives to Finance DRM-CH

Financial resources are critical to develop and implement DRM plans and actions for CH sites and assets, but the lack of budgets specifically dedicated to this does not need to mean a lack of chance to carry out DRM for CH. Throughout this chapter, some resources to help develop DRM actions to protect CH in the face of disasters have been presented. They can be summarised around three main opportunities:

1 *Fundraising Campaigns and Public-Private Alliances.*
    There is a new paradigm in the donors/philanthropists' interests, which seems to be increasing attention to the several risks faced by CH worldwide. Examples of iconic monuments

damaged by disasters, such as the destructive fire in Notre-Dame Cathedral in Paris, show this trend – by April 2022, the donations have reached nearly €844 million from 340,000 donors in 150 countries (France 24, 2022), although this also brought into debate the inequality in terms of support to other CH at risk. Opportunities for fundraising and donations campaigns seem to keep opening to support the protection of CH from crises,[14] and the schemes based on public-private cooperation contribute also to strengthen visibility and support.

2 **Capacity Building and Community Engagement**.

Increasing the network of DRM-CH professionals is a good investment for the discipline itself, reflected in positive cost-benefit results. Likewise, investing in local community resilience around CH areas through trainings, drills, and preparedness activities, have also a positive impact. Theories and calculations about the cost of doing nothing (IFRC, 2019) to reduce risk and be prepared to face a crisis, contribute to show the importance in economic terms of prevention (ICCROM, 2022a). Training courses for professionals, games, and tools for communities, and methodologies and other resources, some mentioned in this chapter, can help develop DRM-CH at low cost.

3 **Technical Support and Knowledge Exchange**.

Collaborations between sectors are critical, as well as international support to establish linkages and connections between different countries that can share experiences and solutions to be adapted to local contexts. Several examples around the world show the added value of these actions, such as a recent congress on experiences in risk management for CH in Ibero-America (ICCROM, 2021), which brought together Ibero-American professionals and institutions working in DRM including CH, to share experiences and knowledge, identify existing challenges and needs, and discuss potential common strategies and concrete actions.

An important opportunity to finance DRM-CH comes through the integration of the discipline into projects using new technologies, such as satellite imagery. For example, the EU Copernicus (2018) program has been developing natural hazards and climate change activities, and connecting with CH, such as through specific programs like the HEritage Resilience Against CLimate Events on-Site (HERACLES) and the PROTection of European Cultural HEritage from GeOhazards (PROTHEGO) projects (see references section). Likewise, the partnership established between UNESCO and the Operational Satellite Applications Programme (UNOSAT) under the UN Institute for Training and Research (UNITAR), brings additional opportunities to protect CH with the use of geo-spatial technologies, benefitting several countries and opening new possibilities to trigger additional investments from other disciplines (Copernicus, 2016).

Furthermore, some innovative ideas that have been raising recently could be applied to finance DRM-CH, such as the trans-disciplinary research project on *Circular models Leveraging Investments in Cultural heritage adaptive reuse* (CLIC) (see references section) that explores the development of *'circular' financing*, business and governance models to be applied to the reuse of CH, demonstrating economic, social, and environmental convenience. And the use of *cryptocurrencies* applied to CH, as presented by the Global Heritage Fund (GHF) *Non-Fungible Token* (NFT) (Sycip, 2021). This consists of converting some CH aspects into digital form, entering the blockchain public ledger and allowing CH to be opened to this new economic sector. Cryptocurrency owners could actively participate in the preservation and conservation. However, this still needs to find a way of being more eco-friendly due to the huge amount of electricity consumed by bitcoins (Lacey, 2022).

Another innovative solution to contribute to the financing on DRM-CH could be establishing popular campaigns such as **adopting monuments** (COE, 2022; Adopt a Heritage) to involve people in their protection through risk reduction activities. In this regard, creating awareness on DRM-CH and linkages with the tourism sector (UNEP, 2008; GFDRR *et al*., 2014), aiming to use the industry to help financing and strengthen the resilience of places and travellers, is another promising opportunity to explore.

In conclusion, even if it results challenging to find direct finance to develop DRM for CH, there are plenty of opportunities to support the integration of both disciplines. Understanding disaster risk and creating awareness about the importance of investing in CH protection is the first step to fund DRM-CH, reduce risk, and increase the resilience of people and their heritage.

Key points

- The financing of DRM for CH faces a double challenge in terms of finding funding for a sector recently formed by two areas already struggling to find financing.
- Disaster risk finance and economics of CH conservation seem running in parallel, facing their own challenges while opening an opportunity to cooperate in the development of DRM-CH financing.
- Several international programs support DRM-CH through direct or indirect financial mechanisms while fostering opportunities for additional support.
- There are some good examples of practices at regional, national, and local levels, which can serve as model or inspiration for others to adapt and replicate in their own contexts.
- Fostering public-private alliances, developing fundraising campaigns, and using new technologies to create awareness and attract interest and support, could help to raise funding for DRM-CH.
- Investing in cross-sectoral partnerships, collaborations, knowledge exchange, capacity building, and community engagement and training can be very useful to support DRM-CH development.
- Lack of dedicated budgets does not need to mean lack of chance to carry out DRM for CH; several resources are available to support professionals while strengthening the international DRM-CH network.

Notes

1 German Corporation for International Cooperation agency (GIZ, 2022).
2 For a comprehensive compilation of definitions from UNESCO, see UNESCO Culture for Development Indicators (UNESCO, 2014).
3 As in the Regional Forum UR2020 for Central America: understandrisk.org/sesiones-tecnicas-ur-centroamerica/ or the UR2018: understandrisk.org/publication/ur2018-proceedings/ (pp. 14–19).
4 A preceding presentation of this tool took place during the *VII Regional Platform for DRR in the Americas and the Caribbean* through a Learning Lab on 4 November 2021: https://rp-americas.undrr.org/2021/cultural-heritage-management-drr-introducing-scorecards-new-addendum-make-cities-resilient.html
5 The guidance for this assistance may be found in the Operational Guidelines' Annex 9: https://whc.unesco.org/en/guidelines/
6 CER has recently become an independent NGO: culturalemergency.org/
7 Following the definition used by the PDNA methodology: **damage** would refer *to total or partial destruction of physical assets existing in the affected area*; while **losses** would include *the changes in flows of goods and services—diminished revenues and/or additional costs, expressed in current values—caused by the disaster that may extend throughout the rehabilitation and reconstruction periods*.
8 Defined by the UN General Assembly as *the use of the recovery, rehabilitation and reconstruction phases after a disaster to increase the resilience of nations and communities through integrating disaster risk reduction measures into the restoration of physical infrastructure and societal systems, and into the revitalisation of livelihoods, economies and the environment (*UNGA, 2016*)*.

9 Episodes from Nepal, Brazil, India, Pakistan, are already available through their website, as well as information on how to contribute (icorp-ontheroad.com/).
10 The system is based in the identification and designation of Cultural Properties (CP) of different importance, classified into seven categories: Tangible CP (structures), Tangible CP (fine arts and crafts), Intangible CP, Folk CP, Monuments, Cultural Landscapes, and Preservation Districts for Groups of Historic Buildings (ACA, no date).
11 Understood in this context as regions, prefectures, and municipalities (Newman *et al.*, 2020, p. ii).
12 For description see Newman *et al.* (2020, p. 75); and GFDRR (2018, p. 18).
13 For more information on the Environmental Water Supply System (EWSS) see GFDRR (2018a, p. 21); and Newman *et al.* (2020, pp. 47–49) for summaries, and Toki, K. and T. Okubo, 'Protection of Wooden Cultural Heritage from Earthquake Disaster', in Proceedings of Meetings on Cultural Heritage Risk Management (Kyoto: World Conference on Disaster Reduction), 94–102.
14 Efforts to support the protection of CH in Ukraine are good example of this: europanostra.org/ukraine-crisis/

References

ACA (no date) *Agency for Cultural Affairs*. Available at: bunka.go.jp/english/; *Introduction to Cultural Properties*. Available at: bunka.go.jp/english/policy/cultural_properties/introduction/; *Awareness Raising on Protection of Cultural Properties*. Available at: bunka.go.jp/english/policy/cultural_properties/awareness_raising/index.html [Accessed 20 August 2022].
Adopt a Heritage (no date) India. Available at: adoptaheritage.in/ [Accessed 20 August 2022].
Agency for Cultural Affairs (2018) *Policy of Cultural Affairs of Japan*, ACA, Tokyo. Available at: bunka.go.jp/english/report/annual/pdf/r1394357_01.pdf [Accessed 20 August 2022].
Anticipation Hub (no date) *Disaster Risk Financing for Humanitarian Action*. Available at: anticipation-hub.org/learn/emerging-topics/disaster-risk-financing/ [Accessed 20 August 2022].
Aymerich, M. (2015) *Towards an Integrated Approach to Funding Cultural Heritage for Europe*, EIB. Available at: europanostra.org/wp-content/uploads/2017/04/2015-FundingCulturalHeritage-EIB.pdf [Accessed 20 August 2022].
Bagan Disaster Risk Management Plan (DRMP) (2018) Washington, DC: World Bank Group. Available at: https://documents1.worldbank.org/curated/en/671391544633861934/pdf/132874-WP-P162815-PUBLIC-Bagan-DRM-Eng.pdf [Accessed 20 August 2022].
Burnham, B. (2019) *A Cultural Heritage Investment Fund as a Strategy for Conserving our Cultural and Natural Legacies* in 2018 US/ICOMOS Symposium 'Forward Together: A Culture-Nature Journey towards More Effective Conservation in a Changing World', November 13–14, 2018, San Francisco, California. Available at: https://www.icomos.org/en/focus/culture-nature/41457-call-for-papers-forward-together-a-culture-nature-journey [Accessed 20 August 2022].
CADRI (no date) *Capacity for Disaster Reduction Initiative*. Available at: cadri.net/ [Accessed 20 August 2022].
CER (2022) *Prince Claus Fund's Cultural Emergency Response*. Available at: princeclausfund.org/cultural-emergency-response [Accessed 20 August 2022].
CHiFA (no date) *Cultural Heritage Finance Alliance*. Available at: heritagefinance.org/ [Accessed 20 August 2022].
Chmutina, K., and J. Von Meding (2019) *Dilemma of Language: 'Natural Disasters'* in academic literature. *International Journal of Disaster Risk Science*, Vol. 10, 283–292. Available at: doi.org/10.1007/s13753-019-00232-2 [Accessed 24 November 2022].
Chmutina, K., J. Von Meding, J.C. Gaillard, and L. Bosher (2017) *Why Natural Disasters aren't All That Natural*. OpenDemocracy. Available at: opendemocracy.net/en/why-natural-disasters-arent-all-that-natural/ [Accessed 20 August 2022].
Clarke, D., A. de Janvry, E. Sadoulet, and E. Skoufias (2015) *Disaster Risk Financing and Insurance: Issues and Results*, Ferdi. Available at: ferdi.fr/en/publications/disaster-risk-financing-and-insurance-issues-and-results [Accessed 20 August 2022].
COE (2022) *Adopt a Monument*. Council of Europe. Strategy 21 – Good Practices. Available at: coe.int/en/web/culture-and-heritage/-/adopt-a-monument [Accessed 20 August 2022].

Copernicus (2016) *UNESCO and UNITAR-UNOSAT Cooperate to Protect Cultural Heritage with Geo-Spatial Technologies*. Available at: copernicus.eu/en/unesco-and-unitar-unosat-cooperate-protect-cultural-heritage-geo-spatial-technologies [Accessed 20 August 2022].

Copernicus (2018) *Copernicus for Cultural Heritage: Satellites to Preserve the Legacy from Our Past!* Available at: copernicus.eu/en/copernicus-cultural-heritage-satellites-preserve-legacy-our-past [Accessed 20 August 2022].

De Masi, F., and D. Porrini (2021) *Cultural Heritage and Natural Disasters: The Insurance Choice of the Italian Cathedrals*. Journal of Cultural Economy, 45, 409–433. doi.org/10.1007/s10824-020-09397-x [Accessed 20 August 2022].

European Commission (no date) *Risk Management for Cultural Heritage*. Available at: https://culture.ec.europa.eu/cultural-heritage/cultural-heritage-in-eu-policies/risk-management-for-cultural-heritage [Accessed 20 August 2022].

European Commission (2018) *European Year of Cultural Heritage 2018*. Available at: https://culture.ec.europa.eu/cultural-heritage/eu-policy-for-cultural-heritage/european-year-of-cultural-heritage-2018 [Accessed 20 August 2022].

European Commission (2021a) Work Plan for Culture 2019–2022 – *Workshop on Complementary Funding for Cultural Heritage – Background Paper and Selected Good Practices.* Available at: europanostra.org/wp-content/uploads/2021/04/202104-EC-Background-Paper-Complementary-Funding-for-Heritage.pdf [Accessed 20 August 2022].

European Commission (2021b) *Funding Opportunities for Cultural Heritage*. Available at: https://culture.ec.europa.eu/cultural-heritage/funding-opportunities-for-cultural-heritage [Accessed 20 August 2022].

EIB Institute (2022) *What We Do: Cultural Heritage*. Available at: https://institute.eib.org/whatwedo/arts/cultural-heritage/ [Accessed 20 August 2022].

EUCPM (2022) *EU Civil Protection Mechanism*. Available at: https://civil-protection-humanitarian-aid.ec.europa.eu/what/civil-protection/eu-civil-protection-mechanism_en [Accessed 20 August 2022].

Financial Protection Forum (2015) *What Is Disaster Risk Finance (DRF)?* Available at: financialprotectionforum.org/what-is-disaster-risk-finance-drf [Accessed 20 August 2022].

France 24 (2022) *In Pictures: Notre-Dame Cathedral Three Years after the Fire*. Available at: france24.com/en/france/20220415-in-pictures-notre-dame-cathedral-three-years-after-the-fire [Accessed 20 August 2022].

GFDRR (no date) *Damage, Loss and Needs Assessment – Tools and Methodology*. Available at: gfdrr.org/en/damage-loss-and-needs-assessment-tools-and-methodology [Accessed 20 August 2022].

GFDRR (2015) *Guide to Developing Disaster Recovery Frameworks*. Available at: gfdrr.org/en/publication/guide-developing-disaster-recovery-frameworks [Accessed 20 August 2022].

GFDRR (2017) *Technical Deep Dive on Resilient Cultural Heritage and Tourism*. Available at: gfdrr.org/en/publication/technical-deep-dive-resilient-cultural-heritage-and-tourism [Accessed 20 August 2022].

GFDRR (2018a) *Assessing and Communicating Risk to Cultural Heritage: The Future of Preserving the Past* in *The Understanding Risk 2018 Proceedings.* World Bank, Washington DC (pp. 14–19). Available at: understandrisk.org/wp-content/uploads/UR2018_Proceedings_Publication.pdf [Accessed 20 August 2022].

GFDRR (2018b) *Technical Deep Dive on Resilient Cultural Heritage and Tourism*, World Bank, Washington DC. Available at: gfdrr.org/en/publication/technical-deep-dive-resilient-cultural-heritage-and-tourism [Accessed 20 August 2022].

GFDRR (2020) *Uzbekistan Resilient Cultural Heritage and Sustainable Tourism Development*. Available at: https://documents1.worldbank.org/curated/en/833831602615313609/pdf/Uzbekistan-Resilient-Cultural-Heritage-and-Sustainable-Tourism-Development.pdf [Accessed 20 August 2022].

GFDRR (2022) *Disaster Risk Management – Fragility, Conflict and Violence Nexus*. Available at: gfdrr.org/en/drm-fcv [Accessed 20 August 2022].

GFDRR, UNDP, European Union, and World Bank (2014) *Post-Disaster Needs Assessments Guidelines: Volume B – Tourism*. Available at: https://recovery.preventionweb.net/publication/post-disaster-needs-assessments-guidelines-volume-b-tourism [Accessed 20 August 2022].

GFDRR, UNDP, European Union, and World Bank (2017) *Post-Disaster Needs Assessments Guidelines: Volume B – Culture*. Available at: documents1.worldbank.org/curated/en/306991493101225270/pdf/114520-WP-PUBLIC-ADD-SERIES-pdna-guidelines-vol-b-culture.pdf [Accessed 20 August 2022].

Giuliani, F., R.G. De Paoli, and E. Di Miceli (2021) *A Risk-Reduction Framework for Urban Cultural Heritage: A Comparative Study on Italian Historic Centres*, Journal of Cultural Heritage Management and Sustainable Development, Vol. 11 No. 4, pp. 499–515. doi.org/10.1108/JCHMSD-07-2020-0099 [Accessed 20 August 2022].

GIZ (2022) *Deutsche Gesellschaft für Internationale Zusammenarbeit*. Available at: giz.de/en/html/index.html [Accessed 20 August 2022].

GHF (2022) *Global Heritage Fund's Environmental Resilience Program*. Available at: globalheritagefund.org/issues/environmental-resilience/ [Accessed 20 August 2022].

Hallegatte, S., J. Rentschler, and B. Walsh (2018) *Building Back Better: Achieving Resilience through Stronger, Faster, and More Inclusive Post-Disaster Reconstruction*, World Bank, Washington DC. Available at: gfdrr.org/sites/default/files/publication/Building%20Back%20Better.pdf [Accessed 20 August 2022].

Hashem, Y., and J. Ambani (2021) *A Story of Change*. ICCROM FAR. Available at: iccrom.org/publication/story-change [Accessed 20 August 2022].

HEF (2021) *UNESCO Heritage Emergency Fund*. Available at: https://www.unesco.org/en/culture-emergencies/heritage-emergency-fund [Accessed 20 August 2022].

ICCROM (2018) *First Aid and Resilience to Cultural Heritage in Times of Crisis Handbook and Toolkit*. Available at: iccrom.org/sites/default/files/2018-10/fac_handbook_print_oct-2018_final.pdf and iccrom.org/index.php/publication/first-aid-cultural-heritage-times-crisis-toolkit [Accessed 20 August 2022].

ICCROM (2020) *inSIGHT: A Participatory Game for Enhancing Disaster Risk Governance*. Available at: iccrom.org/index.php/publication/insight-participatory-game-enhancing-disaster-risk-governance [Accessed 20 August 2022].

ICCROM (2021) *Congress on Experiences in Risk Management for Cultural Heritage in Ibero-America*. Available at: iccrom.org/news/congress-experiences-risk-management-cultural-heritage-ibero-america [Accessed 20 August 2022].

ICCROM (2022a) *First Aid and Resilience to Cultural Heritage in Times of Crisis (FAR) Program*. Available at: iccrom.org/programmes/first-aid-and-resilience-times-crisis-far [Accessed 20 August 2022].

ICCROM (2022b) *Risk Management for Preventive Conservation*. Available at: iccrom.org/section/preventive-conservation/risk-management-preventive-conservation [Accessed 20 August 2022].

ICORP (2021) *ICORP on the Road*. Available at: icorp-ontheroad.com/ [Accessed 20 August 2022].

IFRC (2019) *The Cost of Doing Nothing. The Humanitarian Price of Climate Change and How It Can Be Avoided*. International Federation of Red Cross and Red Crescent Societies, Geneva. Available at: ifrc.org/sites/default/files/2021-07/2019-IFRC-CODN-EN.pdf [Accessed 20 August 2022].

ISS (no date) *When Disaster Meets Conflict*. International Institute of Social Studies. Erasmus University Rotterdam. Available at: iss.nl/en/research/research-projects/when-disaster-meets-conflict [Accessed 20 August 2022].

Jeggle, T., and M. Boggero (2018) *Post-Disaster Needs Assessment: Lessons from a Decade of Experience*. European Commission, GFDRR, UNDP, and the World Bank, Washington DC. License: CC BY-NC-ND 3.0 IGO. Available at: https://openknowledge.worldbank.org/entities/publication/6eae4f5d-0a11-5d1a-9bc9-a8878ba6af5e [Accessed 20 August 2022].

Jigyasu, R., and V. Arora (2013) *Disaster Risk Management of Cultural Heritage in Urban Areas: A Training Guide*, Ritsumeikan University, Kyoto. Available at: preventionweb.net/files/44208_trainingguide1.pdf [Accessed 20 August 2022].

Lacey, R. (2022) *Everything You Need to Know about Eco-Friendly Cryptocurrencies*. The Times. Available at: thetimes.co.uk/money-mentor/article/eco-friendly-cryptocurrencies/ [Accessed 20 August 2022].

Lejtreger, R., and B. Minguez Garcia (2020) *Gestión del Riesgo de Desastres y Patrimonio Cultural en Centroamérica y República Dominicana: Oportunidades para Fortalecer la Resiliencia del Patrimonio y del Turismo Sostenible*, World Bank Group, Washington, DC. Available at: gfdrr.org/en/publication/gestion-del-riesgo-de-desastres-y-patrimonio-cultural-en-centroamerica-y-republica [Accessed 20 August 2022].

Licciardi, G., and R. Amirtahmasebi (2012), *The Economics of Uniqueness: Investing in Historic City Cores and Cultural Heritage Assets for Sustainable Development*, World Bank, Washington DC. doi: 10.1596/978-0-8213-9650-6. License: Creative Commons Attribution CC BY 3.0.

MCR 2030 (no date) *Making Cities Resilient*. Available at: mcr2030.undrr.org/; *The Ten Essentials for Making Cities Resilient*. Available at: https://mcr2030.undrr.org/ten-essentials-making-cities-resilient [Accessed 20 August 2022].

Meenan, C., J. Ward, and R. Muir-Wood (2019) *Disaster Risk Finance – a Toolkit*. GIZ. Bonn and Eschborn. Available at: indexinsuranceforum.org/sites/default/files/Publikationen03_DRF_ACRI_DINA4_WEB_190617.pdf [Accessed 20 August 2022].

Mena, C. (2020) *There Are No Natural Disasters*. Available at: apolitical.co/solution-articles/en/there-are-no-natural-disasters [Accessed 20 August 2022].

Minguez Garcia, B., J. Newman, and D. Tshering (2018) *From Japan to Bhutan: Improving the resilience of cultural heritage sites*. World Bank Blogs. Available at: https://blogs.worldbank.org/endpovertyinsouthasia/japan-bhutan-improving-resilience-cultural-heritage-sites [Accessed 20 August 2022].

Mizutori, M. (2020) *Time to Say Goodbye to 'Natural' Disasters*. PreventionWeb. Available at: preventionweb.net/blog/time-say-goodbye-natural-disasters [Accessed 20 August 2022].

Newman, J., B. Minguez Garcia, K. Kawakami, and Y. Naito Akieda (2020) *Resilient Cultural Heritage: Learning from the Japanese Experience*. World Bank Group, Washington, DC. Available at: https://documents.worldbank.org/en/publication/documents-reports/documentdetail/131211602613832310/resilient-cultural-heritage-learning-from-the-japanese-experience [Accessed 20 August 2022].

Organisation for Economic Co-operation and Development (OECD) (2015), *Disaster Risk Financing: A Global Survey of Practices and Challenges*, OECD Publishing, Paris. dx.doi.org/10.1787/9789264234246-en

R-DMUCH (no date) *Disaster Risk Management of Cultural Heritage in Urban Areas: A Training Guide*. Available at: r-dmuch.jp/eng/project/itc/training_guide/index.html [Accessed 20 August 2022].

R-DMUCH (2022) *UNESCO Chair for Cultural Heritage and Risk Management*. Available at: rdmuch-itc.com/ [Accessed 20 August 2022].

Reconstruction Agency (2020) Japan. Available at: reconstruction.go.jp/english/ [Accessed 20 August 2022].

Romão, X., and E. Paupério (2019) *An Indicator for Post-Disaster Economic Loss Valuation of Impacts on Cultural Heritage*. International Journal of Architectural Heritage, Vol. 15 No. 5. doi: 10.1080/15583058.2019.1643948

SCRI (no date) *Smithsonian Cultural Rescue Initiative*. Available at: culturalrescue.si.edu/ [Accessed 20 August 2022].

SEI (2022) *Political Capabilities for Equitable Resilience (POL-CAPS)*. Stockholm Environment Institute. Available at: sei.org/projects-and-tools/projects/political-capabilities-for-equitable-resilience/ [Accessed 20 August 2022].

Sycip, G. (2021) *Crypto, Non-Fungible Tokens and the Potential of the Metaverse in Cultural Heritage Preservation*. Global heritage Fund. Available at: globalheritagefund.org/2021/12/08/crypto-nft-metaverse-cultural-heritage-preservation/ [Accessed 20 August 2022].

Throsby, D. (2012) *Heritage Economics: A Conceptual Framework* in Licciardi, G., and R. Amirtahmasebi, *The Economics of Uniqueness: Investing in Historic City Cores and Cultural Heritage Assets for Sustainable Development*. World Bank, Washington DC (pp. 45–75).

UIS (2021) *Tracking Investment to Safeguard the World's Cultural and Natural Heritage. Results of the 2020 UIS Survey on Expenditure on Cultural and Natural Heritage*, UIS, Montreal. Available at: uis.unesco.org/sites/default/files/documents/uis_culture_and_heritage_report_2021_web.pdf [Accessed 20 August 2022].

UKRI (2022) *Global Challenges Research Fund*. UK Research and Innovation. Available at: ukri.org/what-we-offer/international-funding/global-challenges-research-fund/ [Accessed 20 August 2022].

UNDRR (2015) *Sendai Framework for Disaster Risk Reduction 2015–2030*. Available at: undrr.org/publication/sendai-framework-disaster-risk-reduction-2015-2030 [Accessed 20 August 2022].

UNDRR (2021) *Journalists Told 'No Such Thing as a Natural Disaster'*. Available at: undrr.org/news/journalists-told-no-such-thing-natural-disaster [Accessed 20 August 2022].

UNDRR (2022a) *History*. Available at: undrr.org/about-undrr/history [Accessed 20 August 2022].

UNDRR (2022b) *Build Back Better*. Available at: undrr.org/terminology/build-back-better [Accessed 20 August 2022].

UNDRR (2022c) *Boosting Disaster Resilience through Indigenous Knowledge-Sharing among UK, Thailand, Nepal*. Available at: undrr.org/news/boosting-disaster-resilience-through-indigenous-knowledge-sharing-among-uk-thailand-nepal [Accessed 20 August 2022].

UNDRR-MCR2030 (2022) *Disaster Resilience Scorecard for Cities: Cultural Heritage Addendum*. Available at: https://mcr2030.undrr.org/cultural-heritage-scorecard [Accessed 24 November 2022].

UNEP (2008) *Disaster Risk Management for Coastal Tourism Destinations Responding to Climate Change*. Available at: preventionweb.net/files/13004_DTIx1048xPADisasterRiskManagementfo.PDF [Accessed 20 August 2022].

UNESCO (2010) *Managing Disaster Risks for World Heritage*. Available at: https://whc.unesco.org/en/managing-disaster-risks/ [Accessed 20 August 2022].

UNESCO (2014) *UNESCO Culture for Development Indicators. Methodology Manual*. Heritage Dimension (pp. 129–138). Available at: https://en.unesco.org/creativity/sites/creativity/files/cdis_methodology_manual_0_0.pdf [Accessed 20 August 2022].

UNESCO (2015) *Resiliency and Cultural Heritage Discussed at the World Conference on Disaster Risk Reduction*. Available at: https://whc.unesco.org/en/news/1255 [Accessed 20 August 2022].

UNESCO (2021) *2020 Heritage Emergency Fund. Annual Progress Report*. UNESCO, Paris. Available at: https://en.unesco.org/sites/default/files/en_hef_annual_report_2020_0.pdf [Accessed 20 August 2022].

UNESCO (2022) *Reducing Disasters Risks at World Heritage Properties*. Available at: https://whc.unesco.org/en/disaster-risk-reduction/#resources [Accessed 20 August 2022].

UNESCO, and World Bank (2018) *Culture in City Reconstruction and Recovery*, UNESCO and World Bank, Paris and Washington DC. License: CC BY-SA 3.0 IGO. Available at: https://openknowledge.worldbank.org/entities/publication/f465176a-3d30-5440-9af9-da9dd3fedf34 [Accessed 20 August 2022].

UNISDR (2017) *Build Back Better in Recovery, Rehabilitation, and Reconstruction. Consultative Version*. Available at: unisdr.org/files/53213_bbb.pdf [Accessed 20 August 2022].

United Nations General Assembly (UNGA) (2016) *Report of the Open-Ended Intergovernmental Expert 2 Working Group on Indicators and Terminology Relating to Disaster Risk Reduction*, Seventy-First Session. Item 19(c). A/71/644. Available at: preventionweb.net/files/50683_oiewgreportenglish.pdf [Accessed 20 August 2022].

UNU-EHS (2021) *Climate and Disaster Risk Financing Instruments: An Overview*. Available at: climate-insurance.org/wp-content/uploads/2021/05/Climate-and-Disaster-Risk-Financing-Instruments.pdf [Accessed 20 August 2022].

UR (2022) *Understanding Risk*. Available at: understandrisk.org/ [Accessed 20 August 2022].

Vecvagars, K. (2006) *Valuing Damage and Losses in Cultural Assets after a Disaster: Concept Paper and Research Options*, UN Publication, Mexico DF. Available at: https://digitallibrary.un.org/record/588652?ln=en [Accessed 20 August 2022].

World Bank (no date) *World Bank Tokyo Disaster Risk Management (DRM) Hub*. Available at: https://www.worldbank.org/en/programs/tokyo-drm-hub [Accessed 20 August 2022].

World Bank (2017) *The CityStrength Diagnostic: Promoting Urban Resilience*. Available at: worldbank.org/en/topic/urbandevelopment/brief/citystrength [Accessed 20 August 2022].

World Bank (2018) *CityStrength Diagnostic. Optional Sectoral Module: Cultural Heritage*. Available at: https://documents1.worldbank.org/curated/en/996471525721935888/pdf/125991-WP-P150083-PUBLIC-CityStrength-Guidebook-2018.pdf [Accessed 20 August 2022].

World Bank (2019) *Lebanon – Cultural Heritage and Urban Development Project*. Independent Evaluation Group, Project Performance Assessment Report 140539. Washington, DC. Available at:

ieg.worldbankgroup.org/sites/default/files/Data/reports/ppar_lebanoncultural.pdf [Accessed 20 August 2022].

World Bank (2020) *Jordan – Cultural Heritage, Tourism, and Urban Development Project*. Independent Evaluation Group, Project Performance Assessment Report 147006. Washington, DC. Available at: https://ieg.worldbankgroup.org/sites/default/files/Data/reports/ppar_jordanculturalheritage.pdf [Accessed 20 August 2022].

World Bank (2022) *Disaster Risk Financing and Insurance (DRFI) Program.* Available at: worldbank.org/en/programs/disaster-risk-financing-and-insurance-program [Accessed 20 August 2022].

WHC (2022) *World Heritage Convention*. Available at: https://whc.unesco.org/en/convention/; *WHC International Assistance*. Available at: https://whc.unesco.org/en/intassistance [Accessed 20 August 2022].

WMF (2022) *World Monument Fund (WMF)'s Crisis Response Program*. Available at: wmf.org/crisis-response-program [Accessed 20 August 2022].

Suggested Further Reading

Financing Disaster Risk Reduction: [Accessed 20 August 2022].

Centre for Disaster Protection – Pre-Agreed Disaster Risk Finance: The Agenda Women's Advocates Should Be Influencing. Available at: disasterprotection.org/blogs/pre-agreed-disaster-risk-finance-the-agenda-womens-advocates-should-be-influencing

European Commission – Disaster Risk Financing: Main Concepts & Evidence from EU Member States. Available at: ec.europa.eu/info/sites/default/files/economy-finance/dp150_en.pdf

GFDRR – Disaster Risk Financing and Insurance Program. Available at: gfdrr.org/sites/default/files/publication/thematic-note-diaster-risk-financing-and-insurance-program.pdf

GFDRR – Global Partnership on Disaster Risk Financing Analytics: Results and Achievements. Available at: reliefweb.int/report/world/global-partnership-disaster-risk-financing-analytics-results-and-achievements

Penn Today: Preparing, and Paying for, Climate Change-Induced Disasters: penntoday.upenn.edu/news/preparing-and-paying-climate-change-induced-disasters

OECD – Developing the Elements of a Disaster Risk Financing Strategy. Available at: oecd.org/pensions/insurance/Developing-elements-of-disaster-risk-financing-strategy-May-2018-conference-outcomes.pdf

OECD – Disaster Risk Financing. Available at: oecd.org/daf/fin/insurance/OECD-Disaster-Risk-Financing-a-global-survey-of-practices-and-challenges.pdf

Financing Cultural Heritage Protection: [Accessed 20 August 2022].

ALIPH – International Alliance for the Protection of Heritage in Conflict Areas. Available at: aliph-foundation.org/

ARCH. Available at: savingculturalheritage.eu/

Circular Models Leveraging Investments in Cultural Heritage Adaptive Reuse (CLIC). Available at: clicproject.eu/

Cultural Heritage Finance Alliance (CHiFA). Available at: heritagefinance.org/

Emergency Funding for Nature in Times of Crisis. Available at: rapid-response.org/

EuropaNostra – Guide for Fundraising Heritage. Available at: europanostra.org/wp-content/uploads/2018/02/Learning-Kit-Fundraising-for-Heritage-CSOs.pdf

Financing Impact on Regional Development of Cultural Heritage Valorisation (FINCH). Available at: https://projects2014-2020.interregeurope.eu/finch/

Heritage Emergency National Task Force (HENTF). Available at: culturalrescue.si.edu/hentf/

HEritage Resilience against CLimate Events on-Site (HERACLES). Available at: heracles-project.eu/

Horizon 2020. Available at: ec.europa.eu/info/research-and-innovation/funding/funding-opportunities/funding-programmes-and-open-calls/horizon-2020_en

HYPERION. Available at: hyperion-project.eu/

Learning from Disaster: Building City Resilience through Cultural Heritage in New Orleans. Available at: https://publications.iadb.org/publications/english/viewer/Learning-from-Disaster-Building-City-Resillience-through-Cultural-Heriage-in-New-Orleans.pdf

Protecting Cultural Heritage from the Consequences of Disasters (PROCULTHER). Available at: proculther.eu/

PROTection of European Cultural HEritage from GeO-hazards (PROTHEGO). Available at: prothego.eu/

STORM. Available at: storm-project.eu/

The Economic Challenge and the Use of Innovative Financial Schemes. Available at: interreg-central.eu/Content.Node/ForHeritage/04.-Use-of-innovative-financial-schemes.pdf

Workshop and Background Paper on Complementary Funding for Cultural Heritage. Available at: europanostra.org/europa-nostra-participates-in-european-commissions-workshop-on-complementary-funding-for-cultural-heritage/europanostra.org/wp-content/uploads/2021/04/202104-EC-Background-Paper-Complementary-Funding-for-Heritage.pdf

Section II
Understanding the Context

5   Heritage and Peacebuilding

Elke Selter

The heritage sector, internationally led by UNESCO, tends to forefront the peacebuilding potential of heritage. This chapter explores how, historically, the idea has grown that there is a direct link between built heritage and peace. As a result, whenever heritage is affected, damaged, or destroyed, the international community, the heritage sector, but also local actors rally to restore, restitute, or rebuild that heritage. Underpinning all these efforts is a seemingly simple concept that more heritage means more peace. Yet, heritage is also often at the centre of conflicts, which seems to imply the opposite, that more peace would require less heritage. Where does the idea that more heritage would mean more peace, come from? Are there ways to make sure that working with heritage yields positive results in a peacebuilding field? This chapter focuses on international efforts to build peace, and thus on the work of UNESCO because UNESCO is the UN organisation specifically created to work with heritage to help the UN build peace. Hence, looking at how UNESCO, which has over time also become a global authority on heritage matters, interprets and implements this mandate tells us a lot about whether and how heritage can build peace.

What Is Heritage and What Is Peacebuilding?

Before exploring the recent history of heritage and international peacebuilding, let us first look at these two concepts. This chapter uses a broad concept of 'heritage' that includes its tangible as well as its intangible aspects (which are discussed in detail in Chapter 2). The preservation of that heritage has, in recent history, especially since the 1990s, become a global norm. In other words: the protection and preservation of heritage are usually considered *the right thing to do*, and the destruction of heritage – wilful or not – is usually seen as 'wrong', even portrayed as pure barbarism. States also find it important for their heritage to be internationally recognised, through platforms like UNESCO, but also by using it as a tool for cultural diplomacy (Ang et al. 2015). Since UNESCO started devising international conventions for heritage, most commonly, heritage tends to be divided into tangible heritage, including both moveable and immovable heritage, and intangible heritage. These subdivisions are artificial, as much as practical, but hold little relevance at a local level, where tangible and intangible values, the built and the moveable aspects of heritage tend to be strongly intertwined. See Chapter 2 for more details.

It is also important to note what Smith called the 'Authorized Heritage Discourse' (AHD) (2006). This is a 'professional discourse' by which heritage is constructed through a Western, material, elitist lens. The AHD refers to the globally dominant heritage discourse, based on an expert-driven approach to conservation that originated in 19th-century Europe and that focuses mostly on the preservation of tangible heritage's inherent values and material authenticity. It corresponds to UNESCO's original approach to heritage, and because of UNESCO's central

DOI: 10.4324/9781003293019-8

role in inspiring global heritage law and policies, it is also reflected in most heritage policies, laws, and management systems around the world. These are often based on identifying specific values and then cementing them in preservation methods that are all about ensuring that these values do not change. That we have all been made to think that there is one proper way of looking at heritage and preserving it is a key element in the complex relationship between heritage and peacebuilding.

This chapter links this global thinking about heritage to 'peacebuilding'. The term 'peacebuilding' was first used by UN Secretary-General Boutros-Ghali in his An Agenda for Peace (1992). In *Peaceland*, Autesserre defines peacebuilding as including 'any and all elements identified by local and international stakeholders as attempts to create, strengthen and solidify peace' (2014: 21). The most commonly used approach by international organisations like UNESCO, tends to be referred to as 'liberal peacebuilding', which aims to restore state control and the rule of law in post-war context by installing liberal, global norms – norms like democratic elections and human rights, but also the protection of heritage (Mac Ginty & Richmond 2007). Paris called it a 'specific kind of social engineering', which combined peace with various elements of physical, social, and economic recovery (2004: 5). Like the AHD, liberal peacebuilding also tends to be seen as a way to impose Western norms elsewhere (Sabaratnam 2011). Other critiques relay that the concepts is very state-centred, focusing on one desired end result (such as a free-market economy or democracy), which does not work everywhere and ignores local realities (Richmond 2011), and scholars like Autesserre (2014) have shown how this results in local communities resisting such peacebuilding efforts. We will see how this is not any different for heritage, where in zones of war, project often help authorities restoring control by bringing back the heritage, without much real consideration for local realities or dissonant local voices. A way to bring in these local voices is referred to as 'transformative justice' (Gready & Robins 2014; Lambourne 2013). Such a transformative approach envisages long-term peace, which requires political, social, and economic transformation alongside more traditional justice needs. The transformative approach to peace brings a variety of local, informal systems, thereby also countering some of the critiques of peacebuilding (and transitional justice) being too Western in its approach and imposing norms that are perhaps not desirable locally.

Heritage and Peace: A Short History[1]

When UNESCO was created in 1945, this was done with the specific aim to assist the UN with achieving global peace through soft sectors like education, science, and culture. The constitution of the organisation that came to be the world's leading authority on heritage matters starts with reference to its core objective: 'building peace in the minds of men' (1945). A key tenet of UNESCO's efforts has been to approach (certain forms of) heritage as a universal good that has the potential to unify people. Its most popular international conventions – the World Heritage Convention (1972), for instance – promoted this idea of a 'heritage of humankind' that surpasses national or local interests. The same ideas were integrated into UNESCO's early operational projects, like the *Nubia campaign* that allowed for the Aswan dam construction to go ahead while Western powers concentrated on saving the heritage that would be flooded by the dam (Betts 2015). Based on this experience in Nubia, until the late 1980s, UNESCO increasingly engaged in projects that were to devise technical solutions to ensure that heritage could be preserved forever, even if wars went on around it. Already here, the foundational principle of the heritage sector's approach to peace can be observed: projects focus on saving heritage while aspiring to build peace but without any specific peacebuilding efforts. Underpinning this idea is the simple binary already noted earlier: if wars mean less heritage, then more heritage must mean (more) peace.

Towards the 1990s, as the Cold War ended, global peace efforts were given a second chance. Despite emerging wars in the Balkans and Rwanda, there was optimism. UN Secretary-General Boutros-Ghali's *Agenda for Peace* (1992) introduced the idea of post-conflict peacebuilding and reshaped peacekeeping missions from a tool to prevent violence towards one that could also build peace. UNESCO saw this as an opportunity to re-invent its original peacebuilding idea by launching a new type of field operations in which heritage was to provide a neutral ground for peacebuilding. You can see this as heritage's contribution to the liberal peace project. This was piloted in Cambodia, where Angkor had served to legitimate the authority of whomever was in charge – whether the Khmer Rouge, the royal family, or the elected government. While this worked to some extent in Cambodia, peacebuilding in the post-Cold War era was a very different undertaking and the certain political neutrality that Angkor had offered was hard to find in the Balkans, for instance, where heritage had been an integral part of an ethnicized war.

Throughout the 1990s, the war in the Balkans set an example of how ethnicized wars could gravely affect heritage. With wars targeting civilians, heritage, too, turned into a common target. The heritage sector had a hard time devising adequate responses. Practically, it tried to do this by mirroring aspects of humanitarianism. For instance, in 1991, the Council of Europe and UNESCO sent observers to Dubrovnik, though their presence could not prevent the historic town from being attacked. Heritage's peace project was thus also struggling in the post-Cold War conflicts. Hence, the sector focused on reconstruction projects that were more in line with the Nubia campaign and the Cambodia program, e.g. rebuilding the old bridge in Mostar.

Also, the idea of international military engagement beyond the traditional 'fighting wars' was further expanded when, in 1999, NATO put the 'right to interfere' – military intervention on humanitarian grounds – into practice in Kosovo, using it to bypass Security Council approval. The same NATO mission decided that it should not just protect civilians but also heritage, marking the start of an intensified involvement of the military with heritage. Safe zones for heritage were created, requiring military protection. They could not entirely prevent destruction during renewed violence in 2004, and up until today, these zones continue to isolate heritage from the people. While such cooperation with militaries is often seen as an important way to better protect heritage, it also meant that the idea of heritage as neutral ground for peace became harder to hold.

The 1990s also saw a turn towards development, linked to the post-Cold War 'peace dividend'. Development NGOs and budgets burgeoned. For the heritage sector, this was a useful advance that allowed it to focus on its technical expertise and on the economic aspects of the liberal peace project. Yet, focusing on development did not do away with the reasons to fight over heritage. On the contrary. For example, the near global ratification of the World Heritage Convention by the early 1990s, the rapid increase in sites on the World Heritage List, and their connection with an unprecedented growth in international tourism made heritage a major global commodity, which was both economically and politically valuable. Not only did states see heritage as a way to enhance economic growth, but they had also learnt that controlling heritage was a way to exercise control over their people. This was done nationally *and* by using UNESCO's regime to endorse these domestic policies internationally. Two examples are Sri Lanka's nomination of a series of Buddhist sites to the World Heritage list in the aftermath of the Tamil Tigers' suicide bombings near the Temple of the Tooth in Kandy (Wijesuriya 2000), or India using the World Heritage label to displace people in Hampi (Bloch 2016). Local contestation grew, and so did the use of heritage in wars. The development turn in the heritage field was further stimulated by the failure to spare Bamiyan (Afghanistan) in 2001. The Taliban's destruction of the rock-hewn Buddha statues followed the most substantive diplomatic campaign UNESCO ever carried out to spare heritage. Yet, it failed, showing that not much can be done when faced with the readiness to attack heritage despite legal provisions for its protection.

A new shift in heritage's relation with peacebuilding arrived with the Global War on Terror. There, the trend of peace and security converging, which had started in the 90s, continued, and heritage tried to come on board again. In the peacebuilding field, the turn of the century was marked by a shift towards security: international engagement was to make the world safer. Strengthening state control and preventing state failure came to dominate, and soft sectors were to either go along or become obsolete. This was very apparent in the humanitarian field (Barnett 2011). Heritage was not immune to that either.

In Iraq and Afghanistan, the US and its European allies were at once financing the military operations, and, as the main global donors, they also funded the aid organisations that were to respond to the damage the military caused, e.g. supporting refugees, medical aid, rebuilding collapsed infrastructure. Because of this dual engagement of the Western powers, they largely defined who received what kind of support in the countries where they were fighting. It turned the use of foreign aid into an alternative means of exercising control or to make up for military engagement. Heritage, too, became a force multiplier. Slow progress in Afghanistan and Iraq, and contentious events like the 2003 looting of the Baghdad Museum that meant negative publicity for the US military, resulted in what can be seen as a 'cultural turn' in US military operations (Clemis 2010), and later also in those of other NATO members. Powerful actors understood that protecting heritage could help their mission by helping to build relations with the local population. But in these asymmetric wars, the less powerful, too, knew that attacking heritage would bring global visibility.

By 2011, during the NATO-campaign in Libya, lessons learned from Afghanistan and Iraq allowed the Western forces to largely spare heritage amidst local threats to target archaeological sites. This experience helped enhance cooperation between the heritage sector and the military. Agencies like UNESCO cooperating with the military as a way to spare heritage, while still limited, marked an important shift from the sector-led diplomatic endeavours in Bamiyan a decade earlier.

The deliberate attacks on heritage that we have seen at least since the 1990s took on new proportions after the war in Libya, when World Heritage-listed monuments in Timbuktu, Mali, were destroyed, followed by Syria, Iraq, and Yemen. The openly publicised destructions by extremist groups alarmed people around the world that war could so openly destroy what was considered a priceless heritage. Contrary to the Balkans, where most tried to blame someone else for the attacks, the attackers now seemed proud to advertise their actions and invited the media along. The attacks also exposed the difficulties the heritage sector experienced when engaging during active conflict, unable to protect or negotiate. Showing that the heritage sector was still relevant called for cooperation with the security sector, like UN peacekeeping, but also for a reframed discourse that made heritage sufficiently appealing to these new actors. As a result, UNESCO started warning against 'cultural cleansing' – a form of ethnic cleansing marked by doing away with that heritage which is thought to represent a particular group of people – and linking heritage trafficking to counterterrorism. Thereby, heritage protection was moved from being a tool to build peace to being a matter of keeping people – locally and within an abstract global community – safe from 'barbarians'. Heritage thereby became fully subject to a process called 'securitisation' (Buzan et al. 1998). This was not only a major discursive turn but also an operational one because in the name of 'security' the heritage sector sought closer cooperation with the military, peacekeepers, and other security sector actors, and initiated a debate on the Responsibility to Protect (R2P) and heritage.

Reconstruction, Shared Heritage, and Peacebuilding

In practice, working with built heritage in post-war settings has often taken the form of reconstruction. Recent examples were the rebuilding of mausoleums in Timbuktu (Mali) and the ongoing reconstruction in Mosul (Iraq). In the heritage field, 'reconstruction' refers to 'a technical process for the restitution of destroyed or severely damaged physical assets and infrastructure' (UNESCO 2018a), which is not the same as 'reconstruction' in the broader humanitarian and peacebuilding fields, where it refers to rebuilding a state, including activities to undo the damage done and rebuilding 'better' by promoting liberal norms (Mac Ginty & Richmond 2007). The heritage sector calls the latter 'recovery' which 'aims at the consolidation of peace and security and at restoring or improving the economic, physical, social, cultural and environmental assets, systems and activities' (UNESCO 2018a).

While from a heritage perspective, it may be desirable to bring back the heritage, as part of a peace process, this approach involves several obstacles. A recurring issue is what stage of a monument is to be rebuilt and how to do that (which can also culturally differ). There are always many 'pasts' to return to. While this may seem like a technical choice, it is not. Such choices leave room for history to be rewritten and thus politically manipulated. Little (2012) demonstrated engaging in a peace process is only politically desirable when it contributes to gaining or retaining power. Narratives of the past play an important role in this, and heritage is a major physical embodiment of such narratives. Reconstruction of heritage – carried out in coordination with the government – is thus particularly vulnerable to such political usages. Indeed, destruction of heritage during wars is often seen as a way to establish new narratives of the past, but so is reconstruction. Sometimes reconstruction can even be seen as removing traces of what happened, and thus contributing to a story of denial. International organisations often try to avoid these political complexities by referring to the reconstructed heritage as 'shared' and thus a common basis for building peace. There are broadly three ways in which such 'sharedness' is understood, and all of these are prone to manipulation.

First, heritage can be considered 'neutral'. Here, heritage is seen as not predominantly associated with a particular warring faction. This can, for example, be heritage dating back to much earlier times for which narratives focus on historically common roots. An example is Bosnia's Stecce (or Stecaks) – medieval monolith tombstones – which are seen as a common heritage but which Kisic showed to be sites of dissonance because of different historiographies appropriating the monuments to different ethnic groups (2016).

Second, heritage can be considered 'shared' because it is 'universal'. Heritage is then rebuilt, not because certain groups associate with it, but because it is to represent all of humanity. Yet, this is a theoretical concept that is very complex in reality, especially in wars where at least one group deliberately attacked that heritage. The reconstruction of Timbuktu's mausoleums (Mali) is an example of this (see case study).

Third is heritage's version of power-sharing or unity governments: each (formerly) warring group gets equal amounts of heritage reconstructed. An example of this is how the Commission to Preserve National Monuments – called the 'Annex 8 Commission' after its reference in the Dayton Peace Agreement – functioned in post-war Bosnia-Herzegovina when lists were drawn up for protection and reconstruction based on the equal representation of the three main ethnic groups, independent of the extent into which each group's heritage had been damaged (Hadzimuhamedovic 2018). UNESCO's current project to 'Revive the Spirit of Mosul', which reconstructs heritage from different religions in the Iraqi town, follows a similar model. However, while trying to balance everything, ethnic affiliations tend to be highlighted and take more prominence (Bennoune 2018). This can be especially counterproductive in situations where targeting of heritage during the war was disproportional.

Rebuilding Heritage in the Balkans

The reconstruction of built heritage in post-war Bosnia-Herzegovina is one of the largest heritage reconstruction campaigns of the post-Cold War era. It allows us to gain some better insight into what works in terms of peacebuilding. This chapter focuses on international efforts, which are particularly complex. Comparing different projects in Bosnia highlights that.

A main example of internationally led projects was the reconstruction of the 'old bridge' in Mostar. The site, which has since become part of the World Heritage List, was destroyed in 1993 and later rebuilt through a UNESCO-World Bank project. From a technical perspective, the reconstruction tends to be considered a success story (e.g. Armaly et al. 2004). However, those who examined the impact on social relations consider it a failure. Calame and Pasic, for instance, concluded that the 'project proved to be relatively unimportant in relation to the ongoing process of long-term social reconciliation in Mostar' (2009: 3). Forde's (2016) ethnographic research revealed how the reconstruction was trying to replace complex reconciliation processes by a simple rebuilding of physical infrastructure, which was insufficient. This was exacerbated by its top-down approach in which international agencies interacted with various levels of government but never truly engaged in a dialogue and needs assessment with the local inhabitants. This is a typical downside of international engagement, which could also be observed in Mali (see Box 5.1).

Box 5.1   Rebuilding Timbuktu's heritage for peace

In 2012, during violent conflict in Northern Mali, World Heritage-listed places in Timbuktu were destroyed, and its famous manuscript libraries were evacuated. The destroyed sites were mainly Sufi mausoleums and a sacred gate at one of the historic mosques. In the immediate aftermath of the destruction, the international community, with UNESCO in the lead, announced that it was ready to rebuild these sites.

A project was mounted, funded mainly by the European Union and Switzerland, and in 2013, the reconstruction started. Technically, the project was a major success because of the number of studies conducted and the expertise used for the reconstruction. It was also done remarkably fast for an international project. However, the project is also portrayed as a contribution to peace in Northern Mali. But is it? First, the reconstruction project was decided upon by a group of (mainly) representatives of international (heritage) organisations during a French-funded meeting at UNESCO in Paris (UNESCO 2013). While UNESCO reports 'a large delegation from Mali', only 1 in 10 participants was Malian. Moreover, the Malians that could travel to Paris represented a 'heritage elite' (Joy 2012). And since almost none of them were from Timbuktu, their information about the local situation was limited, with the first technical missions only undertaken several months later (Diop 2016). This approach continued throughout the reconstruction, where the main interlocuters were the government, descendants of the Saints who were buried in the mausoleums, and traditional masons' associations. Researchers like Oumar Ba (2020) and Traoré Hadizatou (2018) conducted extensive community interviews indicating that there is extremely little ownership of the project among the broad population of Timbuktu. Another conclusion was that selective engagement of some groups, including rumours about who received money from international organisations, created new divisions in society. Even Malian UNESCO staff admit that people in Timbuktu 'felt prisoners of our approach'.[2] All this is perhaps the most apparent when it comes to the annual mud-cladding of the sites, called *crépissage*. There, what used to be a communal effort, is nowadays seen as an activity to be funded by UNESCO.[3]

A second important step in working towards peace through heritage has been the conviction of Al Mahdi by the International Criminal Court (ICC). The ICC's case against Al Mahdi was the first time the world's apex court tried someone for war crimes committed through the deliberate targeting of heritage. Overall, UNESCO, and other heritage bodies with it, welcomed this as justice finally being done for those who target heritage. Research conducted locally, however, draws a different picture (Ba 2020; Hadizatou 2018), one that mirrors what Clark has called the 'distant justice' model of the ICC (2018). Following the conviction, a reparations order was issued amounting to 2.7 million EUR (ICC 2018). That money is raised through the Trust Fund for Victims (TFV), who then implements a series of reparation measures. In this case, the reparations were both individual and collective, and tried to follow an approach that comes closer to that of transformative justice while still being strongly limited by the ICC court order. For instance, the TFV has struggled to define who is a 'victim' of Al Mahdi, entitled to reparations, and to broaden this from UNESCO's method that worked with male descendants of the Saints, to, for instance, also include women.[4] Again, despite concrete attempts to avoid some of the earlier pitfalls that UNESCO had faced, the fact that reparations involve the distribution of money, and that not everyone is equally 'victimised' by the destructions, is likely to further advance the societal divisions that we saw appear after UNESCO's project.

Of course, the situation in Mali and the Sahel region is far too complex for heritage projects to solve. But one can wonder whether the heritage projects contributed positively or not, and for whom. The projects have been successful in bringing back the heritage. By doing so, a strong message was sent that there is no room for those who cannot tolerate diversity. That is valuable, but it was a message that focused on 'the greater good', something that is valuable for a generic, ill-defined public, more than for the people most affected by the conflict. For the inhabitants of Timbuktu and its surroundings, those people who were at the receiving end (or not) of reconstruction and reparations, that general message was sometimes blurred by a feeling of disempowerment, of being left out, or of justice not being done (for them).

Nationally, the reconstruction of monuments in Bosnia was led by the 'Annex 8 Commission'. The Commission, which had representation of the three main ethnic communities, worked through a list of monuments that aimed to also rebuild heritage of these three communities in equal numbers, although some communities had suffered significantly more damage than others. While this approach has been (rightfully) applauded in terms of national ownership, at the local level, Hadzimuhamedovic (2018) demonstrated that it has the same shortcomings as the international approach. This is because it focuses a lot on the political needs (equal representation) and on generic heritage values (which are the main monuments?) and less so on what a transformative process would require.

Then there are local projects, such as the example of the small town of Stolac (Hadzimuhamedovic 2015). In Stolac, an approach was chosen that foregrounded dialogue and inclusion involving also refugees and diaspora. Thereby, it was reconstructed not only because it was valuable heritage but also to send a message to refugees that they were welcome again. This approach has proven successful locally but remains isolated from having a broader impact.

The example of Bosnia-Herzegovina demonstrates two important things. First, that reconstruction can yield positive results locally, but this becomes vastly more difficult and unlikely to succeed the further the implementer is removed from this local level. At national and international levels, more abstract criteria – like ethnic balancing of the number of monuments to be rebuilt – play a more important role. Efforts are carried out 'for the greater good', which can be because a global community cares (as was also the case in Mali) or because it fits within a political agenda. Yet, this 'greater good' is 'a convenient fiction' (Barnett & Weiss 2008: 38).

Second, contributing to enhancing relations requires an approach that acknowledges past events and is grounded in a joint agreement to move forward. This requires an inclusive process and implies that heritage values are not the main priority. Rather, heritage values are the outcome of a process. The case study on Mali in this chapter shows similar critiques on the reconstruction project in Timbuktu. One way in which this can be improved is by enhancing capacities, and practice, of heritage professionals to engage in conflict analysis, like other sectors involved in peacebuilding would do. A tool recently developed by ICCROM is a first step in this regard (Tandon 2021).

Memorialisation

An important element in linking reconstruction to peacebuilding is how past events are acknowledged. Also here, the heritage sector often finds itself in a difficult position where a prioritisation of heritage values can be at odds with the local needs for commemoration (or forgetting). From a peacebuilding perspective, bringing valuable heritage back when the affected population wishes so, can be a major part of dealing with the traumatic past. However, this would require for the reconstruction to focus not so much on (shared) heritage values but on the way in which rebuilding valuable, symbolic structures can be part of a restorative process. That means linking reconstruction to dealing with the past, through memory projects or by honouring a community's right to forget (Lennon 2009). However, according to a recent study by the International Organization for Migration (IOM), the heritage sector 'suffers from a lack of awareness and engagement in post-conflict communities' (Hay 2019: 47). Adding that it 'appears that memorialization in general has been overlooked … with little consideration given to developing a useful approach of acknowledgment and recognition of post conflict memory of trauma' (Ibid.: 50).

This 'fear' becomes even more explicit when it comes to sites of memory (see Chapter 17). For UNESCO, dealing with sites of memory has long been difficult because it is so openly political, creates strong divisions between member states, and, understandably, raises concerns of an even greater politicisation of bodies like the World Heritage Committee. It is not that sites of memory are not part of the World Heritage List. Auschwitz (Poland) and Hiroshima (Japan) were included on the World Heritage List, as were other 'negative' sites relating to slavery, for example. Mostar (Bosnia-Herzegovina) and Bamiyan (Afghanistan) were inscribed after being destroyed, with specific references to their wanton destruction as part of the inscribed value.

Yet, recent years have seen a hesitation to inscribing more such sites. An effort led by Belgium and France to include sites relating to the First World War, combined with an intent from Rwanda to nominate sites relating to the 1994 genocide, led UNESCO and ICOMOS to re-open the debate as to whether such 'sites associated with recent conflicts and other negative and divisive memories' (Beazley and Cameron 2020; UNESCO 2021a, 2021b). The name alone chooses to highlight the *negative* past rather than the *positive* message for the future that many of these sites work to convey through their conservation and museographic work. According to ICOMOS, such sites raise 'fundamental issues with regard to the purpose and scope of the World Heritage Convention and its appropriateness to celebrate properties that commemorate aspects of wars and conflicts' (ICOMOS 2018a: 153). The debate resulted in discussion papers (ICOMOS 2018b, 2020), expert meetings (UNESCO 2021b), and a decision by the World Heritage Committee to halt all 'negative heritage' nominations (UNESCO 2018b).[5] Continuing pressure, in particular by African states, but also by European countries like Belgium and France and other post-war states like Cambodia, finally led the World Heritage Committee in 2021 to establish a working group to further discuss the matter (UNESCO 2021b: para.9).

While the debate is ongoing, it is important to understand that the hesitation to fully consider sites of memory as makes the linking of the international heritage project with peacebuilding particularly difficult. After all, in the places where peace needs to be built, heritage will have such negative connotations. This is very explicit in the case of sites of memory, but it would be naïve to think that reconstructed sites like the mausoleums in Timbuktu are not also sites of memory. Likewise, in cases where heritage was used for national or local memorialisation, this has yielded positive results, for instance, in the cases of Cambodia and Rwanda. Just like reconstruction, they are not immune to political interference (Bolin 2019). Yet, if heritage is to contribute to peacebuilding, it is in these places and the processes that create, manage, and reshape them over time that heritage can contribute to transformative processes.

Conclusion

Heritage's relation to international peacebuilding goes back to at least 1945 when the two domains were explicitly linked in UNESCO's constitution. Yet, the sector has always struggled with this peace-objective. The heritage field foregrounds 'heritage', meaning that at the end of their intervention they want more heritage (and preferably also more peace). Meanwhile, peacebuilding is a process in which heritage could be a means to an end but not the end itself. Looking at the sector's main post-war intervention, reconstruction, demonstrates how this essential difference often leads to meagre results for heritage's contribution to peace. While rebuilding can indeed have important symbolic, social, and economic effects, for it to impact on reconciliation and peace, it needs an inclusive process that does not prioritise the heritage itself. When international agencies involve in reconstruction, such processes rarely take place because many other priorities, including political ones, dominate. The case study on the reconstruction efforts in Timbuktu showed that. It is also evident from heritage's complex relation with memorialisation, an essential component of dealing with the past. For heritage in post-war settings to contribute to peace, its recovery cannot systematically be placed above the local needs for transformation, including for memorialising past events. This does not mean that heritage cannot contribute to peace, some local-level projects have shown positive results. However, to do so, requires the sector to engage in truly open processes that let the affected populations decide what is needed for heritage to contribute to a sustainable future. This would require a fundamental change, away from 'more heritage' as the goal and towards working with heritage as a 'tool' for 'more peace'.

Key Points

- Heritage's relation to international peacebuilding goes back to at least 1945 when the two domains were explicitly linked in UNESCO's constitution.
- One of the main ways in which the heritage sector has tried to link heritage affected by wars to a peacebuilding project is by rebuilding that heritage. Reconstruction was to contribute to economic as well as social revival. The latter assumes that physical

reconstruction can bring back social ties from the past and equally assumes that these ties were good.
- Yet, for reconstruction to contribute to peace, it is essential to involve those people who have been most affected by the conflict and to do so in a manner that helps various groups overcome past events or feel like justice was done.
- This is not possible when the recovery of heritage values is systematically placed above the local needs for transformation, including for memorialising (or forgetting) past events.
- For heritage to contribute to peace, a shift is required away from solely considering the technical aspects of heritage and towards a prioritisation of the requirements of a peace process at all stages of heritage interventions. Memorialisation (and or the right to forget) is an essential component of this. Engaging in understanding the conflict dynamics and building the necessary capacities for this within the heritage field is an essential first step.
- This also requires stepping away from the technical and disciplinary dogmas of heritage preservation and towards truly open processes involving affected populations, letting them decide what is needed for heritage to contribute to a sustainable future for them.

Notes

1 Based on Selter (2021).
2 Author interview with a UNESCO staff in Bamako, November 2018 [translated from French].
3 Ibid.
4 Author interview with a TFV staff, online, October 2020.
5 In comparison, the *Memory of the World Register* has systematically added collections associated with mass atrocities, at times leading to political complexities as was the case with the Nanjing Massacre documentation in 2015.

References

Ang, I., Isar, Y.R., Mar, P., 2015. Cultural Diplomacy: Beyond the National Interest? International Journal of Cultural Policy 21(4), 365–381.
Armaly, M., Blasi, C., Hannah, L., 2004. Stari Most: Rebuilding More than a Historic Bridge in Mostar. Museum International 56, 6–17.
Autesserre, S., 2014. Peaceland: Conflict Resolution and the Everyday Politics of International Intervention. Cambridge University Press, Cambridge and New York.
Ba, O., 2020. Contested Meanings: Timbuktu and the Prosecution of Destruction of Cultural Heritage as War Crimes. African Studies Review 63(4), 743–762.
Barnett, M., 2011. Empire of Humanity. A History of Humanitarianism. Cornell University Press, Ithaca and London.
Barnett, M., Weiss, T.G. (Eds.), 2008. Humanitarianism in Question: Politics, Power, Ethics. Cornell University Press, Ithaca and London.
Beazley, O., Cameron, C., 2020. Study on Sites Associated with Recent Conflicts and Other Negative and Divisive Memories. UNESCO, Paris.
Bennoune, K., 2018. Report of the Special Rapporteur in the Field of Cultural Rights on her Mission to Serbia and Kosovo (No. A/HRC/37/55/Add.1). Human Rights Council, Geneva.
Betts, P., 2015. The Warden of World Heritage: UNESCO and the Rescue of the Nubian Monuments. Past Present 226, 100–125.
Bevan, R., 2016. The Destruction of Memory: Architecture at War. Reaktion Books, London.
Bloch, N., 2016. Evicting Heritage: Spatial Cleansing and Cultural Legacy at the Hampi UNESCO Site in India. Critical Asian Studies 48, 556–578.

Bolin, A., 2019. Imagining Genocide Heritage: Material Modes of Development and Preservation in Rwanda. Journal of Material Culture 25, 1–24.

Boutros-Ghali, B., 1992. An Agenda for Peace. Preventive Diplomacy, Peacemaking and Peace-Keeping (No. A/47/22-S/24111). UN, New York.

Buzan, B., Wæver, O., de Wilde, J., 1998. Security: A New Framework for Analysis. Lynne Rienner Publishers, Boulder and London.

Calame, J., Pasic, A., 2009. Post-Conflict Reconstruction in Mostar: Cart before the Horse. Divided Cities/Contested States. Working Paper 7. https://www.yumpu.com/en/document/read/37902662/post-conflict-reconstruction-in-mostar-cart-before-the-horse

Clark, P., 2018. Distant Justice: The Impact of the International Criminal Court on African Politics. Cambridge University Press, Cambridge.

Clemis, M.G., 2010. The 'Cultural Turn' in U.S. Counterinsurgency Operations. Doctrine, Application, and Criticism. Army History 74, 21–28.

Diop, A., 2016. Mali: Reconstruction De Mausolées à Tombouctou après la crise de 2012: le rôle des communautés, in: Machat, C., Ziesemer, J. (Eds.), Heritage at Risk. World Report 2014–2015 on Monuments and Sites in Danger. ICOMOS, Berlin, 52–57.

Forde, S., 2016. The Bridge on the Neretva: Stari Most as a Stage of Memory in Post-Conflict Mostar, Bosnia–Herzegovina. Cooperation & Conflict 51, 467–483.

Gready, P., Robins, S., 2014. From Transitional to Transformative Justice: A New Agenda for Practice. The International Journal of Transitional Justice 8, 339–361.

Hadizatou, T., 2018. Le bien culturel de Tombouctou, une patrimonialisation discutée ? Conditions locales de réception de la notion de patrimoine de type UNESCO. L'Année du Maghreb 19, 99–114.

Hadzimuhamedovic, A., 2018. Participative Reconstruction as a Healing Process in Bosnia, in: Culture in City Reconstruction and Recovery. UNESCO and World Bank, Paris and Washington DC.

——— 2015. The Built Heritage in the Post-War Reconstruction of Stolac, in: Walasek, H. (Ed.), Bosnia and the Destruction of Cultural Heritage, Heritage, Culture and Identity. Routledge, London and New York, 259–284.

Hay, M., F., 2019. Everyday Sites of Violence and Conflict. Exploring Memories in Mosul and Tal Afar, Iraq. IOM, Bagdad.

ICC, 2018. Public redacted version of 'Decision on Trust Fund for Victims' Draft Implementation Plan for Reparations', 12 July 2018 (No. ICC-01/12-01/15-272- Red). ICC Trial Chamber, The Hague.

ICOMOS, 2020. Second Discussion Paper. Sites Associated with Memories of Recent Conflicts and the World Heritage Convention. Reflection on Whether and How These Might Relate to the Purpose and Scope of the World Heritage Convention and Its Operational Guidelines. ICOMOS, Paris.

——— 2018a. Discussion Paper. Evaluations of World Heritage Nominations Related to Sites Associated With Memories of Recent Conflicts. ICOMOS, Paris.

——— 2018b. Evaluations of Nominations of Cultural and Mixed Properties. ICOMOS report for the World Heritage Committee. 42nd ordinary session, Manama, 24 June–4 July 2018 (No. WHC-18/42.COM/INF.8B1). ICOMOS, Paris.

Joy, C., 2012. The Politics of Heritage Management in Mali from UNESCO to Djenné. Routledge, Oxon and New York.

Kisic, V., 2016. Governing Heritage Dissonance. Promises and Realities of Selected Cultural Policies, Cultural Policy Research Award. European Cultural Foundation, Amsterdam.

Lambourne, W., 2013. Transformative Justice, Reconciliation and Peacebuilding, in: Buckley-Zistel, S., Beck, T.K., Braun, C., Mieth, F. (Eds.), Transitional Justice Theories. Routledge, Oxon and New York, 19–39.

Lennon, J.J., 2009. Tragedy and Heritage: The Case of Cambodia. Tourism Recreation Research 34, 35–43.

Little, A., 2012. Disjunctured Narratives: Rethinking Reconciliation and Conflict Transformation. International Political Science Review/Revue Internationale de science politique 33, 82–98.

Mac Ginty, R., Richmond, O., 2007. Myth or Reality: Opposing Views on the Liberal Peace and Post-War Reconstruction. Global Society 21, 491–497.

Paris, R., 2004. At War's End: Building Peace after Civil Conflict. Cambridge University Press, Cambridge and New York.

Richmond, O., 2011. A Post-Liberal Peace. Routledge, London.

Sabaratnam, M., 2011. The Liberal Peace? An Intellectual History of International Conflict Management, 1990–2010, in: Campbell, S., Chandler, D., Sabaratnam, M. (Eds.), A Liberal Peace?: The Problems and Practices of Peacebuilding. Zed Books, London, 13–30.

Selter, E., 2021. Fighting for Civilization. The International Politics of Heritage at War. PhD dissertation. SOAS, University of London, London.

Smith, L., 2006. Uses of Heritage. Routledge, New York.

Tandon, A. (Ed.), 2021. PATH: Peacebuilding Assessment Tool for Heritage Recovery and Rehabilitation, Toolkit on Heritage for Peace and Resilience. ICCROM, Rome.

UNESCO, 2021a. Reflection on Sites Associated with Memories of Recent Conflicts and Other Negative and Divisive Memories (No. WHC/21/44.COM/8). UNESCO, Paris.

——— 2021b. Outcomes of the Expert Meeting on Sites Associated with Recent Conflicts and Other Negative and Divisive Memories (Paris, 4–6 December 2019) (No. WHC/21/44.COM/INF.8.1). UNESCO, Paris.

——— 2018a. Warsaw Recommendation on Recovery and Reconstruction of Cultural Heritage. Online Document. UNESCO, Paris. URL https://whc.unesco.org/document/169671 (accessed 03 November 2020).

——— 2018b. Funerary and Memorial Sites of the First World War (Western Front) (Belgium, France) (No. Decision 42 COM 8B.24). UNESCO, Paris.

——— 2013. Final Report and Action Plan for the Rehabilitation of Cultural Heritage and the Safeguarding of Ancient Manuscripts in Mali. UNESCO, Paris.

Wijesuriya, G., 2000. Conserving the Temple of the Tooth Relic, Sri Lanka. Public Archaeology 1, 99–108.

Suggested Further Reading

Bevan, R., 2016. The Destruction of Memory : Architecture at War. Reaktion Books, London.

Smith, L., 2006. Uses of Heritage. Routledge, New York.

Tandon, A. (Ed.), 2021. PATH: Peacebuilding Assessment Tool for Heritage Recovery and Rehabilitation, Toolkit on Heritage for Peace and Resilience. ICCROM, Rome.

Walasek, H. (Ed.), 2015. Bosnia and the Destruction of Cultural Heritage, Heritage, Culture and Identity. Routledge, London and New York.

6 Cultural Heritage, Climate Change and Disaster Risk Management

Will Megarry

Introduction

Chapter Outline and Introduction

Climate change is the fastest growing global threat to heritage sites, many of which – natural, cultural, and mixed – are already being affected. These impacts can vary considerably from direct hazards, including rising sea levels, increased precipitation, and aridification, to indirect impacts like climate migration, conflict, and changing land-use strategies. While many of these hazards are global, they are not experienced equally. Lower income countries and communities (and their heritage) are disproportionately affected and indigenous communities are especially vulnerable. The risk to heritage is both immediate and escalating and there is a pressing need within disaster risk management to urgently respond. The International Council on Monuments and Sites (ICOMOS) stated in 2019, 'The impacts of changes are already damaging infrastructure, ecosystems and social systems – including cultural heritage – that provide essential benefits and quality of life to communities' (ICOMOS 2019). This chapter will explore how we can respond to these impacts. It will begin with a section on climate justice and equity which must underpin all climate action. It will then explore climate impacts on cultural heritage globally, the concept of vulnerability, and some attempts to assess and respond to it within a management framework, which is understood as the relationship and interactions of three main factors: hazards, exposure, and vulnerability. The focus will then switch to some potential solutions. While culture and heritage are impacted by climate change, they are also a source of resilience for communities and an asset in our response. This section will focus on the communicative power of culture and its importance in climate adaptation planning and carbon mitigation. Finally, a case study which connects these themes will be presented. This chapter aims to both stress urgency and increase ambition within the heritage sector to respond to the climate crisis.

This increased ambition is critical as the urgency of this situation cannot be understated. The Intergovernmental Panel on Climate Change (IPCC) has predicted that global warming is likely to reach 1.5°C between 2030 and 2052. IPCC Working Group 1 has recently noted that 'The scale of recent changes across the climate system as a whole and the present state of many aspects of the climate system are unprecedented over many centuries to many thousands of years' (IPCC 2021). Even more recently, Working Group 2 stressed that 'The rise in weather and climate extremes has led to some irreversible impacts as natural and human systems are pushed beyond their ability to adapt' (IPCC 2022). There is an urgent need to apply disaster risk management policies and strategies to protect and reduce impacts to cultural heritage, both tangible and intangible.

Table 6.1 Glossary of key terms

Adaptation	In human systems, the process of adjustment to actual or expected climate and its effects, in order to moderate harm or exploit beneficial opportunities.
Exposure	The presence of people; livelihoods; species or ecosystems; environmental functions, services, and resources; infrastructure; or economic, social, or cultural assets in places and settings that could be adversely affected.
Hazards	The potential occurrence of a natural or human-induced physical event or trend that may cause loss of life, injury, or other health impacts, as well as damage and loss to property, infrastructure, livelihoods, service provision, ecosystems, and environmental resources.
Impacts (consequences, outcomes)	The consequences of realised risks on natural and human systems, where risks result from the interactions of climate-related hazards (including extreme weather/climate events), exposure, and vulnerability.
Mitigation	A human intervention to reduce emissions or enhance the sinks of greenhouse gases.
Risks	The potential for adverse consequences for human or ecological systems, recognising the diversity of values and objectives associated with such systems.
Vulnerability	The propensity or predisposition to be adversely affected. Vulnerability encompasses a variety of concepts and elements including sensitivity or susceptibility to harm and lack of capacity to cope and adapt.

Source: IPCC (2021).

Language and Definitions

Language remains an ongoing challenge when exploring the intersection between climate change and cultural heritage. Terms including climate hazards and risks, and impacts and vulnerability tend to be used uncritically or even interchangeably, muddying the waters of an already opaque multidisciplinary pond. Unless explicitly stated, this chapter adheres to the IPCC glossary from the recent *Climate Change 2021: The Physical Science Basis* report (IPCC 2021). For expediency, this includes certain key terms, as outlined in Table 6.1.

Climate Justice and Equity

The topic of climate justice and equity may seem like an unusual place to start this chapter. Long lists of impacts and gloomy predictions presented on a par with other topics. This is unfortunate as the inequalities inherent in climate change dialogues permeate through every aspect of our understanding on the subject from climate impacts to how we respond to them (for a good and recent overview, see Simpson et al. (2022)). This means that any discussion on climate change and culture must recognise equity and justice as fundamental to climate action and this includes cultural heritage perspectives, which offer a unique and undervalued asset in climate action and response.

As mentioned above, recent IPCC reportage notes that climate impacts will be felt across all aspects of society including cultural heritage. It also notes that while 'some development and adaptation efforts have reduced vulnerability. Across sectors and regions the most vulnerable people and systems are observed to be disproportionately affected' (IPCC 2022). While much ink has been spilled on the subject of management planning and adaptation, we remain unprepared for the potential scale of loss and this loss is and will be more acutely felt in lower and middle-income countries where cultural heritage remains more vulnerable to the impacts of climate change. Acknowledging and better understanding this vulnerability imbalance must be at the heart of our societal and professional response. This chapter, therefore, begins with this dedicated section on climate justice and equity, exploring both the origins of the climate crisis as a Western industrialised phenomenon and how our understanding of climate change, its impacts, and our response are heavily based on Western knowledge systems and power dynamics (Simpson et al. 2022).

Climate Change and Justice: An Uncomfortable Truth

The uncomfortable truth is that climate change is indirectly continuing a colonial legacy of Western cultural oppression and destruction, most acutely felt by poorer and marginal communities. While climate change has and continues to be caused primarily by the cumulative greenhouse gas emissions of wealthier Higher Income Countries (HICs), it disproportionately affects poorer Low and Middle-Income (LMIC) ones. Conversely, these LMIC countries also have less capacity and fewer resources to adapt. This dynamic impacts all aspects of society, including cultural heritage which is usually a lower priority in terms of adaptation planning. Added to this are the disproportionate impacts on indigenous traditions and communities within these countries (Boyd et al. 2021). This dynamic, where the power to create meaningful change remains in the hands of the perpetrators, has a knock-on effect on other areas of climate change studies and knowledge including climate science.

Overcoming Existing Barriers to Climate Change Knowledge and Action: Inclusivity and the Need for a Plurality of Knowledge Systems

For the sake of this chapter, it is helpful to categorise knowledge into three broad categories: indigenous knowledge, local knowledge, and scientific knowledge. All include both intangible and tangible elements and while all three are crucial to responding to climate change, they have not all been equally valued to date. Within the climate science world, greater value has traditionally been associated with scientific knowledge systems to the exclusion of the other two. Two examples are discussed here. These explore the importance of scientific knowledge as presented in climate change publications and research, and the exclusion of alternative knowledge systems – including local and indigenous knowledge – from our responses, specifically their value to adaptation planning and carbon mitigation.

Most climate science is funded by and undertaken by researchers from wealthy countries. This creates a knowledge imbalance and also results in a lack of local institutional knowledge and capacities (Simpson et al. 2022). This bias is inherent in systems and processes which rely on academic outputs, including organisations like the IPCC which is tasked with synthesising knowledge and producing policy-relevant but not policy-prescriptive assessments of climate science. How it achieves this aspirational task has been the subject of much criticism in recent decades as the abovementioned biases inevitably create an incomplete picture of global impacts and responses. As Chakraborty and Sherpa (Chakraborty & Sherpa 2021) note,

> The IPCC still remains fairly biased towards epistemic tools emerging from reductionist, constituent ideologies; recognizes nation-states as the most important spatial units and fails to address the wider stakeholder community of non-state actors; and remains unable to conceptualize climate-society relationships that defy stark human-nature binaries.

While there is an increasing desire within the IPCC to acknowledge and respect different knowledge systems, the processes of knowledge synthesis and analytics pathways remain highly quantitative and restrictive (Vardy et al. 2017; Beck & Mahony 2018; Whyte 2020b;

Chakraborty & Sherpa 2021). High-level decision-making has created both a lack of dialogue and a lack of trust with local communities which are not conducive to meaningful action as they retain and perpetuate a lack of qualities central to collaboration including trust, consent, and accountability (Whyte 2020a). Even heritage organisations are built on dualities incompatible with respecting indigenous knowledge, further restricting dialogue. This includes a Western empirical division between culture and nature, and between tangible and intangible cultural heritage, as enshrined in the World Heritage Convention and its subsequent operational guidelines, which are not shared by many traditional knowledge holders (UNESCO 1972, 2021a). Conversely, unpacking and deconstructing this nature-culture duality lies at the heart of our response.

Any conversation about climate change and action must therefore start with an acknowledgment that both the history of climate change and the contemporary responses of governments and organisations are based on Western knowledge systems which prioritise empirical processes over qualitative approaches and favour high-level social engagement over community interaction. This dynamic is reflected in every aspect of climate literature from our understanding of impacts and loss and damage to adaptation planning and carbon mitigation. So how do we respond to this inequality? As Whyte has noted, reflecting the position of many Indigenous Peoples, we cannot solve the problem using the same tools which have caused it (Whyte 2020a). There is a need for a more inclusive and interdisciplinary response which respects a plurality of knowledge systems and engages equally with all stakeholders. Cultural heritage is ideally suited to lead this response, and to act as a bridge between traditionally disparate groups and communities.

Hazards, Risks, and Impacts

Numerous studies have shown a bias in publication and study foci which reflects both national concerns and colonial power structures, where authors from Europe and North America tend to write about places with post-colonial links to these countries (Sabbioni et al. 2010; Orr et al. 2021; Simpson et al. 2022). This means that we lack a complete understanding of the impacts of climate change on global cultural heritage. This is not to say that significant and valuable work has not been undertaken in the area. Climate change impacts every aspect of our tangible and intangible culture and a considerable corpus of literature on the topic has been produced in recent decades. While a complete summary is not appropriate here, excellent reviews and syntheses by Orr et al. (2021) and Sesana et al. (2021) are worthy of note. Also of value is the ICOMOS (2019) *Future of our Pasts: Engaging Cultural Heritage in Climate Action* report, which provides an excellent summary table of heritage typologies and impacts. This includes impacts to moveable heritage, archaeological resources, buildings and structures, cultural landscapes, associated and traditional communities, and intangible cultural heritage from both direct and indirect climate hazards and risks. Many impacts are quite straightforward. For example, the threats from rising sea levels, coastal erosion, and increased precipitation are well known (for some examples, see Figure 6.1). Other impacts are more complicated with nuanced modalities including risks due to indirect factors like displacement or changing land-use patterns. Understanding and quantifying these risks, hazards, and impacts remain challenging and the remainder of this section will explore some previous attempts and methodologies, focusing on studies into World Heritage (WH) at risk, regional assessments, impacts on specific cultural heritage typologies, and impacts on national inventories.

Figure 6.1 Infographics showing various direct climate impacts from well-known WH properties including: (A) Coastal erosion at Rapa Nui, (B) Efflorescence due to saline intrusion and rising aquifers in Bagerhat, Bangladesh, (C) Sea-level rise at Kilwa Kisiwani, Tanzania, and (D) Increased precipitation at the Old and New Towns of Edinburgh, Scotland.

Source: Images from the Heritage on the Edge Project, Google Arts and Culture 2020.

Impacts on Global Heritage

Studies which have received the most international attention have tended to focus on better-known and iconic WH properties. Some of these studies have taken a very broad and high-level approach, exploring impacts across large areas and over long time periods (Colette & Augustin 2013; Marzeion & Levermann 2014). Others have favoured more regional approaches. These include the impacts of sea level rise (SLR) and coastal erosion on 49 coastal cultural WH properties in the Mediterranean which found that 37 were currently at risk from flood based on a 1-in-100 year flood event, and 42 from coastal erosion. These risks will increase significantly by 2100 (Reimann et al. 2018). Across Africa, Vousdoukas et al. (2022) found that 56 coastal World Heritage properties are currently at risk from a similar coastal event based on moderate or high greenhouse gas emissions scenarios (RCP 4.5 and 8.5); however, this will rise to 200 by 2050. This threat was similarly identified during the Third Cycle of Periodic Reporting in the Africa Region which identified climate change and severe weather events as one of the top six factors affecting WH properties in the region (UNESCO 2021b). The focus on WH properties can draw attention away from the day-to-day reality faced by lesser-known sites and properties; however, they can be seen as representative of this wider threat. They are also powerful assets in raising awareness and promoting climate action globally (Megarry & Hadick 2021).

More inclusive regional studies include Hollesen et al.'s (2018) study of around 180,000 sites in the Arctic, which identified specific impacts including the intensification of permafrost and increased coastal erosion. It also identified specific challenges including the monitoring and vulnerability assessment of so many sites across such a large area. Across the world, climate change impacts on insular cultures in the Pacific have been the subject of numerous studies. Examples have focused on heritage impacts as non-economic losses (McNamara et al. 2021) and also on the importance of intangible heritage, identity, and place-based responses (Frediani et al. 2013; McNamara et al. 2021). These more specific regional or even national assessments often focus on specific climate change hazards. This includes the impacts of SLR on national heritage sites in Puerto Rico where 27 coastal sites are currently inundated at high tide, rising to 140 by the end of the century (Ezcurra & Rivera-Collazo 2018), or the effects of increased precipitation on mud architecture in Yemen (Al-Masawa et al. 2018). In other cases, they focus on specific cultural heritage categories. This includes underwater heritage and shipwrecks (Perez-Alvaro 2016; Wright 2016) and museum collections and indoor repositories (Stengaard Hansen et al. 2012; Bertolin 2019). A common theme in all assessments is that while the direct impacts of climate change may seem quite straightforward, understanding the actual vulnerability of structures, sites, properties and collections is more complex. As Viles (2002) noted over two decades ago, 'Although it is conceptually quite simple to envisage the impact of climate change on individual processes, the difficulty comes in trying to weigh up the importance of different impacts'.

Climate Stressors, Impacts, and Understanding Vulnerability

Understanding the vulnerability of cultural heritage to climate change is different from assessing the potential impacts of climate hazards on sites and properties. Since the third IPCC Assessment cycle in 2001, the emphasis has been on assessing vulnerability rather than risk. While risk is the potential for adverse consequences at sites, vulnerability is the predisposition to be adversely affected, taking other parameters into account. Within this framework, vulnerability is understood as a function of a range of factors including climate hazards, risks, impacts, and resilience or adaptive capacity (Daly 2014). Vulnerability acknowledges the agency and capacities of both heritage science and local communities to respond to climate risk and reduced impact and potential risk (Hennessy et al. 2007).

Various tools have been developed to assess the vulnerability of cultural heritage to climate change. These include Daly's (Daly 2014) six-step vulnerability assessment framework which explored vulnerability as a function of exposure, sensitivity, and adaptive capacity, using two WH properties in Ireland as case studies. Guzman et al.'s (2020) landscape-based approach seeks a more sustainable and ongoing dynamic which works in tandem with the conservation and management of site values at WH properties through the periodic reporting mechanism. The Climate Vulnerability Index, which will be explored in more detail later in this chapter, is a further tool developed to provide a rapid assessment of vulnerability at WH properties and is based on balancing potential impacts to the Outstanding Universal Value (OUV) and other key property values (KPV) against the adaptive capacity of sites and their communities (Day et al. 2020). Figure 6.2 contains a schematic chart which identifies key steps in assessing vulnerability.

Key to all these approaches is that they need to be responsive, reflexive, and repeatable as part of ongoing efforts to continually monitor (and not just assess) vulnerability. Site values, climate models and predictions, and adaptive capacities can and will often change so assessments need to be ongoing and inclusive, hearing from a wide range of stakeholders and inputting into conservation and adaptation plans (Daly 2019). This is especially important in contexts where other factors may influence or contribute to community vulnerability including resource access, changing subsistence patterns and social structures, and personal aspirations (Singh et al. 2019). Vulnerability assessment should also respect different and plural knowledge systems. These should include traditional knowledge systems which are rooted in more holistic observations of natural processes and cycles, urban and rural non-indigenous local knowledge systems which focus more on personal and collective experiences and can overlap with indigenous

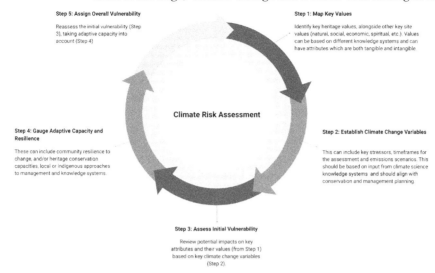

Figure 6.2 The vulnerability cycle.

knowledge systems, and climate science which has evolved from western empirical processes of experimentation and observation (Orlove et al. 2022). Central to this is acknowledging that while marginal and indigenous communities are often the most exposed to climate impacts, they can also be more resilient and resourceful when responding to its impacts (Zhongming et al. 2012). Many steps, including values mapping or gauging adaptive capacity benefit from a more bottom-up approach while understanding climate hazards, risk, and the timeframes of impacts require engagement with climate science. This latter is often hindered by a poor level of climate literacy within the heritage sector.

This approach to assessing vulnerability, while inclusive, remains both problem-oriented and quite hierarchical, necessitating a degree of facilitation. Within natural heritage and ecosystems, alternative tools have been developed which are more proactive and adaptation driven rather than focusing on only vulnerability, focusing on incorporating climate impacts rather than fighting against them (Múnera & van Kerkhoff 2019); however, these do not take into account the adaptive capacity of communities.

Responses

There are many ways that cultural heritage is an asset for climate action. These include contributions to each of the three pillars of climate action, as identified by the UNFCCC: adaptation, mitigation, and finance, through loss and damage and cultural heritage as non-economic loss. This section will briefly address each of these under two subheadings. These explore the value of culture and heritage in climate communication and stressing urgency, and the contribution culture can make to both adapting to climate change and to greenhouse gas mitigation.

Placing People at the Centre of the Response – Climate Stories

Heritage and heritage places are uniquely situated to communicate and stress urgency about climate change and build community resilience (Dawson et al. 2020). Loss and damage of sites are common themes which often include a focus on tourism operators stressing concepts like 'last chance tourism', where people visit special places before they are lost forever (Dawson et al. 2011; Hall 2021). While the destruction of emotive and evocative places is acutely felt, this communicative potential is not limited to loss and damage. Heritage places can also speak to human values and resilience or communicate lessons on adaptation and preservation. This wider communicative remit is best summed up by the phrase, 'every place has a climate story', which seeks to make broader linkages and connections between culture and climate (Rockman & Maase 2017). As with impacts, the focus tends to be on more iconic WH properties and many of these properties are using their platform to increase climate literacy and promote sustainability (Lafrenz Samuels & Platts 2020). Good examples include projects like Heritage on the Edge which explored a range of climate change related themes including carbon mitigation at The Old and New Towns of Edinburgh, Scotland, nature-based solutions and adaptation at the Ruins of Kilwa Kisiwani and Ruins of Songo Mnara, Tanzania, and the importance of new tools and methodologies to record heritage before it is lost at Rapa Nui. Each of these examples illustrates how 'Cultural heritage is also a lens through which we can explore many of the complex issues associated with climate change in a place-based and people-centred way, sympathetic to differing scales and experiences' (Megarry & Hadick 2021).

Knowledge Systems and the Role of Traditional Ecological Knowledge

As previously mentioned, vulnerability assessments are most valuable when they feed into longer term conservation and adaptation planning processing. In addition to the IPCC definition, Adamson et al. (2018) have described adaptation as 'an ongoing process that is managed over time by committing to shorter term actions embedded within a clear long term vision'. The importance of climate change adaptation (CCA) policies for cultural heritage cannot be overstated. Many national heritage organisations have begun developing sectoral plans (for good examples, see Harkin et al. 2020; HEG 2020) and there are also good emerging examples of culture being incorporated into national CCA plans (Kiribati 2014; Sesana et al. 2018; Daly et al. 2020); however, there remains an enormous need for the integration of adaptation planning into standard monitoring mechanisms at site, regional and national level (Daly 2019). Crucially, such plans should not only focus on reducing harm but also on identifying and promoting any opportunities afforded by effective adaptation.

While adaptation plans are needed for vulnerable cultural heritage, cultural heritage can and should also feed into wider adaptation planning. The role of local and traditional knowledge in responding to climate change has been extensively discussed, building on a growing acknowledgement of its value to adaptation planning from the climate science community (Carmichael et al. 2018; Petzold et al. 2020). Often referred to as traditional ecological knowledge (TEK), their utility is often promoted for adaptation. The best known example is the Paris Agreement which states that 'adaptation action should … be based on and guided by the best available science and, as appropriate, traditional knowledge, knowledge of indigenous peoples and local knowledge systems' (UNFCCC 2015). More recently, the Glasgow–Sharm el-Sheikh work programme, drafted during the 26th Conference of Parties in Glasgow in 2021, states that activities of the work programme should 'draw on a variety of sources of information and inputs, including national adaptation plans and adaptation communications, take into account traditional knowledge, knowledge of indigenous peoples and local knowledge systems' (UNFCCC 2021).

This emerging focus on the importance of adaptation is both timely and welcome as the reality of the climate crisis becomes apparent, but this should not overshadow the centrality of carbon mitigation to climate action, and the important role the heritage sector must play in achieving this. Mitigation remains the only cost-effective and definitive tool to combat climate change. Within the heritage sector, the emphasis has often been on reducing the carbon footprint of the built heritage sector (Hambrecht & Rockman 2017; Sesana et al. 2019), but the value of culture to wider actions should also be acknowledged. The linkages between local and traditional knowledge or TEK and adaptation are widely acknowledged; however, the equally important contributions of these knowledge systems to mitigation and decarbonisation are often overlooked. Much intangible heritage, including traditional and local knowledge, pre-dates the most recent two centuries when fossil fuel combustion and extractive land-use change have underpinned economic development, and this knowledge can point the way to post-carbon living. Such strategies should be accompanied by regular evaluation to ensure that they remain both effective in climate action and valuable to communities involved. Examples may include more energy efficient buildings and structures (Flores 2019), using historical mills to generate green energy (Fatorić & Egberts 2020) and promoting traditional land-use strategies as a means to store and sink carbon (Vierros 2017).

It is critically important that adaptation and mitigation strategies which seek to use and benefit from traditional or local knowledge systems do not repeat the mistakes of the past. Communities are not resources or laboratories for analysis, and research must not be extractive and exploitative. There is a risk that, as the climate crisis becomes more urgent, the ends may be seen to justify the means but dialogue, co-creation and informed consent must be at the heart of these efforts. Box 6.1 includes an example from two sites to explore this important co-creation with communities and discuss a range of themes including equity and justice, impacts, risk and vulnerability, and the value of local knowledge systems.

Box 6.1 Values-based climate change risk assessment: Piloting the climate vulnerability index for cultural heritage in Africa

a. Case Study Introduction

The benefits of using case studies to better understand these concepts have been noted elsewhere, both as a way of focusing on discrete threat profiles and impacts (Ford et al. 2010) and as a way of communicating about climate change and heritage (Rockman & Maase 2017). It will explore attempts to understand climate vulnerability at two WH properties, undertaken as part of the *Values-based Climate Change Risk Assessment: Piloting the Climate Vulnerability Index for Cultural Heritage in Africa* (hereafter CVI Africa) project. These are the Ruins of Kilwa Kisiwani and Ruins of Songo Mnara (RKK RSM) WH property in Tanzania and the Sukur Cultural Landscape (SCL) in Nigeria.

The CVI Africa project explored the utility of an existing vulnerability assessment tool – the Climate Vulnerability Index (CVI) – at two WH properties in Africa. The CVI is a systematic and rapid assessment tool that is values-based, science-driven, and community-focused, which builds upon the vulnerability framework approach described in the 4th Assessment Report of the Intergovernmental Panel on Climate Change (IPCC 2007). The vulnerability of both the property OUV and also socioeconomic values are determined by assessing the exposure (the degree of contact), sensitivity (the degree to which an attribute is affected) and adaptive capacity with respect to determined climate stressors (Day et al. 2020). Prior to the project, the CVI had only been applied to sites in Australia (Heron et al. 2020a), Scotland (Day et al. 2019), and the North Sea (Heron et al. 2020b), and the CVI Africa was the first time it was applied in Africa.

The RKK and RSM and SCL WH properties were chosen for three main reasons. Firstly, they represented very different heritage typologies. One is a coastal archaeological site while the other is an inland cultural landscape. Secondly, the threat profile and potential impacts were assessed to be very different at both sites. The RKK and RSM property had a long history of adapting to coastal hazards including coastal erosion (Ichumbaki & Mapunda 2017), while drought and temperature changes were deemed to pose a greater risk to the SCL (Ogunrinde et al. 2021). Finally, both sites had established local and indigenous communities whose lives were intertwined with the properties and who have rich traditions manifest in local and indigenous knowledge systems.

b. The CVI Methodology for the CVI Africa Project

The CVI broadly follows the steps outlined in Figure 6.2. A core steering committee of key stakeholders works with a wider group to identify key values for the properties. These are primarily focused on the OUV; however, other significant property values are also identified and mapped to key tangible and intangible attributes. In-country climate scientists are then employed to identify key climate hazards and generate downscaled climate models for both properties based on current IPCC Representative Concentration Pathways (RCPs). A multi-day workshop is then held with relevant stakeholders from the community, the heritage and tourism sector, representatives from local government and site management to define a timeframe to assess vulnerability, choose a likely RCP, and to identify key climate hazards impacting the site. Working in small and mixed groups, these stakeholders also assess the levels of exposure and sensitivity of key attributes, and the adaptive capacity of the site and community. Key to this process is creating a forum for open and inclusive discussion and debate. All these variables and considerations feed into an overall assessment of both OUV and community vulnerability. Both workshops were held in-country, but local factors and the global health situation necessitated some changes to this process in both cases. This included moving the workshop venues away from the sites and utilising a hybrid remote and in-person format to facilitate greater inclusion. More detailed information can be found in reports for RKK and RSM (Heron et al. 2022) and SKL (Day et al. 2022).

c. Results

The core of the CVI process is a workshop where stakeholders meet and agree on key inputs into the process. At both RKK and RSM, the values mapping exercise began with the WH property's statement of OUV where key values and their associated attributes were identified, examples of which are included in Figure 6.3. Other key property values including social, economic, and spiritual values were identified. Climate change factors and parameters were presented based on commissioned in-country models and reports and, as suspected, the key hazards for both sites were quite different. At RKK and RSM, key hazards included coastal threats including rising sea levels and erosion. Increased precipitation events were also identified as a key hazard. At SCL, key hazards were drought, temperature change, and changes in storm intensity and frequency. Both sets of stakeholders felt that RCP 8.5 was the most realistic and both also selected quite

Figure 6.3 Images from Kilwa Kisiwani WH property including: (A) eroding shoreline showing archaeological deposits, (B) protective wall above the beach below the Malindi Mosque, (C) protective wall in from of the Gereza Fort, and (D) damage to the Gereza Fort from coastal erosion (foreground) and mangrove forests protecting the structure (background).

short timeframes of 20 years (RKK and RSM) and 30 years (SKL). In both cases, these timeframes were chosen to reflect existing conservation concerns or align with site management plans.

Once climate inputs were established and key values and attributes identified, workshop stakeholders discussed their vulnerability by assessing their exposure and sensitivity. Based on these discussions, the sites were assigned an initial vulnerability rating of low, moderate, or high. The adaptive capacity of the property or community was then assessed and this was used to assign separate OUV and community vulnerability assessments for selected timeframes based on RCP 8.5. The OUV vulnerability for RKK and RSM was assessed as being high; however, ongoing efforts for the last three decades to protect the site including the construction on gabion walls and the planting of mangroves along the coast reduced this to moderate (Figure 6.3). The OUV vulnerability for SCL was defined as moderate; however, the adaptive capacity of the community was also deemed to be moderate resulting in an overall assessment of low. At both sites, the community vulnerability was also assessed to be low. Local communities were deemed to be highly adaptive and resilient, devising ways to reduce and adapt to impacts (Sukur) or even benefitting from adaptation efforts (Kilwa).

Table 6.2 CVI Africa summary table with key variables and outcomes

Ruins of Kilwa Kisiwani and Ruins of Songo Mnara, Tanzania	Sukur Cultural Landscape, Nigeria
Archaeological site	Cultural landscape
Statement of Outstanding Universal Value The islands of Kilwa Kisiwani and Songo Mnara bear exceptional testimony to the expansion of Swahili coastal culture, the Islamisation of East Africa and the extraordinarily extensive and prosperous Indian Ocean trade from the medieval period up to the modern era	**Statement of Outstanding Universal Value** The Sukur Cultural Landscape, with the Palace of the Hidi (Chief) on a hill dominating the villages below, the terraced fields and their sacred symbols, and the extensive remains of a former flourishing iron industry, is a remarkably intact physical expression of a society and its spiritual and material culture
Attributes include: The standing ruins including the Gereza and the Great Mosque and unique construction techniques	**Attributes include:** Dry stone structures including the Hidi palace complex, terraces and paved walkways, sacred trees and shrines, and agricultural features
Key climate hazards: 1. Intense precipitation events 2. Sea level rise 3. Coastal erosions	**Key climate hazards:** 1. Drought 2. Temperature trend 3. Storm intensity and frequency
Representative Concentration Pathway (RCP): 8.5	**Representative Concentration Pathway (RCP):** 8.5
Time scale: 2040	**Time scale:** 2050
Adaptive Capacity: Low - Moderate	**Adaptive Capacity:** Moderate
OUV Vulnerability: Moderate	**OUV Vulnerability:** Low
Community Vulnerability: Low	**Community Vulnerability:** Low

The CVI Africa project illustrated a number of key points. It emphasised the importance of inclusivity and discussion to amplify local voices, both to understand the values of properties and why they matter, but also to evaluate how they are being impacted and what is and can be done to reduce these impacts. Central to this is the importance of local (RKK and RSM) and indigenous (SCL) knowledge in both understanding impacts and adaptive capacity. Conversely, it also illustrated the importance of scientific knowledge systems and engaging with climate science to model hazards and climate change over different timeframes and using different scenarios.

Conclusions

This chapter has explored the intersection between climate change and cultural heritage. It has explored a range of themes including climate justice and equity, noting the disproportionate impacts on poorer and indigenous communities and emphasising the need for a broad and inclusive understanding of climate change which respects a plurality of knowledge systems. It then reviewed efforts to quantify the impacts of climate change on heritage, focusing on World Heritage, different climate hazards, and heritage typologies before exploring the subject of vulnerability, presenting a model based on a range of factors including climate hazards, risks, impacts, and resilience or adaptive capacity. The subsequent section explored two ways that culture can contribute to climate action. These are climate communication and the power of culture to tell stories about climate change and stress urgency, and the contribution cultural heritage can make to climate adaptation planning and mitigation strategies. It concluded with a case study focusing on a vulnerability assessment carried out at two African WH properties, the Ruins of Kilwa Kisiwani and Ruins of Songo Mnara in Tanzania and the Sukur Cultural Landscape in Nigeria. It has presented a broad overview of some of the key intersections between climate change and cultural heritage, with an overall focus on justice and equity.

In 2019, Greta Thunberg famously told the world that 'Our house is on fire' (Thunberg 2019; Thunberg et al. 2020). This is an appropriate analogy for those working in cultural heritage. As a sector engaged in the preservation and conservation of our past and present for future generations, we have a duty to save what we can from the flames. But the role of cultural heritage in climate action is not passive. It is not just a resource which needs to be saved. It is also a tremendous asset in our response. So while we work together to save our culture, we should also acknowledge and promote the exciting prospect that our culture may, in fact, save us.

Key Points

- Climate change is the fastest growing threat to cultural heritage and many sites have already suffered serious impacts.
- All climate action must be firmly embedded in concepts of justice and equity. This includes an acknowledgement that existing knowledge systems and modes of interaction have marginalised those most impacted by the climate emergency.
- There is a need for more inclusive approaches to understand concepts like risk, impacts, and vulnerability which include a plurality of both knowledge systems and values and place people at the centre of our climate responses.
- Key to this preparing sites for climate impacts is informed adaptation planning based on a full understanding of vulnerability, including local and Indigenous Knowledge systems.

References

Adamson, G.C.D., Hannaford, M.J. & Rohland, E.J. 2018. Re-thinking the present: The role of a historical focus in climate change adaptation research. *Global Environmental Change: Human and Policy Dimensions*, 48, pp. 195–205.

Al-Masawa, M.I., Manab, N.A. & Omran, A. 2018. The effects of climate change risks on the mud architecture in wadi Hadhramaut, Yemen. In A. Omran & O. Schwarz-Herion, eds. *The Impact of Climate Change on Our Life: The Questions of Sustainability*. Singapore: Springer Singapore, pp. 57–77.

Beck, S. & Mahony, M. 2018. The IPCC and the new map of science and politics. *Wiley Interdisciplinary Reviews. Climate Change*, 9(6), p. e547.

Bertolin, C. 2019. Preservation of cultural heritage and resources threatened by climate change. *Geosciences Journal*, 9(6), p. 250.

Boyd, E. Chaffin, B.C., Dorkenoo, K., Jackson, G., Harrington, L., N'guetta, A., Johansson, E.L., Nordlander, L., De Rosa, S.P., Raju, E. and Scown, M. 2021. Loss and damage from climate change: A new climate justice agenda. *One Earth*, 4(10), pp. 1365–1370.

Carmichael, B. Wilson, G., Namarnyilk, I., Nadji, S., Brockwell, S., Webb, B., Hunter, F. and Bird, D. 2018. Local and indigenous management of climate change risks to archaeological sites. *Mitigation and Adaptation Strategies for Global Change*, 23(2), pp. 231–255.

Chakraborty, R. & Sherpa, P.Y. 2021. From climate adaptation to climate justice: Critical reflections on the IPCC and Himalayan climate knowledges. *Climatic Change*, 167(3–4), p. 49.

Colette, A. 2013. *Case Studies on Climate Change and World Heritage*. UNESCO Publishing.

Daly, C. 2014. A framework for assessing the vulnerability of archaeological sites to climate change: Theory, development, and application. *Conservation and Management of Archaeological Sites*, 16(3), pp. 268–282.

Daly, C. 2019. Built & Archaeological Heritage Climate Change Sectoral Adaptation Plan. Available at: http://eprints.lincoln.ac.uk/id/eprint/38705/ [Accessed July 21, 2021].

Daly, C. et al. 2020. Climate change adaptation planning for cultural heritage, a national scale methodology. *Journal of Cultural Heritage Management and Sustainable Development*, 11(4), pp. 313–329.

Dawson, J. Johnston, M.J., Stewart, E.J., Lemieux, C.J., Lemelin, R.H., Maher, P.T. and Grimwood, B.S., 2011. Ethical considerations of last chance tourism. *Journal of Ecotourism*, 10(3), pp. 250–265.

Dawson, T. Hambly, J., Kelley, A., Lees, W. and Miller, S., 2020. Coastal heritage, global climate change, public engagement, and citizen science. *Proceedings of the National Academy of Sciences of the United States of America*, 117(15), pp. 8280–8286.

Day, J.C. Heron, S. F., Markham, A., Downes, J., Gibson, J., Hyslop, E., Jones, R. and Lyall, A. 2019. *Climate Risk Assessment for the Heart of Neolithic Orkney World Heritage Site*. Edinburgh: Historic Environment Scotland.

Day, J.C., Heron, S.F. & Markham, A. 2020. Assessing the climate vulnerability of the world's natural and cultural heritage. *Parks Stewardship Forum*, 36(1). Available at: https://escholarship.org/uc/item/92v9v778 [Accessed March 1, 2022].

Day, J.C., Heron, S.F., Odiaua, I., Downes, J., Itua, E., Abdu, A.L., Ekwurzel, B., Sham, A. & Megarry, W. 2022. An application of the Climate Vulnerability Index for the Sukur Cultural Landscape. ICOMOS-Nigeria, Abuja, Nigeria.

Ezcurra, P. & Rivera-Collazo, I.C. 2018. An assessment of the impacts of climate change on Puerto Rico's cultural heritage with a case study on sea-level rise. *Journal of Cultural Heritage*, 32, pp. 198–209.

Fatorić, S. & Egberts, L. 2020. Realising the potential of cultural heritage to achieve climate change actions in the Netherlands. *Journal of Environmental Management*, 274, p. 111107.

Flores, J.M. 2019. An investigation of the energy efficiency of traditional buildings in the Oporto world heritage site. *Restoration of Buildings and Monuments*. Available at: http://dx.doi.org/10.1515/rbm-2017-0010 [Accessed March 15, 2022].

Ford, J.D. Keskitalo, E.C.H., Smith, T., Pearce, T., Berrang-Ford, L., Duerden, F. and Smit, B. 2010. Case study and analogue methodologies in climate change vulnerability research. *Wiley Interdisciplinary Reviews. Climate Change*, 1(3), pp. 374–392.

Frediani, P., Frediani, M. & Rosi, L., 2013. Climate change and its impact on tangible cultural heritage: Challenges and prospects for small island nations. In P. Frediani, M. Frediani, & L. Rosi, eds. *Focus on Civilizations and Cultures: Cultural Heritage Protection, Developments and International Perspectives*. New York: Nova Science Publishers, pp. 293–307.

Guzman, P., Fatorić, S. & Ishizawa, M. 2020. Monitoring climate change in world heritage properties: Evaluating landscape-based approach in the state of conservation system. *Climate*, 8(3), p. 39.

Hall, D.R., 2021. *Tourism, Climate Change and the Geopolitics of Arctic Development: The Critical Case of Greenland*, CABI.

Hambrecht, G. & Rockman, M. 2017. International approaches to climate change and cultural heritage. *American Antiquity*, 82(4), pp. 627–641.

Harkin, D. Davies, M., Hyslop, E., Fluck, H., Wiggins, M., Merritt, O., Barker, L., Deery, M., McNeary, R. and Westley, K., 2020. Impacts of climate change on cultural heritage. *MCCIP Science Review*, 16, pp. 24–39.

HEG CC Subgroup 2020. Historic Environment and Climate Change in Wales. Sector Adaptation Plan.

Hennessy, K.B. Fitzharris, B., Bates, B.C., Harvey, N., Howden, M., Hughes, L., Salinger, J. and Warrick, R. 2007. Australia and New Zealand: Climate change 2007: Impacts, adaptation and vulnerability: Contribution of Working Group II to the Fourth Assessment Report of the Intergovernmental Panel on Climate Change. Available at: https://trid.trb.org/view/1154810. [Accessed February 28, 2022].

Heron, S.F. Day, J.C., Cowell, C., Scott, P.R., Walker, D.I. and Shaw, J., 2020a. *Application of the Climate Vulnerability Index for Shark Bay, Western Australia*. Perth: Western Australian Marine Science Institution.

Heron, S.F. Day, J.C., Zijlstra, R., Engels, B., Weber, A., Marencic, H. and Busch, J.A. 2020b. Workshop Report: Climate Risk Assessment for Wadden Sea World Heritage Property. Available at: https://rijkewaddenzee.nl/wp-content/uploads/2020/12/2020_CVI-Workshop-Report-1.pdf. [Accessed 28 February 2022].

Heron, S.F., Day J.C., Mbogelah, M., Bugumba, R., Abraham, E., Sadi, M.B., Pauline, N., Khamis MS, Madenge, S. & Megarry, W. 2022 Application of the Climate Vulnerability Index for the Ruins of Kilwa Kisiwani and the Ruins of Songo Mnara, Tanzania. CVI Africa Project, Dar es Salaam.

Hollesen, J. Callanan, M., Dawson, T., Fenger-Nielsen, R., Friesen, T.M., Jensen, A.M., Markham, A., Martens, V.V., Pitulko, V.V. and Rockman, M., 2018. Climate change and the deteriorating archaeological and environmental archives of the Arctic. *Antiquity*, 92(363), pp. 573–586.

Ichumbaki, E.B. & Mapunda, B.B. 2017. Challenges to the retention of the integrity of world heritage sites in Africa: The case of Kilwa Kisiwani and Songo Mnara, Tanzania. *Azania: Archaeological Research in Africa*, 52(4), pp. 518–539.

Intergovernmental Panel on Climate Change (IPCC). Working Group 1. *Climate Change* 2007: *Impacts, Adaptation and Vulnerability*.

Intergovernmental Panel on Climate Change (IPCC). Working Group 1. *Climate Change* 2021: *The Physical Science Basis Summary for Policymakers*.

Intergovernmental Panel on Climate Change (IPCC). Working Group 2. *Climate Change* 2022: *Impacts, Adaptation and Vulnerability Summary for Policymakers*.

International Council on Monuments and Sites (ICOMOS) 2019. T*he Future of Our Pasts: Engaging Cultural Heritage in Climate Action*, July 1, 2019. Paris: ICOMOS.

Kiribati 2014. *Kiribati Joint Implementation Plan for Climate Change and Disaster Risk Management (KJIP) 2014–2023*. Kiribati: Government of Kiribati.

Lafrenz Samuels, K. & Platts, E.J. 2020. An ecolabel for the world heritage brand? Developing a climate communication recognition scheme for heritage sites. *Climate*, 8(3), p. 38.

Marzeion, B. & Levermann, A. 2014. Loss of cultural world heritage and currently inhabited places to sea-level rise. *Environmental Research Letters*, 9(3), p. 034001.

McNamara, K.E., Westoby, R. & Chandra, A. 2021. Exploring climate-driven non-economic loss and damage in the Pacific Islands. *Current Opinion in Environmental Sustainability*, 50, pp. 1–11.

Megarry, W. & Hadick, K. 2021. Lessons from the edge: Assessing the impact and efficacy of digital technologies to stress urgency about climate change and cultural heritage globally. *The Historic Environment: Policy & Practice*, 12(3–4), pp. 336–355.

Múnera, C. & van Kerkhoff, L. 2019. Diversifying knowledge governance for climate adaptation in protected areas in Colombia. *Environmental Science & Policy*, 94, pp. 39–48.

Ogunrinde, A.T. Oguntunde, P.G., Akinwumiju, A.S. and Fasinmirin, J.T., 2021. Evaluation of the impact of climate change on the characteristics of drought in Sahel Region of Nigeria: 1971–2060. *African Geographical Review*, 40(2), pp. 192–210.

Orlove, B., Dawson, N., Sherpa, P., Adelekan, I., Alangui, W., Carmona, R., Coen, D., Nelson, M., Reyes-Garcia, V., Rubis, J., Sanago, G. & Wilson, A. 2022. ICSM CHC White Paper I: Intangible Cultural Heritage, Diverse Knowledge Systems and Climate Change. Contribution of Knowledge Systems Group I to the International Co-Sponsored Meeting on Culture, Heritage and Climate Change. Charenton-le-Pont & Paris, France: ICOMOS & ICSM CHC.

Orr, S.A., Richards, J. & Fatorić, S. 2021. Climate change and cultural heritage: A systematic literature review (2016–2020). The *Historic Environment: Policy & Practice*, 12(3–4), pp. 434–477.

Perez-Alvaro, E. 2016. Climate change and underwater cultural heritage: Impacts and challenges. *Journal of Cultural Heritage*, 21, pp. 842–848.

Petzold, J. Andrews, N., Ford, J.D., Hedemann, C. and Postigo, J.C. 2020. Indigenous knowledge on climate change adaptation: A global evidence map of academic literature. *Environmental Research Letters*, 15(11), p. 113007.

Reimann, L. Vafeidis, A.T., Brown, S., Hinkel, J. and Tol, R.S., 2018. Mediterranean UNESCO world heritage at risk from coastal flooding and erosion due to sea-level rise. *Nature Communications*, 9(1), p. 4161.

Rockman, M. & Maase, J., 2017. Every place has a climate story: Finding and sharing climate change stories with cultural heritage. In T. Dawson, C. Nimura, E. Lopez-Romero, & M-Y. Daire, eds. *Public Archaeology and Climate Change*. Oxford: Oxbow Books, pp. 107–114.

Sabbioni, C., Brimblecombe, P. & Cassar, M. 2010. *The Atlas of Climate Change Impact on European Cultural Heritage: Scientific Analysis and Management Strategies.* London: Anthem Press.

Sesana, E. Gagnon, A.S., Bertolin, C. and Hughes, J., 2018. Adapting cultural heritage to climate change risks: Perspectives of cultural heritage experts in Europe. *Geosciences Journal*, 8(8), p. 305.

Sesana, E. Bertolin, C., Gagnon, A.S. and Hughes, J.J., 2019. Mitigating climate change in the cultural built heritage sector. *Climate*, 7(7), p. 90. Available at: http://dx.doi.org/10.3390/cli7070090.

Sesana, E. Gagnon, A.S., Ciantelli, C., Cassar, J. and Hughes, J.J., 2021. Climate change impacts on cultural heritage: A literature review. *Wiley Interdisciplinary Reviews. Climate Change*. Available at: https://onlinelibrary.wiley.com/doi/10.1002/wcc.710. [Accessed 28 February 2022].

Simpson, N.P. Clarke, J., Orr, S.A., Cundill, G., Orlove, B., Fatorić, S., Sabour, S., Khalaf, N., Rockman, M., Pinho, P. and Maharaj, S.S., 2022. Decolonizing climate change–heritage research. *Nature Climate Change*. Available at: http://dx.doi.org/10.1038/s41558-022-01279-8 [Accessed: 28 February 2022].

Singh, C. Tebboth, M., Spear, D., Ansah, P. and Mensah, A. 2019. Exploring methodological approaches to assess climate change vulnerability and adaptation: Reflections from using life history approaches. *Regional Environmental Change*, 19(8), pp. 2667–2682.

Stengaard Hansen, L. Åkerlund, M., Grøntoft, T., Ryhl-Svendsen, M., Schmidt, A.L., Bergh, J.E. and Jensen, K.M.V., 2012. Future pest status of an insect pest in museums, Attagenus smirnovi: Distribution and food consumption in relation to climate change. *Journal of Cultural Heritage*, 13(1), pp. 22–27.

Thunberg, G. et al. 2020. *Our House Is on Fire: Scenes of a Family and a Planet in Crisis*. UK: Penguin.

United Nations Educational, Scientific and Cultural Organization (UNESCO) 1972. *Convention Concerning the Protection of the World Cultural and Natural Heritage*. UNESCO, World Heritage Centre.

United Nations Educational, Scientific and Cultural Organization (UNESCO) 2021a. *Operational Guidelines for the Implementation of the World Heritage Convention*. UNESCO, World Heritage Centre.

United Nations Educational, Scientific and Cultural Organization. 2021b. *Report on the Results of the Third Cycle of Periodic Reporting Exercise in Africa, World Heritage Committee, Extended 44th Session, Fuzhou (China)/Online, 2021*. UNESCO.

United Nations Framework Convention on Climate Change (UNFCCC) 2015. Paris Agreement. In *Report of the Conference of the Parties to the United Nations Framework Convention on Climate Change (21st Session, 2015*: Paris).

United Nations Framework Convention on Climate Change (UNFCCC) 2021. Glasgow–Sharm el-Sheikh work programme on the global goal on adaptation. Available at: https://unfccc.int/documents/311181

Vardy, M. Oppenheimer, M., Dubash, N.K., O'Reilly, J. and Jamieson, D., 2017. The intergovernmental panel on climate change: Challenges and opportunities. *Annual Review of Environment and Resources*, 42(1), pp. 55–75.

Vierros, M. 2017. Communities and blue carbon: The role of traditional management systems in providing benefits for carbon storage, biodiversity conservation and livelihoods. *Climatic Change*, 140(1), pp. 89–100.

Viles, H.A. 2002. Implications of future climate change for stone deterioration. *Geological Society, London, Special Publications*, 205(1), pp. 407–418.

Vousdoukas, M.I. Clarke, J., Ranasinghe, R., Reimann, L., Khalaf, N., Duong, T.M., Ouweneel, B., Sabour, S., Iles, C.E., Trisos, C.H. and Feyen, L., 2022. African Heritage sites threatened as sea-level rise accelerates. *Nature Climate Change*. Available at: http://dx.doi.org/10.1038/s41558-022-01280-1

Whyte, K., eds. 2020a. Against crisis epistemology. *Routledge Handbook of Critical Indigenous Studies*. UK: Routledge, pp. 52–64.

Whyte, K. 2020b. Too late for indigenous climate justice: Ecological and relational tipping points. *Wiley Interdisciplinary Reviews. Climate Change*, 11(1). Available at: https://onlinelibrary.wiley.com/doi/10.1002/wcc.603

Wright, J. 2016. Maritime archaeology and climate change: An invitation. *Journal of Maritime Archaeology*, 11(3), pp. 255–270.

Zhongming, Z. Nakashima, D., Galloway McLean, K., Thulstrup, H., Ramos Castillo, A., Rubis, J. and Traditional Knowledge Initiative, 2012. Weathering uncertainty: Traditional knowledge for climate change assessment and adaptation. Available at: http://resp.llas.ac.cn/C666/handle/2XK7JSWQ/10723

7 Cultural Heritage and Urbanization in the Context of Disaster Risk Management in Istanbul

Ebru A. Gencer

Protecting cultural heritage to entrust to future generations is deemed a responsibility for civilisations. Academicians and practitioners have long argued that historical environments should be preserved to 'provide a functional efficiency' and to 'use the most important elements of the past culture to generate future theories by interpreting the data of the past with contemporary methods' (Okyay 1976).[1] As bridges from past to future, heritage sites also form the cultural identities of our communities, cities, and nations, supporting cultural continuity, especially in the context of national and ethnic conflicts worldwide.

Historic preservation discipline was established in Europe at the end of 19th century. However, it was after the First World War that the Western World started to enact legislations that were inclined towards preserving the architectural heritage with historical and cultural importance, and to establish institutions related with preservation. This awareness gathered speed following the Second World War when international foundations such as ICOMOS (International Council of Monuments and Sites) were established with the understanding that cultural and historical heritage not only belong to the countries they are located in but that they are also a common heritage of the world. The 1964 Venice Charter was 'a milestone in changing the definition of cultural heritage and extension of the scope of protection' to include civil structures and immovable cultural properties (Unal 2014, 20).

Despite the increased importance of 'site preservation' as a concept, implementation was less successful due to the modernisation and rapid urbanisation processes following the Second World War. While trying to find answers to 'what' is to be preserved, and 'why' and 'how', many cities with historical importance went through a rapid urbanisation process that led to the permanent destruction of important cultural heritage sites before 'historic preservation' could be institutionalised, and preservation actions could even begin. Within the last decades, in addition to rapid urban development, other crises in the urban context have added to the destruction of cultural heritage sites. Among such crises are human-induced conflicts such as the Balkan wars that led to the disintegration of Former Yugoslavia, those induced by natural hazards such as the 2015 earthquake in Nepal, and the ongoing and future potential impacts of climate change on cultural heritage sites.

Therefore, today there is a greater stress on urban cultural heritage sites requiring a more systematic approach to urban risk management and attention to cultural heritage preservation. This chapter will examine urban risk management approach to cultural heritage through the case study of Istanbul's Historical Peninsula which has been exposed to conflicts, rapid urbanisation, earthquakes, and, more recently, the increasing impacts of climate change related disasters. This case study will review how the concept of cultural heritage preservation has changed through time and how it has been contemplated within the context of planning of the Istanbul Metropolitan Area.

DOI: 10.4324/9781003293019-10

Istanbul's Historical Peninsula: Cultural Heritage Preservation and Urbanisation in the Context of Disaster Risk Management[2]

The City of Istanbul has been exposed to multiple hazards and crises throughout its history. The city was invaded and rebuilt several times and suffered from major earthquakes, fires, and tsunamis. Existing historical records reveal 'mosques and churches collapsing, giant waves forming and ships colliding at the Marmara Sea, thousand dying or living outdoors from fear or due to the fires following earthquakes' (see Fig. 7.1) (Gencer 2007, 305).

Most of these disasters impacted the area of the city within the city walls known as the *Historical Peninsula*, where Istanbul's historical development largely took place; as it evolved from a Greek colonial settlement in 700 BC into the capital of the Eastern Roman Empire, Constantinople, in the fourth century. The city's spatial structure mostly remained unchanged until the Ottoman conquest in 1453 AD and reached its monumental development in the 16th century, leading to the concentration of one of the most valued cultural heritage sites in the world within the Historical Peninsula.

Figure 7.1 Topographical Istanbul during the Byzantine period.

Source: Main map source: R. Janin, Constantinople Byzantine. Developpement urbain et repertoire topographique. By Cplakidas.

While the heritage sites important to the city were rebuilt several times during the centuries due to invasions, fires, and earthquakes, it was in 1696 that a first provision to prevent disasters was declared as the Istanbul head officer prohibited the use of timber architecture to prevent fires for new construction as several fires during the 17th century caused immeasurable damage in the Historical Peninsula. Among them, in a single fire disaster, 20% of the city had perished, in others, many important architectural structures, including a big part of the Topkapi Palace and Eyup and Spice Bazaars burnt down. The ensuing fires and the modernisation movement led to the first post-disaster planning legislation in 1848 with the aim to reduce future fire disasters by applying a grid planning system in the burnt residential areas. This was also coinciding with the Historical Peninsula's first plan– a 1/2500 scaled map drawn by Helmuth Van Moltke during 1836–1837 in order to provide a continuous street network in the walled city fulfilling modernisation efforts by enlarging streets, opening park areas, and carrying out masonry construction.

Meanwhile, the first conscious attempts for the preservation of the historical heritage in the Ottoman Empire also started in the second half of the 19th century. While the first legislation on heritage only considered archaeological heritage, the ensuing ones expanded the context to include Turkish-Islamic art, Greek and Roman ruins, and immovable objects such as buildings with significant cultural heritage. In addition, new decrees also prohibited damage to such objects and their surroundings,[3] leading to the protection of cultural heritage sites from fires and other disasters.

After the establishment of the Turkish Republic, preservation activities which were generally under the responsibility of the central government, became a part of local governments' responsibilities. In 1944, the Advisory Commission of Historical Structures argued for the necessity of conserving historical structures within urban development plans. Modern planning activities focused on giving importance to Istanbul's cultural heritage and the natural environment.

From 1936 to 1950, the French planner Henri Prost's activities left an important impact in Istanbul and on its cultural heritage. Prost's 1939 Historical Peninsula Plan focused on preserving the silhouette of Istanbul by limiting construction heights and preserving its natural beauty and important heritage sites: the Topkapi Palace walls, the Archaeological Park, and Byzantine and Eastern Roman palaces region including the Hagia Sophia and Hippodrome area. On the other hand, despite preserving and amplifying the beauty of the most significant historical sites, Prost's plans also eliminated the old city fabric by way of building large transportation arteries.

These development operations not only led to the loss of civic heritage but also to the loss of the housing stock together with this loss of housing, ensuing population increase in the 1950s with rural to urban migration, led to 'gecekondu' (squatter) building activity in Istanbul (Tekeli 1994), initiating vulnerability and risk accumulation in the city.

After 1939, Henri Prost's Historical Peninsula Plan, the first master planning activity for the Historical Peninsula took place in 1990 with a 1/50,000 scaled Historical Peninsula Preservation Development Master Plan by Professor Gunduz Ozdes.

In addition to this Master Plan, another large-scale plan for Istanbul was the 1/50,000 scaled the Metropolitan Istanbul Sub-Region Master Plan which was prepared by the Greater Istanbul Master Planning Office in 1994. The 1994 plan proposed the preservation of the Historical Peninsula to transform it into an area of tourism, commerce, and culture, and moving population and industrial pressures out of the Peninsula. The 1994 plan analysis involved a series of geophysical studies, including a first-time review of past earthquakes in Istanbul, and explained that 'in Istanbul, the possibility of a damaging earthquake with a magnitude six or more' was quite high (IBB 1995, 52). The plan proposed to 'identify priorities in critical areas due to

Istanbul's earthquake risk and develop alternative mass housing projects for squatters in residential development areas', as well as 'to protect geologic and archeological sites north of the Küçükçekmece Lake'[4] (ibid. 329). In addition to these first-time analysis and proposals reflecting geophysical conditions of the city, the Metropolitan Municipality established *Geotechnical and Earthquake Investigation Directorate* in January 1997 (Özerol 2001, 116).

However, these geological studies were already too late for the informally urbanised city and settlements extending to geologically unstable areas. According to a study of Istanbul Governorship in 1992, 400,000 buildings were 'gecekondus' and 750,000 more were previously regularised by amnesty laws (Sonmez 1996, 140; Mortan 2000, 49). When the August 1999 Marmara Earthquake hit Istanbul, the vulnerability of the city revealed itself with destruction and loss of life. According to the Istanbul Governorship records, 981 people lost their lives, 41,180 residences and workplaces were damaged, and 18,162 families needed temporary sheltering in Istanbul following the August 1999 earthquake.[5]

Post-Disaster Planning Activities in the Historical Peninsula

Following the August 1999 earthquake and an ensuing one on October 1999, named together as the 'Marmara earthquakes', and the heightened possibility of a major earthquake in the Istanbul region in the near future, the Istanbul Metropolitan Municipality requested the preparation of the *Istanbul Earthquake Master Plan* (IEMP) by a consortium of four universities. This document laid the basis for identifying activities and responsible authorities, as well as preparing an action plan for disaster risk mitigation and management in Istanbul.

As part of the *Istanbul Earthquake Master Plan, the Strategic Plan for Disaster Mitigation in Istanbul,* prepared by Boğaziçi and Yıldız Technical Universities, aimed at diminishing 'the hazardous effects of a possible earthquake in Istanbul', and 'supported by a secondary goal of improving the quality of the natural and urban environment' (Ökten et al. 2003a, 195). The *Strategic Plan for Disaster Mitigation in Istanbul* (SPDMI) focused on (a) identifying the problems and the potentials of the Istanbul Metropolitan Area, (b) developing a 'road map' with strategies, planning instruments, and priorities, and (c) examining current legal and institutional issues, and proposing recommendations (ibid., 195–214; Ökten et al. 2003b, 225–406).

The SPDMI team suggested a 'three-fold road map' consisting of macro- and mezzo-level strategies and micro-level implementation for risk-based planning and development based on suitable risk reduction strategies in urban areas (ibid.). The team identified priority working areas based on problematic and potential areas related to legal status of their initial development, urban functions, and population densities, as well as hazard probabilities. According to their findings, the highest areas of priority for risk-based planning and development were found to be the historic areas including both the Historical Peninsula and Galata.

For each priority area, the SPDMI team suggested different planning tools such as *Urban Redevelopment Ignition*, *Local Redevelopment, and Land Readjustment* and specific implementation instruments such as the use of reinforcement and micro-renovations, short-term reinforcement and long-term urban renewal projects, renewal and regeneration strategies, and/or relocation based on topographical reevaluations (ibid., 202).

For the Historical Peninsula, which was designated as a first priority area for risk-based planning and development, the SPDMI team discussed the importance of historical and cultural heritage in the framework of the larger vision of making Istanbul a World City. The planning team argued that in addition to the gradual loss of economic and social dynamics in the Historical Peninsula, an emerging danger was the potential earthquake hazard as it had been impacted several times throughout its history. The planning team classified the Historical

Peninsula into eight sections based on soil studies, which also indicated potential liquefaction and landslide areas due to sloping. In addition to this classification based on hazard potential, the planning team also classified the Historical Peninsula based on functions, densities, and land use. Indeed, the Historical Peninsula is now not only home to immeasurable cultural heritage but also a Central Business District and low-quality multi-storey housing of lower-income residents who has mainly migrated into the city since the 1950s. Therefore, while the Historical Peninsula was affected by earthquakes before, the changing character of housing increases risk not only for that housing stock but also the risk of historical and cultural heritage that exists side by side with that housing stock. Therefore, the earthquake risk classification system in the Historical Peninsula included an urban fabric analysis based on construction methods, age of structure, number of floors, overall geological structure, and the changing neighbourhood character. The team discussed that while the presence of masonry buildings in some neighbourhoods increases damage, in some locations, the geological factor predominates, and in others, density is a risk increasing factor. When all these factors are calculated in a risk classification system, the median scores attributed to each neighbourhood may not represent the true risk, as there may be different urban tissues and different densities. Therefore, the urban tissue should become a critical risk-reduction factor in urban areas, particularly in historical urban areas, where there is not a classical zoning and layout (Ökten et al. 2003b, 296–301).

Due to the historical fabric and central business district function of the Historical Peninsula, the Planning Team argued that the risk reduction strategy of the Historical Peninsula along with Galata and Taksim areas should be considered as a whole. The future of the Historical Peninsula should not be planned using the parameters of a classical Master Planning Strategy but in sub-regions, using principals provided and undertaking work and projects at an urban design scale. For instance, in the Hanlar Region, which is home to the famous Grand and Spice Bazaars, planning for risk reduction should be undertaken at an urban design scale, eliminating the area from risky buildings of no historical value, which otherwise increases overall risk for the site. Rehabilitation of this area can be achieved with a partnership of private partners, business cooperations for each Bazaar and the Ministry of Tourism.

The planning team recommended other strategies in sites that are home to monumental structures, such as those around historical mosques arguing that these sites and monuments are a part of the World Cultural Heritage and their rehabilitation should be sponsored by International Organizations.

The planning team also recommended for the rehabilitation of other sites where there is a combination of monumental structures and examples of civic heritage. At such sites, decision for restoration can be done on a building basis that will not only allow the preservation of such heritage but also reduce their earthquake risks. In addition, other design tools can be used include the design of green axes to prevent the spread of fires with strategic tree planting, the development of micro-infrastructure with pressurised water and neighbourhood pools to be used as water tanks.

The planning team also argued that in areas, such as Galata neighbourhood, where rehabilitation cannot be achieved at a neighbourhood scale due to urban fabric and existing densities, special retrofitting techniques should be investigated and applied. Indeed, retrofitting or rehabilitation at a building scale also became a central measure of *Disaster Risk Mitigation Plan for Cultural Assets*, which was later developed as part of the Istanbul Metropolitan Municipality's Seismic Risk Mitigation and Emergency Preparedness Project (ISMEP) that was initiated in 2006. ISMEP project was undertaken within the framework of Risk Mitigation Planning in Istanbul, ensuring institutional coordination between heritage and planning teams.

This collaboration is based on a protocol that was signed in 2006 between Istanbul Project Coordination Unit and the Ministry of Culture and Tourism for developing proactive measures that can be taken in case of an earthquake for cultural heritage structures in Istanbul that is used by or belonging to the Ministry (Unal 2014, 42). One of the main goals of the ISMEP project is 'to integrate historical assets and cultural heritages in Istanbul with an extensive risk mitigation plan' and 'to include cultural resources into planning within the current policies and programs' (ibid., 43). This project included an inventory of Istanbul's cultural heritage buildings and reviewing their multi-disaster and earthquake performances, performing a loss-damage estimation, and preparing disaster risk mitigation plan. Within this scope, monumental buildings; palaces, administrative buildings and hospitals; museums; library and educational buildings; and civil buildings with historical values were evaluated. This project once again underscored that '[u]rban heritage is an indispensable part of the urbanized people and their identity', and '[t]herefore, this heritage needs to be transmitted to the next generations through cultural reference constituting the conscious and content of history and culture' (Unal 2014, 62).

Conclusion

As this chapter presented, the notion of what to preserve in Istanbul and in what modalities largely followed the institutional development in the field of historic preservation in Turkey and in the West. On the other hand, in Istanbul, modernisation and urbanisation processes took place in a more rapid fashion particularly following the First World War and the establishment of the new Turkish Republic. While the preservation field was evolving, planning and preservation processes took place in parallel but not in a coordinated manner.

The City reacted to various disasters by bringing new planning approaches, starting with adopting grid-planning system to reduce fire disasters in the 19th century. On the other hand, this planning approach led to the destruction of historical residential fabric. Similarly, while wanting to preserve the monumental architecture and historical silhouette of the city, planning operations of Henri Prost that lasted from 1936 to 1950 led to the loss of civic historic structures and housing stock and were replaced with informal and low-quality housing following rural to urban migration to Istanbul that started in the 1950s. Disaster risk studies in the context of planning were not accomplished up until the 1994 plan. However, the City had already accumulated high levels of spatial risk leading to losses in the 1999 Izmit Earthquake.

Planning activities following the 1999 Marmara Earthquakes were the first time when all three fields: cultural heritage preservation, urban planning, and disaster risk management started to be considered in unison.

As part of these planning activities, The Strategic Plan for Disaster Mitigation in Istanbul (SPDMI) argued that Istanbul's Historical Peninsula and Galata Districts should be priority areas for planning for disaster risk reduction and proposed specific implementation instruments. The planning team also developed a risk classification system based on geological studies, functions, densities, and land use and argued that historical urban fabric required local solutions and specific instruments for risk reduction that should be tailored to each site. Some of the proposals of the SPDMI team were later executed as part of Disaster Risk Mitigation Plan for Cultural Assets within the framework of Risk Mitigation Planning in Istanbul Metropolitan Municipality's Seismic Risk Mitigation and Emergency Management Project, providing the institutional coordination among all three fields.

This change in approach to develop coherent strategies for disaster risk management, urban development, and cultural preservation underscores the increased awareness, globally, of the significance of cultural heritage for social resilience and sustainable development. In addition, increasing disaster risks, including those of climatological origins, is not only a threat to cultural heritage but also to sustainable urbanisation processes and urban resilience.

As this short example on planning activities in the nexus between preservation and risk reduction presents urban cultural heritage preservation within the scope for disaster risk reduction requires context-based instruments tailored for individual sites and should be considered in unison with risk-based planning activities. The increasing stress on urban cultural heritage from crises and disasters requires such unified and systematic approach to urban risk management ensuring that urban heritage is protected for our cultural continuity and resilience and sustainable development at large.

Key Points

- Following the establishment of the Turkish Republic in 1923, planning and historic preservation approaches in Istanbul mainly followed parallel tracks, and they seldom included disaster risk reduction measures.
- Planning activities following the 1999 Marmara Earthquakes were the first time when all three fields: urban planning, disaster risk management, and cultural heritage preservation were considered in unison: The Strategic Plan for Disaster Mitigation in Istanbul (2002) argued that Istanbul's Historical areas needed to be considered as priority areas for disaster risk reduction and that they required specific instruments based on contexts of their urban fabric.
- Such coherent strategies that take into consideration of disaster risk management of historic sites within the larger context of risk-informed planning can lead to the sustainable preservation and increased resilience of our cultural heritage.

Notes

1 Okyay, İsmet, 'Tarihsel Çevreyi Koruma ve Ulusal Kültür Sorunsalı', *Mimarlık*, 1976, No: 3, p. 38 (author's translation).
2 This case study is based on E.A. Gencer (1996). *Applicability of the Urban Preservation Plans: The Case of the Istanbul Historical Peninsula*. Masters Thesis, Mimar Sinan University, Istanbul; and E.A. Gencer (2007). *Natural Disasters, Vulnerability and Sustainable Development: Global Trends and Local Practice in Istanbul*. PhD Thesis, Columbia University, New York.
3 Zeren, Nuran, 'Kentsel Alanlarda Alınan Koruma Kararlarının Uygulanabilirliliği', Istanbu Technical University, Faculty of Architecture, PhD Thesis, pp. 28–30, Istanbul, 1981 (author's translation).
4 Author's Translation.
5 Istanbul Governorship Disaster Management Center briefing 2002.

References

Gencer, E.A. 1996. *Applicability of the Urban Preservation Plans: The Case of the Istanbul Historical Peninsula*. Masters Thesis, Mimar Sinan University: Istanbul (unpublished).
Gencer, E.A. 2007. *Natural Disasters, Vulnerability and Sustainable Development: Global Trends and Local Practice in Istanbul*. PhD Thesis, Columbia University: New York.
Istanbul Buyuksehir Belediyesi (IBB). 1995. 1/50.000 Olcekli Istanbul Metropolitan Alan Alt Bolge Nazim Imar Plan Raporu. [1/50.000 scaled Istanbul Metropolitan Area Sub-Region Master Plan Report]. IBB, Planlama ve Imar Daire Baskanligi, Sehir Planlama Mudurlugu: Istanbul.
Mortan K. (ed). 2000. *Istanbul bir sosyo ekonomik degerlendirme*. [Istanbul: A Socio Economic Evaluation]. T.C. Istanbul Valiligi (Istanbul Governorship): Istanbul.

Ökten, A., Saglam, B., Sengezer, I., Dincer, G., Batuk, E., Demir, E., Koc, A., Gul, Y., Evren, E., Seckin, T., Cekic, I. and Emem, O. 2003a. 'Urban Planning, Legal Issues, Administration, Finance', in Bogazici University (BU), Istanbul Technical University (ITU), Middle East Technical University (METU), *Earthquake Master Plan for Istanbul*. Geotechnical and Earthquake Investigation Department, Planning and Development Directorate, Istanbul Metropolitan Municipality (IMM): Istanbul.

Ökten, A., Sengezer, I., Dincer, G., Batuk, E., Koc, A., Gul, Y., Evren, E., Seckin, T., Cekic, I. and Emem, O. 2003b. 'Yerlesim Caismalari' [Settlement Studies], in Bogazici Universitesi (BU), Istanbul Teknik Universitesi (ITU), Orta Dogu Teknik Universitesi (ODTU), *Istanbul Icin Deprem Master Plani.* [Earthquake Master Plan for Istanbul.] Zemin ve Deprem Inceleme Mudurlugu, Planlama ve Imar Dairesi, Istanbul Buyuksehir Belediyesi (IBB): Istanbul.

Okyay, İ.1976. Tarihsel Çevreyi Koruma ve Ulusal Kültür Sorunsalı [Preserving historical environment and national culture question]. *Mimarlık*, No: 3, p. 38.

Özerol, U. 2001. Proceeding in Istanbul ve Deprem [Istanbul and Earthquake], 111–20. Istanbul Kultur ve Sanat Urunleri, Istanbul Buyuksehir Belediyesi (IBB): Istanbul.

Sonmez, M. 1996. *Istanbul'un iki yuzu: 1980'den 2000'e degisim.* [Two faces of Istanbul: Transformation from 1980 to 2000]. Arkadas: Ankara.

Tekeli, I. 1994. *The Development of the Istanbul Metropolitan Area: Urban Administration and Planning.* IULA-EMME and YTU: Istanbul.

Unal, Z. 2014. Protection of cultural heritage. Istanbul Seismic Risk Mitigation Project (ISMEP), Istanbul Project Coordination Unit, ISMEP Guide Books, Vol. 6. Istanbul. https://commons.wikimedia.org/w/index.php?curid=5084599

Zeren, N. 1981. 'Kentsel Alanlarda Alınan Koruma Kararlarının Uygulanabilirliği' [Implementation of Preservation Decisions in Urban Areas], Istanbul Technical University, Faculty of Architecture, PhD Thesis, pp. 28–30, Istanbul.

8 Vernacular Built Heritage and Disaster Resilience

Rajendra and Rupal Desai

Around the world, people have learnt to value the priceless heritage from the ancient world, especially the built heritage in the form of temples, palaces, forts, etc., since they link us to the past history, and often leave us in awe of what our forefathers could do without the help of the modern science and technology. On the other hand, the vernacular-built heritage, most commonly in the form of houses, is as fascinating since it demonstrates how closely our lives have been linked to the local context, and is priceless since, being not dependent on the energy consuming modern construction materials, it has been recognised as truly sustainable. Most regions of the earth stretching from the equator all the way to the Arctic have this priceless heritage which is still alive in many areas, and Indian Subcontinent is no exception.

But today for a literate Indian in the urban areas, the word 'house' reminds her of cement, steel, and bricks. An overwhelming majority of buildings in urban and suburban India are getting built out of these materials. It does not matter if it is a single-storey house or a multi-storey one and if it is in Kashmir in North or Kerala in South. In other words, the building construction has nothing to do with the 'local context'.

But the housing scene is markedly different in the rural areas of the country. While thinking about building a house, most house owners, barring a few rich ones, will think of what is most easily found locally in the surrounding area such as stone, mud, timber, bamboo, and bricks (which is nothing but transformed mud).

Still, most everyone does aspire for cement and steel mainly because they are considered as signs of modernity and prosperity. Also, wrongly or rightly, these materials are seen as signs of permanence. On the other end, the traditional local materials are seen as a sign of poverty and impermanence.

Vernacular Architecture and Local Context

Building design, as well as building construction rooted to a particular region, is an integral part of the precious vernacular architecture built heritage of the respective region. Historically people built buildings using materials that were most easily available locally, which are logically termed as 'local materials'. These materials dictated the form of the buildings including the most common sizes for various spaces. Over generations the local climate patterns also contributed in a big way to the form as well as the construction of the building such that the building would perform well in the face of the elements.

It is best to look at some examples of the vernacular-built heritage from different parts of India. For each of these cases, the basic details about the construction and its relationship with the local context as well as reasons for their disaster vulnerabilities and capacities are explained with particularities specific to that heritage.

DOI: 10.4324/9781003293019-11

Marathwada, Maharashtra State, Western India

Marathwada region on the Deccan Plateau is situated right on top of Basalt rock which offers excellent construction material for walls. It is in rain shadow, and hence, gets little rain. As a result, the walls in houses are made of locally found Basalt stone, whereas the roof is made of a thick layer of soil laid on a deck made out of local timber (Figure 8.1).

Skills such as stone quarrying and shaping, carpentry, and soil preparation have been historically easily available locally. Since the wall and roof, both, are thick, they insulate the interior very well, thus keeping it cool on a hot summer afternoon and warm on a cold winter night. The rainwater drains off the soil-covered roof since the soil used for this is impervious. House plan is introverted. In other words, the houses open in an internal courtyard. This creates a well-protected courtyard that offers safe and shielded space for the women and children of the family. The houses have rather formidable appearance, and their historic tales about them. But the Latur earthquake revealed their vulnerability to earthquakes which wasn't known earlier.

It was observed that the walls lacked interlocking of stones that stitch together the wythes or the wall faces. As a result, in the earthquake when the ground shook, under the impact of lateral inertia forces thus generated, the wall faces separated and buckled leading to disintegration of the stone walls. The timber deck supporting the heavy mud roof was found to lack rigidity. Hence, it swayed back and forth under the impact of the lateral seismic forces, in the process hitting the walls to hasten their collapse. As a result, even a low magnitude earthquake resulted in the destruction of over 35,000 houses and death of over 9,500 individuals.

Figure 8.1 Marathwada region on Deccan Plateau of Western India.

Figure 8.2 Tribal area of South Gujarat.

South Gujarat State, Western India

In this area of heavy rainfall there is an abundance of biomass. As a result, biomass provides all basic materials for walling and roofing. Temperatures are moderate. Mud is plastered on the vegetal matter such as locally found cane or even agriculture residue in the walls to seal the walls against wind and creatures (Figure 8.2). So, the thin walls provide adequate protection from the elements and intruders, if any. Roof is four sided and steep with large overhangs. As a result, the wall heights are small and they are well protected against heavy rains. Selected portions of walls are not plastered to create windows for adequate light and ventilation. Thus, these houses are made primarily from local materials using local skills, and hence, are truly sustainable. This type of construction is often vulnerable in high winds and insects.

The high winds damage the roofing tiles since they are not anchored to the support structure. Tiles get blown off in high winds. Historically people had easy access to hard timber for framing and reed for the lattice, which were long lasting. But in recent decades due to increased demand for the industrial use, soft timber and agriculture waste have found their way in house construction. These succumb rapidly to the insect attack, requiring their replacement every few years.

Uttarakhand State, in Foothills of Himalaya, Northern India

Most of Uttarakhand sits on mountains, and that means easy access to rocks of different types. In many areas, rubble or broken stone for masonry walls is available in plenty, often right on the ground. The flat stone suitable for roofing too is found in most areas through mining. Although the state has witnessed much deforestation, fast growing variety like Pine is found in most regions. As a result, Random Rubble Masonry has been the principal wall type, and *Pathal*

Figure 8.3 Stone walls not laced with timber.

(flat stone) supported on timber understructure has been the principal roof type in all the hill areas of the state (Figure 8.3). Thick stone walls provide much needed insulation against the winter cold, and so does the *Pathal* roof. Walls and floors plastered with mud create cozy ambiance for those living in the house. In short, the system is totally based on local materials and skills.

There are examples of remarkably strong houses with earthquake resistant features like element of timber in horizontal and vertical directions within masonry walls and light weight living quarters at the top that not only withstood powerful earthquakes but also provided security against the threat of invaders. But the earthquakes of 1991 and 1999 showed that majority of houses are vulnerable to earthquakes because of dilution of such features.

The stone houses built in the past several decades have done away with the interlaced timber elements in the walls. This resulted in poor wall to wall connection which was manifested by cracking of walls at the wall-to-wall junction, and greatly reduced flexural strength which led to vertical cracking during the earthquake. In addition, with steadily falling skill standards of the local building artisans, the random rubble masonry is now devoid of interlocking of stones as well as 'through' stones. As a result, these earthquakes saw high level of damage to the stone masonry walls.

Kashmir State, Northern India

Parts of Kashmir sit on mountains where stone/rubble is easily available and hence, the walls are most commonly made of random rubble masonry in mud mortar. Flat stone for roofing is available only in a very few places. In the past, wood shingles were used for roofing when hard wood like Deodar was accessible. But since the advent of CGI sheeting, all other options have gone in oblivion because of the advantages of CGI, in spite of being nonlocal. In the mountains

Figure 8.4 Urban (valleys) 'Dhajji Divaari' – timber laced brick or stone masonry walls.

timber too is available, although not in abundance. Hence, roof understructure has been of timber. The other part of Kashmir, namely the Valley, has an abundance of soil appropriate for brick making. Hence, brick walls of different types have been the principal wall type in the region. The most traditional type is called Taq which has thick brick masonry walls made in mud mortar and is found most commonly in urban centres like Srinagar. A timber laced brick masonry called *Dhajji Diwaari* too has been used (Figure 8.4). Most houses have attic which creates an excellent barrier against the cold coming in through the roof.

The brick masonry laced with timber, called Dhajji Diwaari, in urban areas, has been found to be earthquake resistant due to fully braced timber lacing and the light weight brick masonry infill. In Taq type buildings the combination of thick brick shear walls with timber floor diaphragm brings earthquake resilience. In addition, the top storey is made of light weight Dhajji Diwari with light roof to add to the resilience. But stone masonry buildings in the hill regions have been observed to be seismically vulnerable primarily because of the poor workmanship of the masons, and absence of disaster resisting features such as Bands. In 2005 earthquake, these buildings suffered extensive damage in the form of cracks of different types, and also wall collapses.

Ladakh (Himalayan State), Northern India

Ladakh being a high altitude desert has very little precipitation. Soil/earth is available in plenty in most areas. Hence, it provides an excellent roofing material. For walling the adobe or hand moulded sun-dried mud blocks is the most viable walling option (Figure 8.5). In some regions,

Figure 8.5 Adobe/mud block masonry with flat roof.

though, stone is also available. Hence, in such pockets, one also finds random rubble masonry walls. Being arid, timber required for the roof is not most easily available locally. But it is relatively less difficult to procure it as compared to steel, cement, aggregates, and sand required for RC slab. Hence, flat mud-timber roof is very much in use, especially with very little rain. Ladakh experiences extremely cold temperatures for several months. The thick mud and stone walls along with the mud roof provide excellent insulation to keep the houses warm. Mud plastered walls and floors create a cozy ambiance for those living in these houses.

These buildings are truly local, being dependent 100% on local materials and skills. Past earthquakes in Kashmir did not cause any major damage to this type of buildings primarily because of the distance. But this type of buildings lack earthquake resisting features and in addition have heavy roof. So they certainly are vulnerable to future earthquakes. In addition, the recent experiences of heavy precipitation in the form of rain instead of snow, due to climate change, have demonstrated vulnerability of this earth-based construction.

The accumulated snow typically melts and drains out of the spouts projecting out of the roof parapet. Snow, as it comes down, does not erode the mud walls and mud roof. But rain erodes the mud surfaces of the flat roof as well as the unprotected mud walls. In addition, the rain permeates in mud walls making them wet through the body of the walls. As it gets wet it expands and then when it dries it cracks, thus further increasing the damage. Repeated rain results in serious damage to the structure making it vulnerable to future rains as well as earthquakes.

NE States of India – Assam, Meghalaya, Arunachal, Mizoram, etc.

The North East of India has an abundance of biomass including timber, bamboo, and grasses. Hence, the local building systems, commonly knowns as 'Assam Style House' Figure 8.6, with

Figure 8.6 'Assam type' timber framing with cane infill mud plastered walls and pitched roof.

some variations, are totally dependent upon these local materials. High precipitation and high seismic activities among other factors dictate the house form as also the construction. Walls are made of a mix of timber, cane (*Ikra*), and bamboo. In some regions, they are entirely of bamboo where as in others there is timber frame with infill of cane or split bamboo. In Mizoram, cement sheets have become very popular for the infill within timber frame. Historically, thatch of different types was the most predominant roofing. But the advent of CGI sheeting has changed that for the ease of construction and maintenance. In many regions frequently subject to water inundation, the houses are made on stilts. In most cases, these stilts are of timber or bamboo, and they are quite well braced for wind and earthquake. But now reinforced concrete stilts are also entering the scene. The poorly designed interface between the traditional superstructure and the modern concrete stilts has been observed to bring vulnerability during a recent earthquake.

Recent earthquakes in the region have demonstrated the seismic resilience of this 'Assam' type construction. The light weight of the structure, the ductility of timber frames and cane-bamboo infill along with bracings between the stilts, plus anchoring of some sort between various elements ensure earthquake and cyclone resilience. The 2011 Sikkim Earthquake had amply demonstrated this when the RCC had suffered fair amount of damage, while the traditional 'Assam Type' timber laced houses suffered little or no damage. In a number of RCC buildings, the RC structural elements, especially the columns, suffered severe distress that could not be repaired by local people, and some suffered total or partial collapse.

Important Attributes of Vernacular Building Systems

The most important attributes of these vernacular building systems are their (i) low cost, (ii) sustainability, (iii) contribution to local economy, and (iv) ability to generate local jobs.

In each one of the above six cases from India it is evident that they are in harmony with the local climate, use predominantly the local materials which have very small or no carbon footprint and require no long distance polluting transportation, and finally, allow reuse of almost all the construction materials when situation arises. In short, these building technologies are truly sustainable and green.

From the economic angle, it is evident that this type of construction is most affordable since the materials used, be it stone, or timber, or earth, etc., are not energy intensive nor require costly long distance transportation, and hence, accessible to the masses. Only the rich opt for other types of construction that uses expensive materials and requires outside specialists. Every stage of the application of these technologies starting from the production of materials such as stone, timber, reed, adobe, etc., involves local people, and hence, creates much local employment. This in turn results in a high economy multiplier effect. Finally, the simplicity of technologies permits substantial participation of individuals in the construction of their own houses.

Impact of Disasters on Vernacular Buildings

In the past few decades, disasters caused by natural hazards like earthquakes and cyclones in different parts of the Indian Subcontinent caused a lot of destruction of the vernacular buildings (Figure 8.7). This destruction points at the construction quality and construction typology, the two principal factors governing the building performance. In other words, the quality is often found to be compromised and the disaster resisting features that impart special properties to the structures of a particular type to withstand the deformations that the forces of disasters caused by natural hazards generate are inadequate or totally missing. The developments of the past several decades including dilution of the skill and know-how of building artisans, entry of modern

Figure 8.7 Severely damaged rubble masonry building in 2005 Kashmir Earthquake.

materials without the necessary knowledge and loss of easy access to the important local construction materials are the root cause behind this construction scene.

Key Factors for Increasing Disaster Vulnerability of Vernacular Built Heritage?

Six cases of vernacular built heritage from India represent widely different geo-climatic regions. Each of these regions has been hit by one or more disasters caused by natural hazards. These destructive events have taught valuable lessons about why the buildings are vulnerable in the face of such disasters.

It has been observed that, with rare exceptions, those who are constructing do not have the right know-how of the correct use of the local materials, like stone, bricks, etc., in the masonry walls. Most of the time, in masonry construction, the basic rules are violated.

For example …

- The basic rules that ensure the correct placement of the masonry units, like stones, bricks, etc., in the masonry walls to make them strong.
- Basic but critical rule like all vertical joints must be filled with mortar to minimises the movements between the masonry units.
- Right way of ensuring proper connection between adjacent walls which ensures good 'box' action required for disaster resistance.

The other major issues are …

- Wooden Bands incorporated in the construction of masonry walls at different levels that made the walls disaster resistant and which were a common feature in the past have disappeared due to issues like scarcity of materials and their cost. But as a result, the **building artisans have become oblivious of them**.
- Similarly, in highly seismic regions, the buildings had vertical timber elements anchored to the masonry to bring in ductility to the walls. These too have disappeared.
- Anchoring of roof and floors to the wall that they are resting on, help greatly in making buildings disaster resistant. It has been observed that elements like floor joists, rafters, purlins, ridge beams, etc. all are simply placed on the walls without any secure connection to the walls. As a result, during the disaster, there is a lot of movement between them causing more damage to the walls.
- Connections between different structural elements of floors and roofs are found to be inadequate or missing. Often single nails are driven in rather causally which simply pull out during the disasters, resulting in more damage.
- Anchoring of roofing to the understructure is generally inadequate or all together missing. This leads to great damage to roofing as well as to the roof understructure.

Moving away from Vernacular Building Systems

The vernacular buildings that our artisans have been building during the past several decades have seen a gradual drop in the quality, integrity, and disaster resistance of the structures. As a result, the disasters have brought a lot of undue hardships and huge losses to the people. **Every major disaster that caused widespread destruction has shaken peoples' confidence in their building technologies that they have known and used all their life.** Generations ago, when people totally relied on local materials and skills, they had no other option but to work on the local building system to improve them. This is how the vernacular building systems in disaster prone regions became disaster resistant over generations (Figure 8.8). But now, especially with

Figure 8.8 Vertical timber reinforcement anchored to timber band and the wall.

the establishment favouring the modern building materials, people are more or less co-opted into abandoning of the traditional or vernacular building systems in favour of modern non-local systems and materials. Every engineer that visits the site talks about discarding the vernacular local system and recommends bringing in modern building systems, especially **in the name of safety and durability**. As a result, those who have to rebuild after a disaster end up opting for what one could call '**hybrid vernacular system**', unless compelled by some other overpowering issues such as economics or the unavailability of the materials.

Over the past few decades, **these modern materials that essentially are alien to the local areas have secured a strong footing in the local vernacular building typologies**. Unfortunately, the knowledge for the right use of these materials hasn't reached their users. But the blind confidence in them has led to abandoning of the age old rules of construction and special disaster resisting features of vernacular system, thus inviting vulnerability.

This has created a false sense of security in the occupants of such buildings where these materials are used. It is not surprising that these vulnerable hybrid vernacular buildings have suffered a lot of damage in one disaster after another.

Increasing Dilution – Entry of the Corner 'Columns'

In the aftermath of Kutchh earthquake, RC frame elements began to creep in to the vernacular traditional building systems. The onslaught of the RC frame elements, primarily the RC columns, has spread like wildfire all over the country, and anyone who can afford, wants these columns

Figure 8.9 Severely damaged vernacular stone masonry house having RC columns in corners without beams.

incorporated with the traditional vernacular load-bearing masonry building in its corners. In short, the '*column waalaa ghar*' or 'the house with columns' has become one more typology and a norm, without realising that it is an extremely vulnerable aberration of the vernacular masonry buildings (Figure 8.9). This is so since the masonry wall-to-wall connection is disrupted by an alien element that has no bonding with the walls. Sadly, these columns by themselves do not support anything and are a total waste of money. Thus in the past several decades, the Vernacular Building systems have been steadily diluted. In this manner, a day is not far when the vernacular building systems and form may be fully lost.

Loss of Vernacular Architecture – What Does It Mean to Us?

There are a number of issues that need to be examined …

Identity

From an esoteric perspective, it would mean a loss of unique regional identity. Historically in many areas, people have been proud of their traditions, with type of houses and their construction. The arid region of Kutchh in western India is known for its *Bhunga*, which is the name of the local round mud houses which are very much in vogue today. Visitors come to see, among other things, Bhunga. Not too long ago, people of the Marathwada villages in the Western state of Maharashtra in India were proud of their stone-timber-earth houses. But the 1993 Latur earthquake changed all that and the rest is history. The stone and timber Koti Banal architecture of the Himalayan State of Uttarakhand too is another iconic architecture. But the visitors to Latur or Uttarakhand know nothing about them. They have a potential to draw a lot of tourists from around the world. One more example that deserves a mention is the earthen architecture of the high altitude desert areas of Ladakh, and Lahul – Spiti of Himachal Pradesh State of India. Most beautiful houses are built in this cold desert using local earth. So the fame of the vernacular architecture could go a long way in preserving it. But ensuring their disaster resistant character is a pre-requisite to their promotion.

Local Employment and Economy

Lack of employment is one of the biggest problems faced by rural areas, especially those in the more remote areas such as up in the mountains or out in the desert. People are leaving their, often pristine, rural natives in search of jobs, ending up in urban hovels. Since the vernacular building systems depend on the local materials and local skills right from the extraction of the materials to the finishing of the building, they create more local employment and contribute to the local economy than other type of building systems. Wealth of the locals, when spent on construction, goes back in the local economy rather than ending up in the coffers of some industry far away.

Carbon Footprint

There is yet another reason why we should not lose the vernacular building systems. Today in climate change the world is facing the biggest challenge ever faced by humankind. To counter it the search is continuously on for more energy efficient greener systems in all aspects of our living. So the energy efficient housing is the need of the hour. Not only the house should consume less energy in providing comfortable living conditions to the occupants, but it should consume less energy in construction. In other words, the embodied energy of the materials used in construction should be as low as possible. The materials should not require a huge amount of energy for reaching the site. In short the building system should have the smallest possible carbon footprint. Vernacular building systems meet this requirement unlike the other systems.

Water Footprint

Water requirement of the construction is one more important point to be taken into consideration, especially when sweet water availability is dropping rapidly globally. There are very few areas where water availability is not a problem. Hence, greater the water requirement of a building system, greater will be the problems in the execution of that system, and greater will be the unreliability of the outcome of such a construction. Water availability is generally not a crucial factor in the application of the vernacular building systems since they use very little water.

Summing this up, it would be reasonable to say that the loss of the Vernacular Building Systems would be a catastrophy, especially for the rural people, and hence, we cannot afford to lose this invaluable heritage.

So for the locals their own vernacular building system should be the most preferred system. For this, if the common irritants of this system are taken care of then there should not be any reason for people to go for any other system.

In the past several decades, much has been lost on this front. But we still have not reached the point of no return. And at the same time, in India, thanks to the effort of a few, much work has been done in different parts of the country focusing at strengthening of the local vernacular building systems. A continuous, concerted, and systematic countrywide effort is, nonetheless, required to make a lasting impact. It is also required that such an effort is supported by people at large, education institutions and various government agencies. With the current scenario in Latur region in front of us, a look back at our own efforts that we made in the post Latur

Box 8.1 Way Forward

In order to preserve this precious heritage actions are required on a number of fronts, as briefly described here.

Disaster Resilience: Restoring of disaster resistance to the vernacular building system is a very important step since poor performance in the disasters has been one major reason for the people to lose confidence in them and look for an alternate, howsoever expensive. This has to be done on two fronts.

The **new buildings built with the improved vernacular building system** must be capable of withstanding without total collapse the disasters of most likely intensity that are likely to strike the region (Figure 8.10). Secondly, to ensure that the existing vernacular buildings do not succumb to the likely disasters in the region they must be retrofitted. **Retrofitting of the existing vulnerable vernacular buildings** (Figure 8.11), houses and public buildings alike, could go a long way in saving the vernacular building heritage since it costs less, it is fast, people can continue to occupy the buildings partially and it can be done in phases. So (a) specifications and procedures must be evolved; (b) full scale demonstrations must be done for people to see what improvements are required and how they are installed; (c) local building artisans, petty contractors, and government engineers must be trained in this direction; (d) Official Schedule of Rates (SoR) must be prepared and formally inducted in the government system to pave way for the government interventions in this direction; (e) IEC materials must be prepared and put in public domain; and (f) Government must launch schemes to promote these measures through incentives to the people to encourage them to bring long term safety against future disasters.

Figure 8.10 Earthquake resistant house.

Figure 8.11 Retrofitting and restoration completed.

Research to Remove the Irritants and Up-Grade Building System: All building systems, vernacular and others, have irritants. These typically are related to maintenance, or to the introduction of some improvement, etc. Introduction of decentralised timber preservation is the need of the hour to make fast growing tertiary timber like Pine acceptable rather than people opting for RC out of compulsion. All this demands technical research. There is a need to combine the vernacular systems with modern science, technology, and materials. This could help improve performance and help rebuild peoples' confidence.

Main-Streaming of Vernacular Buildings: It is important that people feel proud of these systems and should feel attracted. The public imagination of this type of houses as a thing of past or something only for the poor, or something primitive needs to be eradicated. For this, it is important to demonstrate that such houses too can have all modern conveniences, and that they can be rodent and insect proof like the modern *pukka* houses that people generally aspire for.

Promote Vernacular Building Heritage for Tourism: Pamphlets could be prepared on the vernacular houses of the region, tours could be planned to visit the selected heritage houses in the region. Home-stays in such houses could be aggressively promoted. This will indirectly inspire people to up-grade their vernacular houses rather than replace them with the modern ones. This will for sure help people inculcate pride in them.

Sensitisation of Future Building Professionals: There is a desperate need to educate the students of civil engineering and architecture on the traditional/vernacular building systems including the scientific and technical information to inspire respect for them, to help them relate to a role that the vernacular systems can play to tackle the challenge of climate change. This would be best achieved through inducting this topic in their curriculum.

Earthquake rehabilitation phase over seven long years from 1994 to 2000 has a valuable message. It clearly says that unless the government agencies recognise the vernacular building system as legitimate and worthy of adoption for their own institutional buildings, it is unlikely that this valuable heritage will be preserved for the posterity.

Key Points

- Vernacular-built-heritage of the common people, which is still alive around the world, is just as important and priceless as that from the ancient world.
- Vernacular buildings have performed poorly in recent disasters due to dilution of the proven construction technology.
- In recent decades, in the name of permanence, the building professionals have started rejecting the vernacular technologies in favour of the modern building technologies having high carbon footprint.
- The current trend could lead to extinction of precious vernacular-built-heritage before long.
- But the Vernacular Heritage can be saved if efforts are made to synthesise them with the modern science and materials to improve their performance in the present context.
- Retrofitting of the existing vulnerable vernacular buildings is a fast and affordable option to save the Vernacular Heritage.
- Removal of the common irritants in vernacular building system as also the up-grading of system demands research.
- All this requires the inclusion of the vernacular building systems in the education and research programs by the prime education institutions around the world.

# 9 Traditional Livelihoods for Climate Action and Disaster Risk Management

Sukhreet Bajwa, Tanaya Sarmah, Ranit Chatterjee, and Rajib Shaw

Introduction

Livelihood, in its most basic sense, is a way of living, earning, and sustaining. It has not only an economic focus but also a larger emphasis on capacities, assets and activities required to maintain well-being. Climate induced hazards and disasters have caused a lot of damage and loss to the resources in environment which sustained the livelihoods. Traditional livelihoods refer to the occupations led by indigenous people as well as those which are rooted in our cultural and heritage space. In more simpler term, Traditional livelihoods refer to the means of securing the necessities of life that have been used over generations. These are part of the intangible heritage of the country. The traditional occupations generally include activities like fishing, agriculture, livestock rearing, handicrafts, weaving, and selling non-timber forest-based products. Disasters often destroy the local flora and fauna of the region, which in turn impacts the lifestyle of the communities. While the impact on the tangible heritage is easily apprehensible, the impact of the intangible heritage is difficult due to its interlinkage with the development process and needs much longer engagement. The reliance on the community for cultural continuity is very high in such cases.

Many of these traditional livelihoods are endemic to a certain geography and variations can be experienced from one cluster to the other. The traditional livelihoods become a part of the living heritage (intangible heritage) owing to the skill set, knowledge, and other imbedded characteristics. Needless to say, the traditional livelihoods are a living testimony of evolution and diversification of human race that have withstood the challenges put forth by disasters and climate changes. Many marginalised and indigenous communities in the informal settlements around the world depend heavily on traditional livelihoods for making a living. But with the change in frequency and severity of hazards and aggravated climate change trends coupled with the development pressure to make societies highly technology dependent, traditional livelihoods continuity is in jeopardy.

The practices associated with traditional livelihoods often interact with the environment and are locally evolved and intergenerationally transmitted to help communities follow environmental patterns and maintain livelihood security (Berkes et al., 2000) as cited in (Venugopal et al., 2019). Hence, the continuity of traditional livelihood requires understanding the risk patterns and making necessary modifications to the livelihood approaches. This has resulted in ingraining the risk management and adaptation as a part of their operations. It is good to make a distinction here between traditional knowledge in livelihood and the knowledge base in traditional livelihoods. Here we are more concerned about the latter which is a subset of the first and presents a more focused understanding of the capacity as a part of the traditional livelihoods' evolutions process. The presence of traditional livelihood practices in subsistence communities acts as a social security safety net, so long as the environmental patterns remain consistent and predictable.

DOI: 10.4324/9781003293019-12

Community is at the centre of heritage conservation, climate action, and disaster risk management. The Sustainable Development Goals (SDGs), Sendai Framework for Disaster Risk Reduction, and Paris Climate Agreement all put focus on people as the driving force to bring in change (Hill & Gaillard, 2013; Davidsson, 2020). Tradition and heritage are important to the society as it gives them a sense of identity and belongingness to a place, culture, cuisine, monument, or a ritual. Traditions are rooted in people, and this is their strength. The discourse in disaster risk reduction and resilience emphasises on making risk reduction a part of culture (Rosa et al., 2021). This only goes on to show how important these two tangents are and more so the need for their integration (Ravankhah et al., 2017). As traditional livelihoods are at the centre of the communities' lifestyle and sustenance, this chapter explores the role of communities in protection of traditional livelihoods for climate action and disaster risk management. The chapter would further dwell into understanding the traditional livelihood practices and their sustainability in the face of climate induced hazards.

The chapter focuses on the need for protecting cultural heritage for sustainable livelihoods. The various case studies are undertaken to highlight the existing gaps as well as good practices in this domain. Traditional livelihood is focussed as an important intangible heritage which is self-sustainable as well as includes a scope to teach the younger generations life skills. The various threats to livelihood due to non-integration of disaster risk management practices are also studied. The next section highlights the need for community-led action vis-à-vis the expert-led project implementation for conservation of cultural heritage through disaster risk management. This section provides some concrete actionable strategies which can be taken up to strengthen the role of people. Finally, the chapter concludes summarising the key takeaways in the end.

Traditional Livelihoods in the Context of Culture and Disaster Risk Management

The integration of cultural heritage in DRM has been emphasised in the Sendai Framework for Disaster Risk Reduction (SFDRR) 2015–2030, Priority for Action 1 – 'understanding disaster risk' and in Priority for Action 3 – 'investing in disaster risk reduction for resilience' (UNISDR, 2005). The international organisations, such as UNESCO, identifies that among the two types of cultural heritage (i.e., tangible and intangible). Intangible heritage is more fluid in definition and fragile to protect. The importance of intangible cultural heritage is not the cultural manifestation itself but rather the wealth of knowledge and skills that is transmitted through it from one generation to the next. Intangible heritage is a source of community-based resilience and sustainable development.

Traditional livelihoods, such as traditional craftsmen, depict the most tangible form of culture through their products. Traditional livelihoods are an important part of not only culture but also of society and local economy. Local and indigenous practices of fishing, agriculture, and food are an important part of everyday resilience in tribal and forest dependent communities. However, in a discourse of cultural conservation the focus is mostly on built monuments and tangible part of the culture. A specific example of traditional livelihoods and their interplay in tangible culture is from the Grand Shrine of Ise, Japan. As reported by Munjeri (2004), the Empress of Japan issued a decree (in 690 AD) for the renewal of the wood-and-thatch Grand Shrine through refurbishment in every 20-year period. This custom was termed as the 'shikinen zotai', which entails reconstruction using new material whereas shrine furniture, decorations, sacred treasures, and garments are reproduced and reinstalled. Thus, the craftsmen involved in the reconstruction activities are traditional artisans whose has been trained with the help of knowledge that has been passed down through generations. This intangible cultural heritage of traditional craftsmanship has helped Grand Shrine of Ise preserve its originality and its setting for the past 1,000 years. The process of 'shikinen zotai' is an inseparable part of the Grand Shrine and cannot be overlooked as a cultural heritage. Not only is the knowledge an important component for preservation, but it has also been financially providing for scores of traditional craftsman families over the centuries. The Umaid Bhawan Palace of Jodhpur was built to provide employment to the traditional artisans during the famine of 1929 (Admin, 2019). The construction of a huge palace employed the local farmers and gave them a steady source of income. The mere presence of this heritage monument spurs a growth in the cultural identity of the region over time and which eventually got linked with the livelihood of the residents.

In the event of a disaster in and around a heritage site the linkage between livelihoods and intangible cultural heritage may be affected. For instance, if pilgrimage activities to temples get disrupted due to an unforeseen disaster event, it may not only impact the smaller businesses (related to worship or temple activities) around the temple but also upset the very foundation of the 'temple economy' (Good, 2004) comprising of priests and other ritualistic service providers. It is thus important to explore the dimensions of resilience which caters to the preservation of such traditional livelihoods.

Cultural heritage is a synchronized relationship involving society (that is, systems of interactions connecting people), norms and values (that is, ideas, for instance, belief systems that attribute relative importance). Symbols, technologies, and objects are tangible evidence of underlying norms and values. Thus, they establish a symbiotic relationship between the tangible and the intangible (Bouchenaki, 2003). The intangible heritage should be regarded as the larger framework within which tangible heritage takes on shape and significance. As seen from Figure 9.1 below, while the tangible heritage is at the centre of all interactions, however, to transform a mere monumental/emblematic space into a cultural heritage a series of intangible interactions takes place within the sphere of interaction. Since all interactions stem from the needs of people, all the arrows in the diagram point away from the people. Primary social interactions are partaken surrounding the tangible heritage and may include physical interaction with the space (e.g., visit to a monument), celebration of the space (e.g., hymns or songs surrounding the monument/festival), or contemporary knowledge of the intangible culture (e.g., traditional knowledge of monument construction). Associated financial interactions further stem from livelihoods and the application of traditional knowledge. Finally, over a period, all such interactions

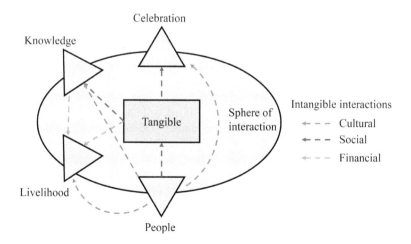

Figure 9.1 Complex interplay between intangible and tangible heritages.[1]

engrave into the cultural fabric of the heritages. The deep interconnections, as depicted by the diagram, further ascertains that intangible heritages form a complex interplay with the tangible heritage and cannot be ignored.

Along with timely and effective emergency management, it is vital to deal with disaster risks proactively, with coordination from multiple stakeholders and interaction among them. Failure to do so may result in delaying the recovery process due to potential conflicts of interest among the stakeholders. Traditional artisans, craftsmen, and those who depend on heritage-based livelihoods are important stakeholders in such conversations, as seen in case study (1) (Box 9.1). To ensure that cultural heritage is considered in DRM planning, raising awareness of heritage values among the associated stakeholders is to be encouraged while increasing their competencies in dealing with cultural heritage in cases of emergency. This will, in the long run, create opportunities for economic coping capacity, social and cultural coherence, reduction of community vulnerability using indigenous knowledge and techniques, etc.

Box 9.1   Case Study #1: Intertwined tangible and intangible heritage: Case of Bungamati in Nepal (Chatterjee, 2018)

Kathmandu valley has evolved as the economic hub of Nepal with the highest investment from the private sector. The 2015 earthquake had severe impacts on the Kathmandu Valley (Adhikari et al., 2015). Bungamati, a peri-urban Newari settlement in the Kathmandu Valley, is currently known for its traditional handicrafts (metal carving, wood carvings), which are highly dependent on the tourism sector for their survival. Bungamati has special reference to 1934 Earthquake in which 99% of the buildings were damaged beyond repair and the recovery process was community based, which suggests existence of strong social ties. A longitudinal study of the changing livelihoods in Bungamati is presented here.

Newari residential buildings have been multi-storey highlighting their urbane characteristics. The allocation of space vertically is based on the notion of purity of function. As the result the space closer to ground as often used for cattle shed and storage while upper floors are used for living and kitchen. Prior to the 1934 earthquake, Bungamati was known for its tie ropes used for fastening wooden rafter of thatch houses. Post 1934, a shift to tiles from thatch as roof material led to people of Bungamati look for alternate source of livelihood. The need for having masons and carpenters to reconstruct the buildings led to evolution of specific skill sets in Bungamati. In addition, farming, livestock rearing, and dairy became the primary livelihood for the majority of the Bungamati households. This change in livelihood in line with the notion of Newari notion of purity within a built space led to creation of two entry doors at the ground floor of the buildings, one for animals and other for humans. Till the early 1990s, Bungamati mainly had been relying on barter system but with economic reforms in Nepal, change in livelihood from dairy to more organised sector jobs and need for money for procuring daily necessities become the focus. Many of the grounds were converted to shops and were operated by the house owners as shops for handcrafts, groceries, etc. This also changed the solid void ratio of the front façade leading to combining of the two doors into one big shop entry and as a result in creation of soft storeys and changing the vertical alignments of openings. Many of these buildings collapsed during the 2015 earthquake in Bungamati.

Traditional Livelihoods for Climate Action

People engaged in traditional livelihoods are custodians of natural resources. They rely on the nature to fulfil their basic needs. In this context, they promote sustainable production and consumption. Livestock and animal husbandry are one of the oldest livelihoods. The traditional livelihood is not of mass production but one where the animals are dependent on pasturelands for grazing. The pasturelands are in turn protected by the communities as they are important for providing the fodder. Other climate adaptation measures such as traditional systems of water management, flood protection all become adopted by communities to save and protect the natural resources which are source of their income. The traditional agricultural practices continue to remain important methods of climate change mitigation and to ensure sustainability. Many aspects such as sustainable agricultural practices and local understanding of weather include the knowledge of risk perception which is ingrained in a culture. One example is from Bangladesh, where people revived the traditional cultivation method called 'dhaap' or 'baira' (DISD, 2021). These are traditional floating vegetable gardens in the form of artificial islands which rise and fall with swelling waters. Such practices provide a major boost to developing resilient ecosystems with support of traditional livelihoods. In India, there is a boost being given to production and consumption of traditional crop such as 'millets' as it is climate resilient. The revival of this indigenous practice is not only to ensure climate adaptation but also to provide nutritional security and enhanced income to small and marginal farmers. Small-scale farming is more environmentally sustainable than industrial farming. Small-scale farmers are also custodians of diverse traditional seed varieties and are more attuned to organic ways. Nature-based solutions which diversify production and encourage the growth of local food processing industries, can support an ecosystem of small-scale producers and a vibrant economic landscape connected with local and national markets (IFAD). As observed in case study (3) (Box 9.3), the traditional agrarian practices of Kuhls for irrigation are more environmentally friendly solutions to provide water to crops and apple orchards.

Box 9.2   Case Study #2: Community-led water harvesting system

The Manas National Park (MNP) is a biodiversity hotspot recognised by UNESCO as a world heritage site. It is in the state of Assam, India. It is home to a diverse range of fauna due to the variety of ecosystems ranging from forested hills, alluvial grasslands, and tropical evergreen forests. The Manas-Beki River system contributes to maintaining the diversity of this ecosystem. The MNP is prone to seasonal flooding, erosion, water scarcity, and siltation. Frequent release of excess water from the Kurishu dam upstream brings debris and damages the MNP due to inundation. Some of the challenges identified in the MNP include inadequate training in DRM for the forest officials and the absence of an early warning system when the river water level rises.

During the frequent dam breakage, the farmlands of the locals also get affected. The forest officials and the local communities come together and help in the restoration process. The forest officials undertake timely awareness meetings with the communities located on the fringes of the MNP. In the monsoon season, the elephants often come to graze and, as a result, destroy the crops. The forest officials in cooperation with the local communities send these elephants back to the forest areas. The problem of erosion due to flooding in the Manas-Beki river system is tackled through the use of boulder-based anti-erosion measures. In this process, stakeholders from allied departments are collaboratively involved.

Here the Dongs, dating back to the 1940s, are the indigenous water harvesting systems which include water supply through canals from a perennial river. This practice was developed in response to the water scarcity issue and to fulfil the water needs for agricultural purpose. Use of Dongs minimizes the impacts of flooding by funnelling the off water into large tanks for recycling. It reduces soil erosion and flood hazards by collecting rainwater and reducing the flow of stormwater to prevent urban flooding. This practice is an example of indigenous knowledge developed to sustain agricultural livelihoods and deal with water scarcity in the area and managed through a community-led system.

Sources: Murti and Buyck (2014); Das, Das and Boruah (2017).

Box 9.3 Case Study #3: Environmental and cultural implications of climate change and the shifting of Apple Orchards in Kinnaur, Himachal Pradesh, India

The Kinnaur district in Himachal Pradesh is the state's hydel power hub with 53 hydroelectric power projects. It is part of the old Tibet road pass and has a rich heritage of religious diversity, cultural practices, and diverse flora and fauna. Kinnaur has been witnessing frequent landslides and it is also geologically sensitive to earthquakes. The changes in land use, deforestation and encroachment of the forests, and loss of biodiversity, are posing serious threats. Global warming has resulted in a change in the rainfall pattern, due to which there is a considerable increase in the amount of rainfall witnessed in the district. Tunnelling, excavation, and blasting are carried out for further hydropower projects making the landscape more fragile. Apple cultivation is one of the major sources of income for the local population. The changing rainfall patterns have led to the shifting of the apple orchards to areas of higher elevation. The areas with higher elevations are steep and tend to have less fertile soil and deplete quicker than areas in lower elevations. The apple orchards in steeper areas are at greater risk of landslide hazards. The inhospitable landscape of Kinnaur thus poses a challenge of low availability of cultivable land for each household.

Kinnaur has a rich heritage of cultivating traditional crops and herbs. It serves as an ideal locale to examine the dynamics of human-environment interactions and to contribute to a body of knowledge on the sustainability of mountain livelihoods. Forest products on which the communities used to depend include fuelwood, chilgoza pine nuts, wild apricots, medicinal herbs, and edible mushrooms, all of which were collected and processed collectively by the local communities. Most villages irrigate their apple orchards through a system of Kuhls (irrigation channels), which are controlled by the force of gravity and are fed by melting glaciers or small rivulets. Traditionally, every household participated in the decision-making process about water-related affairs and was mandated to volunteer to maintain the Kuhls. Today, however, as people have become busy with their orchards, community solidarity and cohesion have diminished. Considering these practices which used to be an integral part of the cultural heritage of the region, the challenge, therefore, is to revive the traditional practices for building relationships for collective decision-making.

Source: Rahimzadeh (2020).

Sustainability of Traditional Livelihoods

While the threads of tradition, climate action, and disaster risk management are best woven together, sustainable livelihood can be an important stitch which brings the best of these worlds together. A livelihood comprises the capabilities, assets, and activities required for a means of living. Sustainable livelihood focuses on the interests of communities and recognizes the complexity of people's daily lives. As per Ashley and Carney (1999), *'A livelihood comprises the capabilities, assets and activities required for a means of living. A livelihood is sustainable when it can cope with and recover from stresses and shocks and maintain or enhance its capabilities and assets both now and in the future, while not undermining the natural resource base'*. For the purpose of this chapter, we will focus on livelihood which is built upon knowledge, skills, and understanding about local culture, values and traditions which are passed on from one generation to another. It also includes livelihood built around heritage monuments and areas. Some of the examples can include pottery, handicrafts, traditional weaves, handlooms, paintings, folk music and dances, agricultural practices, heritage tourism, climate and weather understanding, etc. The tangible heritage, including the monuments and religious sites is an important source of livelihood for the local communities. Ranging from street vendors to tourist guides, these sites are a hub of the formal economy as well as a huge proportion of the informal economy. These tangible heritage sites are also the harbinger of intangible heritage including rituals, traditions, and practices. Cultural heritage properties are repositories of traditional knowledge.

Traditional livelihood practices of interacting with the environment are locally evolved and intergenerationally transmitted to help communities follow environmental patterns and maintain livelihood security (Berkes et al., 2000) as cited in Venugopal et al. (2019). Coming to the sustainability aspect, i.e., resilience to future shocks or threats without undermining the natural resource base. It requires understanding the existing risks or threats to these livelihoods. The presence of traditional livelihood practices in subsistence communities sustains livelihood security, so long as the environmental patterns remain consistent and predictable. As observed in the case of Kinnaur district in Himachal Pradesh, the subsequent warming of the region has led to the communities to shift the apple orchards at a higher elevation (Rahimzadeh, 2020). This has not only led to changes in the environment but has also led to an increase in the risk of landslides due to cutting of hills at the top for apple plantation. Further, during the Covid-19 pandemic, many local artisans suffered due to non-availability of market as well as finances to back them up. Multiple stressors related to climate change, globalisations, and technological change interact with national and regional institutions to create shocks to place-based livelihoods, inspired by Reason (2000) as cited in Olsson et al. (2014). As per IPCC, *'climate-related hazards exacerbate other stressors, often with negative outcomes for livelihoods, especially for people living in poverty'*.

As per the Sustainable livelihood framework, there are five types of assets or capitals upon which livelihoods are built, namely human capital, social capital, natural capital, physical capital, and financial capital (Serrat, 2017). Weather events and climate change affect the natural assets on which certain livelihoods depend directly, such as rivers, lakes, and fish stocks. Further, such events and climate hazards also erode social and cultural assets (Olsson et al., 2014). The informal social networks of the community get disrupted due to climatic and non-climatic stressors. This leads to further problems in resource mobilisation. The impact of climate hazards on financial assets is clear as observed due to job losses and reduced productivity in the agriculture sector, forest-based livelihoods, fisheries, etc. Loss of physical assets may be observed in damages caused to houses, farmlands, etc. The impact on human capital surfaces when the younger generation finds the traditional livelihood to be less lucrative and moves away to cities to work in other sectors.

Overlaying the five categories with heritage and the impact of climate hazards, they can be presented as below in Table 9.1:

Table 9.1 Sustainable livelihood: Influence of heritage and impact of climate change

S. No.	Sustainable livelihood component	Heritage	Impact of climate change induced hazards
1	Human capital	Traditional knowledge and skills	Loss of manpower, younger generation moving away to urban areas, etc. disaster induced displacement
2	Social capital	Shared values, behaviours, beliefs, customs	Disruption of bonding, bridging, and linking capital
3	Natural capital	Natural resources of land, rivers, forests, etc.	Erosion of land, drought, extreme weather events, floods, etc. leading to conflict and decreasing access to resources, higher cost
4	Physical capital	Traditional equipment, tools, hand mills, etc.	Loss of houses, farmlands, etc., community assets, built heritage
5	Financial capital	Formal and informal sources of savings, pool-funding, micro-finances, etc.	Livelihood loss, rise in cost, higher market risk

Role of Communities in Conservation of Traditional Livelihoods

People govern their interaction with the environment based on their socio-cultural understanding, belief, and practices. People are bearers of the culture and the traditional knowledge. Within the evolution of the cultural practices lies the need to prosper and grow. Hence, people play an important role for livelihood sustainability as well as for environmental sustainability. As per Serrat (2017), sustainable livelihood is not only dependent on the five types of assets but is also impacted by policies and processes governing it. Policies and processes can provide an incentive to influence the choices that people make. Recognising the impacts of climate change, the policies need to re-discover the component of heritage and sustainability to provide livelihood opportunities. The evolution may include aspects of adaptation practices which can be integrated within the traditional heritage domain. Further, the risk reduction and resilience components can be added as capacity building exercises.

However, such policy-based initiatives can only be successful if they come up as community-based initiatives and become part of tales of resilience in the cultural space. Hence, it is important to not only link the micro with the macro but also to establish linkages between human-environment interactions which are solely governed by livelihood outcomes. While policies may provide an overall broad incentive, the micro level power dynamics of each place vary. These variations further decide the access to the incentives. Examples of micro level power dynamics include differential access to resources and impacts of climate change based on gender, class, and geographical location. Empowering the communities with adequate knowledge and resources can provide a basis for innovative ways of resilience building, as is seen in case study (4) (Box 9.4).

The community-based initiatives must begin right from identifying the existing challenges and gaps through participatory research and then collectively designing an adaptation strategy to ensure resiliency. The basic set of principles, as identified by Serrat (2017), includes the community-based approach to be inclusive, holistic, multi-level, multi-stakeholder, and dynamic. Further, the aspects of sustainability should be checked thoroughly to be addressing economic,

> **Box 9.4   Case Study #4: Sustainable livelihoods from natural heritage on islands**
>
> Nearly one-third of World Heritage Sites categorized as natural heritage are island sites. These islands frequently experience exacerbated environmental and social vulnerabilities.
>
> Small populations and tight networks can produce a strong sense of community and can be advantageous to island livelihoods. Traditional knowledge coupled with past experiences can provide skills which permit the island dwellers the flexibility of coping with extreme events. A few instances are provided in the paragraph below.
>
> The Charles Darwin Foundation on the Galapagos Islands employs 143 people, almost 90% of whom are Ecuadorian. Another example is a conservation project for a low-land forest on Espiritu Santo, Vanuatu, which involves the community to develop livelihoods. In southern England, Brownsea Island is a haven for red squirrels, one of the UK's most endangered native species, because their natural forest habitat has been cut down. UK's National Trust maintains on-island staff and seasonal workers, which protect Brownsea's natural heritage and is helpful for their livelihood as well. UK's National Trust also manages Rathlin Island, Northern Ireland which is a bird viewpoint. The Trust facilitates the island dwellers while they manage the natural and cultural heritage. Such community-based conservation has become popular on islands because community participation has tended to produce improved approaches for the development of the island as well as the livelihood of the dwellers. Conservation-related tourism can thus support creating many livelihood options such as operating tours, running accommodation and restaurants, and selling souvenirs. At the same time, it must be made pertinent that tourism is sustainable so as to not harm the natural and cultural heritage of the island.
>
> Research centres, such as the Bellairs Research Institute in Barbados, are not tourists per se but contribute to livelihoods by bringing revenue, contributing to conservation, and boosting the importance of natural heritage throughout the island. In many cases, local researchers are also involved in the institute's work and management.
>
> *Source:* Kelman (2007).

social, institutional, and environmental aspects. As observed in case study (2) (Box 9.2) the community-led water harvesting system of Dongs became an important aspect of maintaining the biodiversity in the region.

As per Ma et al. (2021), the traditional culture in both tangible and intangible forms has become an important element of drawing the attention of tourists. The households in Wuyuan County, Jiangxi province, China, have been able to successfully convert it into lucrative commercial activity leading to sustainable livelihood goals. In line with this, in India too, many holiday sites are now promoting rural tourism and providing stays in eco-friendly houses as a showcase of the local culture. This tourism can further be classified as agro-tourism, food-based, and stays close to nature.

Cultural connect and culture-based livelihoods are also important factors in undertaking reconstruction activities after a disaster. There are two important aspects related to it, while one caters to the mental and emotional well-being, the other is more related to financial earning based on livelihood restoration. Hence, people who are cultural torch bearers can also aid in recovery activities after a disaster.

Conclusion and Key Points

It is inferred that traditional livelihoods are an important component for climate action and disaster risk management. With respect to the case studies (1), (2), and (3), people have a central role in ensuring the continuity of culture through traditional livelihood initiatives and make it resilient to various natural hazards. Some of the key steps for conserving the traditional livelihoods are:

- Traditional livelihood practices of each region need to be documented as part of indigenous knowledge resource base for climate adaptation and assign values (qualitative and quantitative) and assess the interdependency of ecosystem and other tangible heritage as a part of system.
- Trainings and capacity building support need to be provided to local government authorities to implement central policies in support of traditional livelihoods. In many cases, the skill sets diminish over decades and hence creation of champions among indigenous communities will ensure continuity.
- Heritage and culture-based tourism needs to be re-looked from the perspective of developing ecotourism as a sustainable livelihood measure for the surrounding communities.
- Policy intervention is needed to develop market linkages in support of traditional products for farmers, artisans, etc., to enhance the lucrativeness of such practices and encourage youth to take up such traditional livelihoods.
- Existing social safety nets need to be re-visited to include environmentally friendly public works such as construction of ponds and rejuvenation of lakes in support local livelihoods.
- Leverage digital and new technology for showcasing the richness, cultural values, and the ecosystem dependency on the traditional livelihoods at local national and global platforms.
- Efforts should be made to restore intangible heritage linked livelihoods after a disaster and access to easy finance be made available to cover the losses incurred and also ensure the skill set deterioration is minimised.

Note

1 The figure is copyright free.

References

Adhikari, L.B., Gautam, U.P., Koirala, B.P., Bhattarai, M., Kandel, T., Gupta, R.M., Timsina, C., Maharjan, N., Maharjan, K., Dahal, T. and Hoste-Colomer, R., 2015. The aftershock sequence of the 2015 April 25 Gorkha–Nepal earthquake. Geophysical Supplements to the Monthly Notices of the Royal Astronomical Society, 203(3), pp.2119–2124.

Admin. (2019, October 14). Umaid Bhawan Palace in Jodhpur. History, timing, attractions, museums, palace, entry fees. Retrieved April 15, 2023, from https://www.visittnt.com/blog/umaid-bhawan-palace-jodhpur/

Ashley, C. & Carney, D. (1999). *Sustainable Livelihoods: Lessons from Early Experience*. Department for International Development. Retrieved from https://www.shareweb.ch/site/Poverty-Wellbeing/resources/Archive%20files/Sustainable%20Livelihoods%20-%20Lessons%20From%20Early%20Experience,%20Caroline%20Ashley,%20Diana%20Carney%201999.pdf

Bouchenaki, M. (2003). The interdependency of the tangible and intangible cultural heritage. *ICOMOS 14th General Assembly and Scientific Symposium*. Zimbabwe : Victoria Falls.

Berkes, F., Colding, J., & Folke, C. (2000). Rediscovery of traditional ecological knowledge as adaptive management. Ecological applications, 10(5), 1251–1262.

Chatterjee, R. (2018). Post Disaster recovery of formal and informal businesses: Case study of Kathmandu Valley after 2015 Nepal earthquake. Available at: https://repository.kulib.kyoto-u.ac.jp/dspace/handle/2433/232443 (Accessed: April 13, 2023).

Das, P.J., Das, N. & Boruah, J. (2017). A compendium on Local practices for mitigating risk of water induced disaster and climate change in the Brahmaputra river basin, Assam, India. Aaranyak Technical Report, June 2017, Draft version. https://www.academia.edu/43804749/The_Dongs_of_Subankhata_Reducing_water_scarcity_using_community_based_traditional_canal_irrigation_system_in_Baksa_District_Assam

Davidsson, Å. (2020). Disasters as an opportunity for improved environmental conditions. *International Journal of Disaster Risk Reduction*, 48, 101590.

DISD. (2021). United Nations. Department of Economic and Social Affairs and Social Inclusion. *Challenges and opportunities for Indigenous peoples' sustainability* | Available at: https://www.un.org/development/desa/dspd/2021/04/indigenous-peoples-sustainability/ (Accessed: April 13, 2023).

Good, A. (2004). *Worship and the Ceremonial Economy of a Royal South Indian Temple*. Lampeter, Wales: Edwin Mellen Press.

Hill, M. & Gaillard, J.C. (2013). Integrating disaster risk reduction into post-disaster reconstruction: A long-term perspective of the 1931 earthquake in Napier, New Zealand. *New Zealand Geographer*, 69(2), 108–119.

IFAD. (n.d.). *Four ways nature-based solutions benefit rural people and communities*. Available at: https://www.ifad.org/en/web/latest/-/four-ways-nature-based-solutions-benefit-rural-people-and-communities (Accessed: April 13, 2023).

Kelman, I. (2007). Sustainable livelihoods from natural heritage on islands. *Island Studies Journal*, 2(1), 101–114.

Ma, X., Wang, R., Dai, M., & Ou, Y. (2021). The influence of culture on the sustainable livelihoods of households in rural tourism destinations. *Journal of Sustainable Tourism*, 29(8), 1235–1252.

Munjeri, D. (2004). Tangible and intangible heritage: From difference to convergence. *Museum International*, 56(1–2), 12–20.

Murti, R. & Buyck, C. (eds.). (2014). *Safe Havens: Protected Areas for Disaster Risk Reduction and Climate Change Adaptation*. Gland, Switzerland: IUCN, xii + 168 pp.

Olsson, L., Opondo, M., Tschakert, P., Agrawal, A., Eriksen, S.H., Ma, S., Perch, L.N. & Zakieldeen, S.A. (2014). Livelihoods and poverty. In *Climate Change 2014: Impacts, Adaptation, and Vulnerability. Part A: Global and Sectoral Aspects. Contribution of Working Group II to the Fifth Assessment Report of the Intergovernmental Panel on Climate Change*. [Cutter, S. & Kaijser, A. (eds.)] Cambridge, United Kingdom and New York, NY: Cambridge University Press, 793–832.

Rahimzadeh, A. (2020). Socio-economic and environmental implications of the decline of chilgoza pine nuts of Kinnaur, Western Himalaya. *Conservation & Society*, 18(4), 315–326.

Ravankhah, M., Chmutina, K., Schmidt, M. & Bosher, L. (2017). Integration of cultural heritage into disaster risk management: Challenges and opportunities for increased disaster resilience. In *Going Beyond*. [Albert, M. T., Bandarin, F., & Roders, A. P. (eds.)] Cham: Springer, 307–321.

Reason, J. (2000). Human error: models and management. *Bmj*, 320(7237), 768–770.

Rosa, A., Santangelo, A. & Tondelli, S. (2021). Investigating the integration of cultural heritage disaster risk management into urban planning tools. The Ravenna case study. *Sustainability*, 13(2), 872.

Serrat, O. (2017). The sustainable livelihoods approach. *Knowledge Solutions*, 21–26. https://doi.org/10.1007/978-981-10-0983-9_5

UNISDR. (2005). Hyogo framework for Action 2005–2015. http://www.unisdr.org/2005/wcdr/intergover/official-doc/L-docs/Hyogo-framework-for-action-english.pdf. Accessed 15 April 2022.

Venugopal, S., Gau, R., Appau, S., Sample, K. & Pereira, R. (2019). Adapting traditional livelihood practices in the face of environmental disruptions in subsistence communities. *Journal of Business Research*, 100, 400–409. https://doi.org/10.1016/j.jbusres.2018.12.023

Section III
Understanding the Challenges

10 All Fired up

The Inseparability of Nature and Culture in Disaster Risk Management

Steve Brown

A crowd of us watched in a numbed silence as houses exploded.

(Vea 2020)

Over 2019–2020, parts of eastern, south-eastern, and south-western Australia experienced the worst wildfire or 'bushfire' season on record, colloquially known as the country's 'black summer'.[1] More than 15,000 fires occurred across the continent, resulting in a combined impact area of up to 19 million hectares or 73,000 square miles (Filkov et al. 2020). The areas impacted included, for example, 80% of the Blue Mountain World Heritage area in New South Wales (NSW) (Smith 2021) and 53% of the Gondwana World Heritage rainforests in Queensland. Between September 2019 and March 2020 at least 3,500 homes and thousands of other buildings were lost and 34 people died in the fires. A report prepared for the World Wildlife Fund estimated that the area burnt in the 2019–2020 fires would have contained almost three billion native vertebrates – comprised of approximately 143 million mammals, 2.46 billion reptiles, 181 million birds, and 51 million frogs (van Eeden et al. 2020).

Amongst the cultural heritage items damaged or destroyed were a large, but unknown, number of Indigenous modified trees,[2] as well as innumerable historic built structures. Many of the buildings lost were farm buildings (e.g., homes, shearing, and other sheds) and other structures (e.g., fences, stockyards, water tank stands), as well as crops and pastures associated with agricultural and pastoral landscapes. Within the Kosciuszko National Park in the Australian Alps region of NSW, which is a cultural landscape inscribed on Australia's National Heritage List (Australian Government 2008), ten historic mountain huts (bush-crafted timber and pressed-iron dwellings constructed by stockmen and prospectors) were destroyed, as was the 1890s Kiandra Courthouse (Figure 10.1). The level of destruction rivalled that of the 2003 bushfires across the Australian Alps in which 48 historic huts and 551 houses were destroyed and an estimated 1.73 million hectares burned. The 2003 bushfires burnt across Victoria, NSW, and the Australian Capital Territory during a drought that ranked as one of the worst in over 100 years of official Australian weather records (Worboys 2003).

It is estimated that the 2003 bushfires cost the local tourist industry up to AUS$121 million in lost income and impacted some 1,000 jobs. In comparison, the Insurance Council of Australia estimated that between November 2019 and the middle of February 2020 (when most fires had been brought under control) bushfire losses were approximately AUS$2.32 billion (US$1.66 billion) in insured claims (38,181 claims were lodged) (Gangcuangco 2020).[3] These losses do not include loss of income and productivity (Centre for Disaster Philanthropy 2020), including, for example, the impacts on burned farmland exacerbated by loss of livestock and stored livestock feed, and the impacts of ash and smoke on soils and water quality.

Figure 10.1 Kiandra Courthouse, Kosciuszko National Park, following extensive restoration in 2010 (left) and following the damage sustained by bushfire (January 2020). The structure consists of the 1890s stone courthouse and associated police quarters and a 1950s chalet. (Source: Wikimedia Commons).

The purpose of recounting these terrifying facts and figures is to highlight the impacts of bushfire disasters on natural and cultural heritage, as well as on livelihoods and economies. The emphasis here is on natural and cultural heritage because disasters, such as the Australian bushfires of 2019–2020 and 2003, had profound and cumulative consequences on both domains. In this chapter, I argue that the separation of natural from cultural heritage is futile and unnecessary in the work of disaster risk reduction and management since both are intimately entwined in the histories, heritages, lives, and identities of local, regional, and national places and communities. That is, the impacts on the whole can be greater than the sum of their parts. Throughout the chapter, I will continue to use fire as a case example because of its prevalence in Australia.

Additionally, the reason for using fire as a case example is my personal experience with the threat of fire. I live on a 60-hectare (150-acre) rural property, much of it covered by Open Eucalypt Forest (Australian National Herbarium 2016). Eucalypt species are vulnerable to fire due to their high oil content,[4] and can be extremely dangerous when hot, dry weather conditions prevail. On my property, there can be runs of days during which maximum temperatures exceed 40° Celsius (104° Fahrenheit). High temperatures, extended dry conditions, eucalypt forest, and undulating terrain (fires burn faster uphill and during such an event can build in intensity and speed) are a potentially deadly combination. When they are large enough, bushfires can generate local weather impacts such as lightning, tornadoes, and fire-storms (CSIRO 2021). Given that I live 8 kilometres along a one-way, unsealed road, the chances that I will be compelled to 'prepare, stay, and defend' rather than 'flee' (or leave early) if fires come from the west (the prevailing wind direction), makes me acutely aware of bushfire risk. In the summer of 2019–2020, two vast bushfires came within 20 kilometres of my property. Although the property was fortunate not to be 'burned out', there were many weeks when substantial amounts of smoke enveloped the property and persisted (see discussion below on the health impacts of bushfire smoke and pollution).

(Un)Natural Disasters

It has become increasingly recognised in the field of disaster management that disasters are not natural (see Chapter 1, this volume; UNDRR 2021), even if the associated hazard is conceptualised as entirely 'natural' – for example, the eruption of the Hunga Tonga-Hunga Ha'apai volcano in the Kingdom of Tonga on 15 January 2022; and the powerful, magnitude 9.1, undersea earthquake that struck off the coast of Sumatra island, Indonesia, and set off the 26 December 2004 'Boxing Day' Indian Ocean tsunami. Rather, the current focus of disaster risk reduction and management is on reducing and managing conditions of *hazard*, *exposure*, and *vulnerability* in order to prevent losses and alleviate the impacts of disasters. As presented by the UNDRR (United Nations Disaster Risk Reduction), addressing 'underlying risk drivers will reduce disaster risk, lessen the impacts of climate change and, consequently, maintain the sustainability of development' (UNDRR 2021).

The linking of natural and cultural factors is a significant and necessary component of disaster risk reduction thinking and practice. The importance of cultural and natural assets is recognised, for example, in the intended outcome of the *Sendai Framework for Disaster Risk Reduction 2015–2030*:

> The substantial reduction of disaster risk and losses in lives, livelihoods and health and in the *economic, physical, social, cultural and environmental assets* of persons, businesses, communities and countries.
>
> (UNISDR 2015, Article 16; italics added)

This outcome, along with many articles in the Sendai Framework, give focus to the impact on and protection of cultural and natural assets (UNISDR 2015, Articles 5, 24[d], 30[d]) rather than the actual role of cultural and natural heritage in contributing to the reduction of disaster risks and recovery from disaster events. While the use of traditional knowledge and practices (typically forms of intangible cultural heritage) is recognised as important in contributing to the design and implementation of policies, plans, and standards (UNISDR 2015, Articles 24[i], 27[h], 36[a(v)]), the role that cultural and natural heritage plays in disaster risk prevention, mitigation, preparedness, response, recovery, and rehabilitation is neither well conceptualised nor sufficiently acknowledged. Furthermore, the need to link nature and culture, natural and cultural heritage, tangible and intangible heritage, is not specifically made apparent in the framework.

In other fields of work developed and promoted by the United Nations – including World Heritage, Climate Change, and Sustainable Development – such linkages are becoming increasingly emphasised. For example, strong and cogent arguments have been made concerning the ways in which cultural heritage can drive the transitions in land use, buildings, and other sectors that are required to meet Paris Agreement and Glasgow Climate Pact targets (ICOMOS 2019). Equally strong, evidence-based arguments have been made to demonstrate the potential of harnessing cultural and natural heritage in achieving sustainable development, as well as mobilisation and localisation of heritage in achieving the Sustainable Development Goals (Labadi 2018; Culture 2030 Goal Campaign 2021; Labadi et al. 2021).

These are significant conceptual shifts, not least of all because they seek to position culture and nature, cultural and natural heritage, as forces for change rather than only positioning heritage places as passive objects of disaster impacts. Consequently, how can disaster risk reduction better recognise and respond to the role of heritage as a force for change? As I emphasise in this chapter, the linking of nature and culture is an important conceptual component in disaster risk management practices, despite the long history of separating them in Western thinking and practice – a history that I briefly background now.

Dualism 1: Nature and Culture[5]

The divide or dualism between nature and culture has marked much of the past work in disaster risk reduction. It is linked to 'a habit of thought which developed in the West from the 17th century but with older antecedents' (Byrne et al. 2013, p. 1). Linked to a Cartesian worldview, the nature-culture divide, in which nature is considered separate from humanity, is typically seen as a product of the Age of Enlightenment, an intellectual era that dominated Europe from the 17th to the 19th centuries. This period is linked to the emergence of rational thinking, ideas of the mind and body as separate, the separation of church and state, and the increasing split between, and secularisation of, science and the humanities. For geographer Lesley Head,

> There is an eerie similarity in the 20th century incarnation of both science and arts: they frame the human as separate to the rest of nature.
>
> (Head 2011)

Dualisms (or separate sets of concerns) became increasingly seen as opposites (or binaries) and include nature-culture, past-present (a dualism grounded in a linear time perspective), material-symbolic (or tangible-intangible), and human-nonhuman. This divisive way of thinking increasingly shaped the way in which disciplines in the natural and social sciences and humanities were conceived and practised. Heritage scholar Denis Byrne views the nature-culture duality as a contrast between constructs of nature as 'people-less' and concepts of cultural heritage places as 'nature-less' (Byrne et al. 2013).

Furthermore, this dualistic philosophy was not confined to Europe but was exported as part of the European colonial project and thereby imposed on Indigenous peoples and other colonised communities. By 1914, it is estimated that European-controlled colonies occupied 84% of the globe (Hoffman 2015, pp. 2–3). Colonialism, which often emphasised a dualism between civilised (European) and primitive (or 'other'; cf. Said 1978 on Orientalism), imposed the idea of nature and culture as separate onto many cultures where a more holistic worldview was the norm. This legacy has continued in the postcolonising era (cf. Moreton-Robinson 2003) and can be seen, for example, in the UNESCO *World Heritage Convention* which, while lauded as the first United Nations' document to include both nature and culture, is underpinned by the separation of cultural from natural heritage (UNESCO 1972, Articles 1 and 2). This separation remains pervasive and, I contend, is evident in the historical challenges it created for disaster management thinking and practice.

To some degree, the Sendai Framework links nature and culture – a consequence of the internationalism involved in its creation. Nevertheless, the idea that people and nature are separate infuses parts of the document (e.g., Articles 6 and 24[d]) and, thus, is a topic area that warrants review and revision in future iterations.

By way of example, the nature-culture dualism is particularly evident in the governance and management of protected areas (national parks, nature reserves, etc.). In the case of the World Heritage listed 'Tasmanian Wilderness' (UNESCO World Heritage Centre 1989), for example, a variety of national and regional legislation exists across natural and cultural domains (including the separation of Aboriginal and 'historic' or non-Indigenous cultural heritage). These include the Commonwealth *Environment Protection and Biodiversity Conservation Act* 1999, and Tasmanian State legislation including: *Nature Conservation Act 2002*, *National Parks and Reserves Management Act 2002*, *Historic Cultural Heritage Act 1995*, and *Aboriginal Heritage Act 1995*. Ideally, the *Tasmanian Wilderness World Heritage Area Management Plan* (DPIPWE 2016) is the place where cultural and natural values are integrated. However, and typical of many such plans, cultural and natural values, and management are treated separately (DPIPWE 2016, parts 2, 3, 5), even though the plan calls for the recognition of the property as an 'outstanding Aboriginal Cultural Landscape' (DPIPWE 2016, 103–104). This call is made, in part, to counter the use of the term 'wilderness' – a term meaning 'empty' and seen as denigrating of more than 45,000 years of Aboriginal occupation (Spence 1999). Thus, in terms of legislative arrangements and management, natural and cultural values are treated as separate with minimal recognition of the entanglement or interconnection of these categories and of significantly different Western and Indigenous cosmologies (on the latter, see *tebrakunna* country & Lee 2019). The treatment of fire is an example – framed as natural in the management plan and not linked to Aboriginal cultural burning as a deep time management practice.

Dualism 2: Western Science and Traditional Knowledge

As much as nature and culture is a dualism in protected area management, so is the split between Western science and traditional knowledge. Science-based knowledge is typically applied in the management of biodiversity and geodiversity, while traditional knowledge is often framed as folklore, spirituality, or superstition and, therefore, not necessarily taken seriously in 'science-based' land management planning (cf. Byrne 2014). Here I am using 'traditional knowledge' to mean the evidence-based and experiential knowledge acquired and transmitted by local communities over many generations; and includes, for example, Indigenous traditional ecological knowledge and customary management systems. I say 'evidence-based' deliberately because, first, evidence-based knowledge is not the prerogative of Western science (cf. McNiven 2016, p. 29), and, second, experience and knowledge of the environment are incorporated in many ways into different cultures – for example, as customary laws (including taboos), stories, and cultural and spiritual practices (cf. Anderson 2013). Additional points to be made are that Indigenous knowledge comprises human and non-human (e.g., sacred natural sites; rivers and mountains attributed personhood; kinship with natural species) – qualities that cut across the tangible/intangible divide (McNiven 2016) and, in its application in anti-colonial contexts, Indigenous knowledge and narratives express the politics of cultural justice (Menzies & Wilson 2020).

Parks Victoria is a protected area agency based in the state of Victoria, Australia, and is well known for implementing a 'healthy parks, healthy people' approach to land management in a way that seeks to link human health and wellbeing with experience of the natural environment (Parks Victoria 2020a). Parks Victoria has also adopted a framework of 'Managing Country Together' (Parks Victoria 2020b). 'Country' is an Australian Aboriginal word that refers to a traditional knowledge system and to a concept with a whole-of-landscape meaning. The 'Managing Country Together' framework recognises and acknowledges the need for Indigenous Traditional Owners and park staff to share their diverse knowledges in managing landscapes. This recognition and acknowledgement are not only about Western science respecting and valuing Indigenous knowledge and vice versa, but also a matter of social and land justice consequential to the dispossession and loss of Traditional Owner ancestral lands, separation of families, and attempted suppression of Indigenous culture across Australia.

Ngootyoong Gunditj Ngootyoong Mara South West Management Plan is a management plan that exemplifies a 'two-way approach'; that is, the combination of Western management approaches to protected areas alongside Gunditjmara Traditional Owners' knowledge and experience of Country (Parks Victoria and Gunditj Mirring Traditional Owners Corporation 2015). 'Ngootyoong gunditj, ngootyoong mara' translates from the Gunditjmara Dhauwurd Wurrung language to 'healthy Country, healthy people'; and recognises that the environmental, cultural, economic, and social benefits of being on and caring for Country leads to healthy people and communities. The Management Plan is a strategic guide for managing and protecting over 130 parks, reserves, and Gunditjmara owned properties (in total covering more than 116,000 hectares) in south-west Victoria. It integrates Gunditjmara Traditional Owners' knowledge into park management.

> The plan respects the community's connections to the planning area, in particular the connections of the Gunditjmara Traditional Owners. The plan recognises that Country means the whole of the environment including nature and heritage, and material and spiritual components. The plan takes a landscape-scale planning approach for protecting natural and cultural values, and recreation and tourism management.
> (Parks Victoria and Gunditj Mirring Traditional Owners Corporation 2015, p. iii)

A key goal of the *Ngootyoong Gunditj Ngootyoong Mara South West Management Plan* relates to fire: 'Gunditjmara participate in fire management and their traditional ecological and cultural knowledge is integrated into fire management planning and practices' (Parks Victoria and Gunditj Mirring Traditional Owners Corporation 2015, p. 53).

Fire is a controversial issue in Australia, and increasingly so following the extreme bushfires that blazed across southern and eastern Australia over the summer of 2019–2020. The catastrophic blazes were exacerbated by global climate change, a cause of increased frequency or severity of fire weather; that is, periods with a high fire risk due to a combination of high temperatures, low humidity, low rainfall, and often high winds (University of East Anglia 2020). Australian Aboriginal people have used fire as a land management tool for tens of thousands of years (Hallam 1975; Gammage 2011; Pascoe 2014; Fletcher 2020; Steffensen 2020). The term 'fire-stick farming' (Jones 1969), now better framed as 'cultural burning', is used to describe traditional Aboriginal land practices of 'caring for Country': a complex notion related both to personal and group belonging and to maintaining and looking after the ecological and spiritual wellbeing of the land and of oneself. Cultural burning was and is universally practiced across Australia but applied in different ways in different regions and different ecological zones by different Aboriginal groups. Cultural burning practices typically involve small-scale burns at the right times of year and in the right places and have been shown to reduce the intensity of bushfires and emissions (Fisher & Altman 2020).

The Western science approach to fire risk management has, for more than a century in Australia, relied on the application of hazard reduction or planned burns – i.e., the controlled use of fire to reduce fuel such as dead wood, leaf litter, bark, and shrubs in forested and grassy landscapes. Although much debated, planned burning has not been adequate, despite the enormous financial costs and human resource requirements of implementation, in stemming extreme fire events (on forest management and wildfires, see Lindenmayer et al. 2020). What is required, according to Indigenous groups such as *Firesticks Alliance*, is increased valuing of Aboriginal knowledge systems of cultural burning as well as greater resourcing and capacity building (Firesticks n.d.). A crossover between Indigenous and mainstream science-based fire management groups is increasingly being called for across southern Australia (cultural burning is better recognised in northern Australia) to reduce the severity of future bushfires and better care for native fauna and flora, and the well-being of all communities. The approach to fire and landscape management being undertaken by Gunditjmara and Parks Victoria seeks to implement such a crossover and knowledge exchange.

I now return to the case example of bushfire in Australia in order to explore issues relevant to linking nature and culture in disaster risk reduction. It is important to note that some of the largest and most intense fires are associated with protected areas – including those in the Blue Mountains and Australian Alps national parks that I referenced at the start of this chapter. My focus is on strategies for disaster risk reduction, preparedness, response, and recovery and the ways in which these have been contested between science-driven approaches and Indigenous/local knowledge and practices.

Linking Nature and Culture in Australia's Bushfire Risk Management

The bushfires of the summer of 2019–2020 in Australia were preceded by long periods of drought in many parts of the country, were largely extinguished following high levels of rainfall (including severe hailstorms and flooding) in February 2020, and were followed by the onset of the Covid-19 pandemic (the latter now into its third year). What this series of events demonstrate is that disasters are not necessarily singular, one-point-in-time events but rather can be a series of multi-year hazards and happenings that have interacting, compounding, and cumulative social (personal and community), environmental, economic, and political consequences. Droughts, hailstorms, floods, and the pandemic exacerbated the terrible losses and huge impacts of bushfires and vice versa.

Each of these hazards was, and continues to be, the consequence of natural and cultural factors. This can be seen, for example, in the case of the Covid-19 global pandemic (a virus that has 'moved freely along the pathways of trade and international capital'; Roy 2020). In the case of pandemics, it is worth reflecting on the relation of humans to animals via viruses. Human bodies are ecosystems made up of cells, bacteria, fungi, etc., as well as being hosts to and inhabited by a large number and diversity of viruses. Viruses can be transmitted from other animals to humans (as well as from humans to other animals). For example, Ross River Virus, which originated in Australia, spread from mammals and birds to humans via mosquitos; and Hendra Virus from flying foxes to horses and on to humans. The spread of these and other pathogens are common in human/animal evolutionary biology (termed 'zoonotic pathogens'; i.e., able to infect other host species), and have become exacerbated through increased human population size as well as by changes in human societies and social organisation – including industrialisation, intensive agriculture, animal domestication, and urbanisation (Woolhouse & Gowtage-Sequeria 2005). This is particularly evident, for example, in the case of the second plague pandemic (a virus transmitted from fleas on rats to humans) which arrived in Europe in 1347 CE via the land and sea trade routes of the Silk Road system (Schmid et al. 2015). Another example is the spread of Leptospirosis, a disease that can be spread from cows (via milk contaminated by the urine of infected animals) to humans; and a disease which has been exacerbated by changes in the technology of dairying from hand milking (from the side of the cow) to milking using rotary milking platforms that milk from the back of the cow. Viruses illustrate that the distinction between humans and non-human species – and humans and viruses – is a false dualism and that humans are also animals. The rise and spread of pandemics illustrate how human and non-human (whether animal or plant species, topographic features, object beings, or spirit beings) are deeply intertwined and form entangled relationships.

Equally, nature and culture are entangled and entwined in bushfires. Fire is both a natural and cultural phenomena. 'Natural' fires are typically associated with lightning strikes although this is not as straightforward as it might seem (Figure 10.2). Consider, for example, human

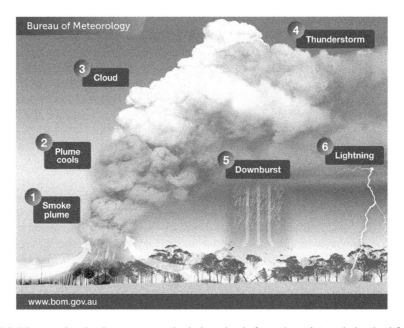

Figure 10.2 Diagram showing how pyrocumulonimbus clouds form above intensely hot bushfires. Pyrocumulonimbus clouds created by bushfires were reported during the Australian bushfires in 2019/ 2020. These events represented the entanglement of local natural and cultural histories, land management practices, and changing climate.

Source: Australian Government, Bureau of Meteorology, with permission.

induced climate change, which is responsible for increasingly warm air and sea temperature conditions and extreme weather events, and, in turn, has increased the number of lightning strikes (Romps et al. 2014).[6] That is, human inaction on climate has increased hot, dry conditions and the likelihood of extreme weather events and, thus, the vulnerability of forests and grasslands to fire started by lightning strikes. Hence, lightning has both 'natural' and 'cultural' origins and dimensions – its distribution and intensity is deeply influenced by past and present human action (Aich et al. 2018).

On the other hand, 'cultural' fires (used in cooking, signalling, warming, land management, ceremony, and socialising) are a mix of a 'natural' element (fire, both flame and smoke) and manipulation of fire behaviour for the benefit of human societies. As noted previously, Australian Aboriginal societies used fire as a tool for tens of millennia in practices of hunting and vegetation management. A consequence of the latter is that Australia's vegetation became shaped by fire management; that is, the skilled application of fire in Indigenous land management promoted certain species and vegetation communities, as well as the distribution of those communities. When militaristic, capitalist, colonial invaders settled in Australia in the late 18th century, they did not immediately recognise the ways in which Indigenous land management had modified and controlled much of Australia's vegetation. Subsequently, without the application of Indigenous fire knowledge and management, much of Australia's vegetation rapidly changed and, along with changing climates from the mid-20th century, became increasingly fire prone and prone to larger and more intense forest and grassland bushfires.

In Australia, it has been estimated by Geoscience Australia that about half of all bushfires are started by natural causes (mostly from lightning), and half by people who ignite them either intentionally (i.e., arson) or accidentally (e.g., cigarettes and sparks from machinery). An example of the latter is, on 27 January 2020, when an Australian Army helicopter, operating in the remote Orroral Valley, sparked a massive bushfire when it landed and the heat of the helicopter's taillight set alight dry grass, following which the helicopter's propeller fanned the sparks. Thus, human misadventure (a cultural activity) led to the ignition of a fire (a 'natural' phenomena) which developed into a bushfire that scorched some 87,500 hectares[7] (the vegetation being a product of long-term natural and cultural creation, as discussed above).

Fire as Natural and Cultural: Implications for DRM

In this section, I consider each of four areas of disaster risk management: risk reduction, preparedness, response, and recovery. In each instance, I highlight aspects that engage with the notion of culture and nature as entangled in relation to bushfires. My aim is not to be comprehensive but rather to provide selected examples that highlight the benefits of integrated approaches and the shortcomings of seeing culture and nature as separate.

Disaster Risk Reduction

Australia has prepared and adopted a *National Disaster Risk Reduction Framework* (Commonwealth of Australia 2018; see also Box 10.1). Although developed and published prior to the 2019–2020 bushfires, the document provides a useful way of examining thinking at a national level. The Framework draws on the Sendai Framework (UNISDR 2015), the 2030 Agenda for Sustainable Development (United Nations 2015a), and The Paris Agreement (United Nations 2015b).[8] The Framework is applied across and between four key environments: built, social, natural, and economic. However, while the Framework is inclusive of social and natural environments, it does not make strong linkages between these domains. For example, fire and

Box 10.1 Australian Government Royal Commission into National Natural Disaster Arrangements Report ('Bushfires Royal Commission')

Australia has a history of more than 240 inquiries about 'natural' disasters. Governments typically establish Royal Commissions following the most severe catastrophes. Their purpose is to inquire into and better understand the causes of and responses to disasters; and to recommend on what should be done to improve emergency management arrangements concerning all phases of disaster risk management: mitigation, preparedness, response, and recovery.

The *Royal Commission into National Natural Disaster Arrangements* was established on 20 February 2020 in response to the extreme and 'unprecedented' bushfire season of 2019–2020 that resulted in devastating loss of life, property, and wildlife, and severe environmental destruction. The central task of the Commission was to inquire into, and report on, national disaster arrangements – that is, arrangements involving all levels of government, the private and not-for-profit sectors, communities, families, and individuals. The Inquiry received extensive evidence – more than 270 witnesses, almost 80,000 pages of tendered documents, and more than 1,750 public submissions. A final report was completed in October 2020.

The report comprises 24 chapters and 600 pages. A number of the chapters address 'heritage' including wildlife and heritage (Chapter 16) and Indigenous land and fire management (Chapter 18). The report notes that six World Heritage properties were affected, including extensive burnt areas across the 'Gondwana Rainforests of Australia' (54% burnt), 'Greater Blue Mountains Area' (82% burnt), and the Budj Bim Indigenous Cultural Landscape (pp. 354, 356); and that over 330 threatened species and 37 threatened ecological communities were in the path of the bushfires (p. 355). Beyond noting that Australia has a 'vast number of heritage places of significant value' (p. 353), the report does not categorise what the cultural heritage places are nor give the numbers of heritage places destroyed and/or damaged by the 2019–2020 bushfires. For example, the loss of culturally modified trees was incalculable since many such Indigenous heritage items had never been officially documented.

The report is notable for incorporating information on Indigenous use of fire. It notes that the application of fire by Australian Indigenous peoples, a practice that extends over tens of thousands of years, includes 'cultural burning' as one component of broader Indigenous land management practices. Cultural burning contributes to improved ecosystem resilience, as well as enhances productive landscapes and supports other customary values. The report recommends: 'Australian, state, territory and local governments should (*sic*) engage further with Traditional Owners to explore the relationship between Indigenous land and fire management and natural disaster resilience' (p. 396).

To be fair, the Australian Government Royal Commission Report tries to present a holistic view of the 2019–2020 bushfire emergencies. It includes, for example, discussion of and recommendations relating to the 'natural' environment and the potential role of Indigenous fire management practices. Nevertheless, the report does not integrate information on heritage places of cultural significance but positions such places as passive objects of disaster impacts rather than as forces for resilience and change. Thus, while inquiries into disasters are an essential mechanism in improving mitigation, preparedness, response, and recovery in disaster risk management, the role of cultural heritage in disaster management arrangements in Australia will benefit from being better recognised and integrated.

'severe fire weather' are framed as 'natural hazards', part of a wider issue of the politics of climate change in Australia.

Australia is one of the world's biggest per capita greenhouse gas emitters.

When emissions from Australia's current coal, oil, and gas exports (3.6% of global total) are added to domestic emissions (1.4% of global total), Australia's contribution to the global climate pollution footprint is about 5%. This is equivalent to the total emissions of Russia, which is ranked the fifth biggest CO_2 emitter globally (Parra et al. 2019). In 2020, the *Climate Change Performance Index* (Burck et al. 2020) ranked Australia last on climate policy out of those 57 countries responsible for more than 90% of greenhouse emissions. At the COP26 Glasgow Summit, Australia was one of the few nations not to strengthen its 2030 targets. The point that I am making here is that the politics of climate science, in which Australian conservative national governments have denied the role of humans in global warming, has not worked favourably in reducing disaster risk with regard to bushfires because climate is conceptualised as purely a 'natural hazard'.[9] By contrast, Australian government scientists have demonstrated that human induced climate change is the overwhelming factor driving the country's ever-more intense bushfires (Canadell et al. 2021) – directly contradicting claims by the country's conservative political leaders.

Disaster Preparedness

In Australia, bushfires can ignite in remote areas (such as the Orroral Valley example presented above), across rural areas (e.g., most commonly grass fires), and through wildland-urban interfaces. In all such instances, the origins of bushfires can be natural (e.g., lightning – a factor exacerbated by human-induced climate change [Romps et al. 2014]) and cultural (arson or accident), but in all cases, are exacerbated by natural-cultural factors.

An amazing story of bushfire preparedness and response relates to the protection of the Wollemi Pine (*Wollemia nobilis*), so called 'dinosaur trees' because they have existed from up to 200 million years ago. The only recorded stand – some 200 trees – is located in a remote sandstone gorge in the Wollemi National Park in the Blue Mountains, about 200 kilometres northwest of Sydney. They were thought extinct by Western science until documented in 1994. The pine grove was threatened by the Gospers Mountain fire, which started by lightning strike in October 2019 and burned more than 500,000 hectares. To protect the rare Wollemi Pine grove, the NSW National Parks and Wildlife Service and NSW Rural Fire Service utilised aircraft tankers to drop fire retardant and specialist firefighters were winched into the gorge by helicopter to set up an irrigation system. As the fire approached, helicopters dropped water on the fire edge to reduce its impact on the trees (Morton 2020). These actions prevented significant loss of mature trees, although many juvenile plants were severely burnt and 'most trees shorter than eight metres showed 100% canopy scorching and only 2% of plants between five centimetres and two metres have begun to resprout' (Cox 2021). This incredible tale of survival is one in which a natural species required a significant level of human intervention and action to ensure its ongoing existence. This dedicated fire management strategy, responding to complex natural-human factors, has been essential to ensure the survival of the species beyond the black summer bushfires.

Disaster Response

Much has been written with regard to response to wildfire, both in Australia and elsewhere. While there are many dimensions to this complex field, I focus here on one aspect of bushfires – bushfire smoke and the health response.

There are many kinds of smoke, with some Indigenous traditional smoking techniques producing smoke with significant positive microbial effects (Sadgrove et al. 2014). Smoke from contained and controlled fires has been used in Aboriginal Australian cultural practices for centuries, including in 'smoking ceremonies' used to welcome people to Country, as well as in ceremonies associated with the repatriation of human remains and other special events. In some places, 'Aboriginal women use smoke from burning selected leaves to protect newborn babies' (Smith et al. 2020).

Then there is the unwelcome smoke of bushfires. During the 2019–2020 bushfires, smoke-related air pollution reached hazardous levels in major metropolitan areas including Canberra, Australia's 'bush capital' (Rodney et al. 2021).[10] On New Year's Day (1st January) 2020, following days of substantial air pollution, Canberra's air quality reached an extreme peak with an Air Quality Index (AQI) reading of 7,700 – exceeding the Australian air quality limit by 77 times, well into the 'Hazardous' AQI level of above 200 (Remeikis 2020). Subsequent studies have demonstrated that prolonged exposure to bushfire smoke can have substantial effects on health – on physical health, mental health, sleep patterns, and the intersections of these (Rodney et al. 2021).[11] Rodney et al. (2021) propose that the response to the health impacts of bushfire smoke include,

> Improved public health messaging ... to address uncertainty about how individuals can protect their and their families health for future events. This should be informed by identifying subgroups of the population, such as those with existing health conditions, parents, or those directly exposed to fire who may be at a greater risk.

Such responses to the health impacts of smoke-related air pollution require considerations of the complex interlinkages of factors that are both natural (smoke including particulate matter) and cultural (such as human lifeways and behaviours, technologies of air purification, and medical evidence). Consequently, any attempts to address the challenges of smoke-related air pollution require territorial approaches – i.e., engagement of multiple sectors across rural-urban areas, implementation by a range of stakeholders across environmental and cultural fields, and secured by multi-level governance.

Disaster Recovery

Australia is still recovering from the 2019–2020 bushfires. Many homes and parts of towns, including infrastructure, are still being rebuilt and, thus, testing the resilience of local communities. Here I want to focus on one form of recovery, which concerns the rebuilding of domestic and community gardens. Gardens are powerful spaces for propagating feelings of place attachment, identity, and connection (Brown 2018). Individual plants can be reminders of friends or family members from whom they were gifted; and parts of gardens can evoke memories of collective endeavour. Feelings for gardens can be experienced at individual, familial, friendship group, and community levels.

Replanting and/or restoring gardens after bushfires can be important in the processes of rebuilding lives, recovering human physical and mental health, remembering loss, reconnecting with family and friends, and reviving landscape. Redesigning a garden and replanting can also be a time to consider the types of plants that can support preparedness against the impacts of future bushfires. For example, the CSIRO (Commonwealth Scientific and Industrial Research Organisation), which is an Australian Government agency responsible for scientific research, has published 'tips' for re-planting after bushfires (Cranney 2020).

They recommend designing a 'fire-smart' garden (based on species selection and landscaping[12]) and choosing more fire resistant and retardant plants (e.g., succulents). The purpose of the latter is that:

> Fire retardant plants can absorb more of the heat of the approaching bushfire without burning than more flammable plants. They can trap burning embers and sparks and reduce wind speeds near your house if correctly positioned and maintained. Fire resistant ground covers can be used to slow the travel of a fire through the litter layer. Fire resistant shrubs can be used to separate the litter layer from the trees above.
>
> (Chladil & Sheridan 2006)

Redesigning and replanting a garden necessary interlinks nature and culture (Head & Muir 2007) and interweaves tangible elements (plants, rocks, garden beds, water systems) as well as intangible associations, including memory, identity, and connection.

Burning Bird and Bogans: Hazards as Complex Natural, Social, Economic, and Political Risks

Two fires that threatened Gozinta, the property on which I live, had unusual beginnings. The source of the Currandooley fire of January 2017, which burned 3.400 hectares, was a low flying bird. The bird, possibly a crow, flew through powerlines causing them to arc and ignite the bird's feathers, following which the alight bird fell to the ground into dry grass where a fire was ignited. High temperatures, dry ground fuel, and strong winds at the time created the dangerous conditions in which this 'fluke' event occurred. In December 2016, a vehicle started a fire, which burned 500 hectares and came within a kilometre of my property. 'Bogans'[13] drove a vehicle recklessly around a paddock, and then either parked or stalled their heated car in dry grass, causing a fire to ignite. Those involved in this event waited some 30 minutes before calling the Rural Fire Service; they thought they could extinguish the fire themselves. Ten fire crews, two water bombing helicopters, and a DC 10 air tanker were required to bring the fire under control.

In the period between September 2016 and the end of March 2017, there were 17 fire incidents in the region where I live: five fires were attributed to burning off vegetation (both legally and illegally); four fires were blamed on 'powerlines' – one being a transformer fire and another being the infamous flaming crow incident; mowers and slashers and one motor vehicle stall caused a further three fires; lightning caused two; a human-caused house fire was another cause; and two fires were listed as undetermined, with one suspected as arson (District Bulletin 2017).

I conclude this chapter with these local stories and statistics in order to emphasise the ways in which fire is both natural and cultural and, therefore, that disaster risk management connected with fire needs to consider the ways in which both domains are entangled. The 17 fire incidents listed above are evidence of the complex ways in which natural and social causal factors lead to bushfire; no matter how hazards and risks are defined.

In the context of disaster risk reduction, 'hazard' generally means 'something that can be dangerous or cause damage'.[14] 'Risk' refers to a chance of something happening that will have a negative effect, and reflects both the likelihood of the unwanted event and the potential consequences of that event. These meanings do not distinguish between nature and culture, natural and cultural heritage, tangible and intangible heritage. And, as I have argued throughout this chapter, nor should the work of disaster risk management since each of these dualisms is artificial and, in reality, are linked and deeply entangled.

Key Points

- Nature and culture are intimately entwined in the histories, heritages, lives, and identities of local, regional, and national places and communities.
- The separation of natural from cultural heritage is futile and unnecessary in the work of disaster risk reduction and management.
- The linking of natural and cultural factors is a significant and necessary component of disaster risk reduction thinking and practice.
- Positioning heritage places as passive objects of disaster impacts fails to recognise that culture and nature, cultural and natural heritage, as forces for change.
- Beware of 'flaming crows' during periods of high temperatures, dry ground fuel, and strong winds.

Notes

1 In Australia, 'wildfire' is more commonly referred to as 'bushfire', where 'bush' is a term meaning both wooded vegetation and areas beyond the major metropolitan centres.
2 Modified trees include trees that were carved or scarred as part of Aboriginal Australian cultural, social, and economic practices. Trees onto which designs were carved are associated with burials, ceremonial grounds, and/or territorial markers. Scarred trees retain evidence of where bark or wood was removed in order to manufacture canoes, shields, or containers; or where footholds were cut into the trunks to gain access to resources such as possums or honey.
3 Taking into account the hailstorms of November 2019 and January 2020, as well as the Australian east coast storms and floods in January–February 2020, the total cost of insurance claims from the four events (bushfires, hailstorms, storms, and floods) exceeded AUS$5.19 billion (US$3.65 billion).
4 The leaves produce a volatile, highly combustible oil, and the ground beneath the trees is covered with large amounts of litter which is high in phenolics, preventing its breakdown by fungi. Wildfires burn rapidly under them and through the tree crowns.
5 In this and the following section, I draw on critiques of dualisms (Brown 2023a, 2023b).
6 For example, Romps et al. (2014) predict that the number of lightning strikes will increase by about 12% for every degree of rise in global average air temperature.
7 The ensuing fire burned about 80% of Namadgi National Park (82,700 hectares), 22% of Tidbinbilla Nature Reserve (1,444 hectares), and 3,350 hectares of rural lands (ACT Government n.d.).
8 Australia is a party to each of these international agreements.
9 In contrast, a recent study concluded that as a result of anthropogenic climate change, the prevalence of days of high-risk bushfire weather has increased, conservatively, by at least 30% since 1900 (van Oldenborgh et al. 2021).
10 The main pollutants of health concern are particulate matter (PM): fine particles of up to 2.5 microns or 10 microns in diameter. These tiny particles are present in smoke generated most commonly by wood heaters, bushfires, vehicle emission exhausts, and dust storms. PM is particularly hazardous to human health since its tiny size enables the particles to penetrate deep into the human system once inhaled, even entering the bloodstream. Health impacts of exposure to particulate matter include an increased risk of cardiovascular disease, respiratory disease, and lung cancer (IQAir 2022).
11 A preliminary evaluation of the air pollution health burden in eastern Australia estimated that bushfire smoke was responsible for 417 deaths, over 3,000 hospitalisations for cardiovascular and respiratory problems, and 1,305 presentations to emergency departments with asthma (Borchers Arriagada et al. 2020). The total smoke-related physical health costs during the 2019–20 bushfire period have been estimated at AU$1.95 billion (Johnston et al. 2021.
12 For example, the Victorian Government's Country Fire Authority (CFA) has developed four principles for garden design in fire-prone areas: (1) Plant to create defendable space; (2) Remove flammable objects from around the house; (3) Break up fuel continuity (keep plants separate and avoid flammable mulches); and (4) Carefully select, position, and maintain trees (CFA 2021).

13 'Bogan' is an Australian slang term for a person whose speech, clothing, attitude, and behaviour, including in relation to driving cars, are considered unrefined or unsophisticated. As well as a term of derision, bogan can be used as a self-identifying term of pride.
14 Oxford Learner's Dictionaries. 'Hazard', https://www.oxfordlearnersdictionaries.com/definition/english/hazard_1.

References

ACT (Australian Capital Territory) Government. n.d. *Ororral Valley Fire Impact Report*. Available from: https://www.environment.act.gov.au/ACT-parks-conservation/bushfire_management/ororral-valley-fire-impact-report (Accessed 26 January 2022).
Aich, V., Holzworth, R., Goodman, S.J., Kuleshov, Y., Price, C., and Williams, E. 2018. Lightning: A new essential climate variable. *Eos*, 99. https://doi.org/10.1029/2018EO104583.
Anderson, M.K. 2013. *Tending the Wild Native American Knowledge and the Management of California's Natural Resources*. California: University of California Press.
Australian Government. 2008. *National Heritage Places – Australian Alps National Parks and Reserves*. Available at: https://www.awe.gov.au/parks-heritage/heritage/places/national/australia-alps (Accessed 23 January 2022).
Australian Government, Bureau of Meteorology. *When Bushfires Make Their Own Weather*. https://media.bom.gov.au/social/blog/1618/when-bushfires-make-their-own-weather/.
Australian National Herbarium. 2016. *Eucalypt Open Forests*. Available at: https://www.anbg.gov.au/photo/vegetation/eucalypt-open-forests.html (Accessed 23 January 2022).
Borchers Arriagada, N., Palmer, A.J., Bowman, D.M., Morgan, G.G., Jalaludin, B.B., and Johnston, F.H. 2020. Unprecedented smoke-related health burden associated with the 2019–20 bushfires in eastern Australia. *Medical Journal of Australia*, 213, pp. 282–283. https://doi.org/10.5694/mja2.50545.
Brown, S. 2018. The nature of attachment: An Australian experience. In B. Verschuuren, & S. Brown, eds. *Cultural and Spiritual Significance of Nature in Protected Areas: Governance, Management, and Policy*, Abington and New York: Routledge, pp. 280–293.
Brown, S. 2023a. From culture and nature as separate to interconnected *naturecultures*. In S. Brown & C. Goetcheus, eds. *Routledge Handbook of Cultural Landscape Practice*, chapter 1.3. Abington and New York: Routledge.
Brown, S. 2023b. From difficult dualisms to entangled complexity. In S. Brown & C. Goetcheus, eds. *Routledge Handbook of Cultural Landscape Practice*, chapter 1.4. Abington and New York: Routledge.
Burck, J., Hagen, U., Höhne, N., Nascimento, L. and Bals, C. 2020. *Climate Change Performance Index: Results 2020*. Bonn: Germanwatch.
Byrne, D. 2014. *Counterheritage: Critical Perspectives on Heritage Conservation in Asia*. London and New York: Routledge.
Byrne, D., Brockwell, S., and O'Connor, S. 2013. Introduction: Engaging culture and nature. In S. Brockwell, S. O'Connor, & D. Byrne, eds. *Transcending the Nature Culture Divide in Cultural Heritage: Views for the Asia Pacific Region*. Terra Australis 36. Canberra: The Australian National University, pp. 35–52.
Canadell, J.G., Meyer, C.P., Cook, G.D., Dowdy, A., Briggs, P.R., Knauer, J., Pepler, A., and Haverd, V., 2021. Multi-decadal increase of forest burned area in Australia is linked to climate change. *Nature Communications*, 12, 6921. https://doi.org/10.1038/s41467-021-27225-4.
Centre for Disaster Philanthropy. 2020. *2019–2020 Australian Bushfires*. Available at: https://disasterphilanthropy.org/disaster/2019-australian-wildfires/ (Accessed 22 January 2022).
CFA (Country Fire Authority). 2021. *Landscaping*. Available from: https://www.cfa.vic.gov.au/plan-prepare/how-to-prepare-your-property/landscaping (Accessed 4 February 2022).
Chladil, M. and Sheridan, J. 2006. *Fire Retardant Garden Plants for the Urban Fringe and Rural Areas*. Hobart: Tasmanian Government.
Commonwealth of Australia. 2018. *National Disaster Risk Reduction Framework*. Canberra: Commonwealth of Australia.

Cox, L. 2021. Wollemi pines given special protected status after being saved from bushfire disaster. *The Guardian*, 15 January 2021. Available from: https://www.theguardian.com/environment/2021/jan/15/nsw-wollemi-pines-given-special-protected-status-after-being-saved-from-bushfire-disaster (Accessed 43 February 2022).

Cranney, K. 2020. Five tips for replanting your garden after bushfires. *CSIROscope*, 31 January 2020. Available from: https://blog.csiro.au/five-tips-for-replanting-after-bushfires/?fbclid=IwAR2dP3PPi_9SMcfO3VHnHLtJYExa-NqvbyK608Bi8cmdr2ZBvQzAcL8fDkg (Accessed 4 January 2022).

CSIRO. 2021. *The 2019–20 Bushfires: A CSIRO Explainer*. Available at: https://www.csiro.au/en/research/natural-disasters/bushfires/2019-20-bushfires-explainer (Accessed 23 January 2022).

Culture 2030 Goal Campaign. 2021. *Culture in the Localization of the 2030 Agenda: An Analysis of Voluntary Local Reviews*. Published in Barcelona, Paris, Abidjan, Montreal, The Hague, and Brussels, in the frame of the 4th UCLG Culture Summit taking place on 9–11 September 2021.

District Bulletin. 2017. What caused most of the recent area bushfires? *The Independent District Bulletin*, 10 April 2017. Available from: https://districtbulletin.com.au/what-caused-most-recent-area-bushfires/ (Accessed 5 February 2022).

DPIPWE. 2016. *Tasmanian Wilderness World Heritage Area Management Plan 2016*. Hobart: Department of Primary Industries, Parks, Water, and Environment.

Filkov, A.I., Ngo, T., Matthews, S., Telfer, S. and Penman, T.D. 2020. Impact of Australia's catastrophic 2019/20 bushfire season on communities and environment. Retrospective analysis and current trends. *Journal of Safety Science and Resilience*, 1, 44–56.

Firesticks, n.d. *Firesticks: Cultural Burning, Healthy Communities, Healthy Landscapes*. Available from: https://www.firesticks.org.au/ (Accessed 20 November 2020).

Fisher, R. and Altman, J. 2020. The world's best fire management system is in northern Australia, and it's led by indigenous land managers. *The Conversation*, 10 March 2020. Available from: https://theconversation.com/the-worlds-best-fire-management-system-is-in-northern-australia-and-its-led-by-indigenous-land-managers-133071 (Accessed 27 January 2022).

Fletcher, M.-S., 2020. This rainforest was once a grassland savanna maintained by aboriginal people – until colonisation. *The Conversation*, 11 May 2020. Available from: https://theconversation.com/this-rainforest-was-once-a-grassland-savanna-maintained-by-aboriginal-people-until-colonisation-138289 (Accessed 27 January 2022).

Gammage, B. 2011. *The Biggest Estate on Earth: How Aborigines Made Australia*. Sydney: Allen and Unwin.

Gangcuangco, T. 2020. Revealed: Insurance bill for 2019–20 summer catastrophes. *Insurance Business Australia*. Available from: https://www.insurancebusinessmag.com/au/news/breaking-news/revealed-insurance-bill-for-201920-summer-catastrophes-223760.aspx (Accessed 26 January 2022).

Hallam, S. 1975. *Fire and Hearth: A Study of Aboriginal Usage and European Usurpation in Southwestern Australia*. Canberra: Australian Institute of Aboriginal Studies.

Head, L. 2011. More than human, more than nature. *Griffith Review*. Available from: https://www.griffithreview.com/articles/more-than-human-more-than-nature/ (Accessed 24 January 2022).

Head, L. and Muir, P. 2007. *Backyard: Nature and Culture in Suburban Australia*. Wollongong: University of Wollongong Press.

Hoffman, P.T. 2015. *Why Did Europe Conquer the World?* Princeton: Princeton University Press.

ICOMOS (International Council on Monuments and Sites). 2019. *The Future of Our Pasts: Engaging Cultural Heritage in Climate Action*. Paris: ICOMOS.

IQAir, 2022. Air Quality in Canberra. Available at: https://www.iqair.com/au/australia/act/canberra (Accessed 2 February 2022).

Johnston, F.H., Borchers-Arriagada, N. Morgan, G.G. et al. 2021. Unprecedented health costs of smoke-related $PM_{2.5}$ from the 2019–20 Australian megafires. *Nature Sustainability*, 4, pp. 42–47. https://doi.org/10.1038/s41893-020-00610-5

Jones, R. 1969. Firestick farming. *Australian Natural History*, 16, pp. 224–231.

Labadi, S. 2018. Historical, theoretical and international considerations on culture, heritage and (sustainable) development. In P.B. Larsen, & W. Logan, eds. *World Heritage and Sustainable Development: New Directions in World Heritage Management*, Abington: Routledge, pp. 37–49.

Labadi, S., Giliberto, F., Rosetti, I., Shetabi, L. and Yildirim, E. 2021. *Heritage and the Sustainable Development Goals: Policy Guidance for Heritage and Development Actors*. Paris: ICOMOS.

Lindenmayer, D.B., Kooyman, R.M., Taylor, C., Ward, M. and Watson, J.E.M. 2020. Recent Australian wildfires made worse by logging and associated forest management. *Nature Ecology Evolution*, 4, pp. 898–900.

McNiven, I.J. 2016. Theoretical challenges of indigenous archaeology: Setting an agenda. *American Antiquity*, 81(1), pp. 27–41.

Menzies, D. and Wilson, C. 2020. Indigenous heritage narratives for cultural justice. *Historic Environment*, 32(1), pp. 54–69.

Moreton-Robinson, A. 2003. I still call Australia home: Indigenous belonging and place in a postcolonising society. In S. Ahmed, A.M. Fortier, M. Sheller, & C. Castaneda, eds. *Uprootings/Regroundings: Questions of Home and Migration*, Oxford: Berg Publications, pp. 23–40.

Morton, A. 2020. 'Dinosaur trees': Firefighters save endangered Wollemi pines from NSW bushfires. *The Guardian*, 15 January 2020. Available at: https://www.theguardian.com/australia-news/2020/jan/15/dinosaur-trees-firefighters-save-endangered-wollemi-pines-from-nsw-bushfires (Accessed 3 February 2022].

Parks Victoria. 2020a. *Healthy Parks, Healthy People*. Available from: https://www.parks.vic.gov.au/healthy-parks-healthy-people [Accessed 27 January 2022].

Parks Victoria. 2020b. *Managing Country Together*. Available from: https://www.parks.vic.gov.au/managing-country-together [Accessed 27 January 2022].

Parks Victoria and Gunditj Mirring Traditional Owners Corporation. 2015. *Ngootyoong Gunditj Ngootyoong Mara South West Management Plan*. Portland: Parks Victoria.

Parra, P.Y., Hare, B., Fuentes Hutfilter, U. and Roming, N. 2019. Evaluating the Significance of Australia's Global Fossil Fuel Carbon Footprint. Sydney: Climate Analytics for the Australian Conservation Foundation.

Pascoe, B. 2014. *Dark Emu: Agriculture or Accident?* Broome: Magabala Books.

Remeikis, A. 2020. Canberra Chokes on world's worst air quality as city all but shut down. *The Guardian*, 3 January 2020. Available at: https://www.theguardian.com/australia-news/2020/jan/03/canberra-chokes-on-worlds-worst-air-quality-as-city-all-but-shut-down [Accessed 2 February 2022].

Rodney, R.M., Swaminathan, A. Calear, A.L., et al. 2021. Physical and mental health effects of bushfire and smoke in the Australian capital territory 2019–20. *Frontiers in Public Health*. https://doi.org/10.3389/fpubh.2021.682402.

Romps, D.M., Seeley, J.T., Vollaro, D. and Molinari, J. 2014. Projected increase in lightning strikes in the United States due to global warming. *Science*, 346(6211), pp. 851–854. doi: 10.1126/science.1259100.

Roy, A. 2020. The pandemic is a portal. Available from: https://www.ft.com/content/10d8f5e8-74eb-11ea-95fe-fcd274e920ca [Accessed 27 January 2022].

Sadgrove, N.J., Jones, G.L. and Greatrex, B.W. 2014. Isolation and characterisation of (−)-genifuranal: The principal antimicrobial component in traditional smoking applications of *eremophila longifolia* (scrophulariaceae) by Australian Aboriginal peoples. *Journal of Ethnopharmacology*, 154(3), pp. 758–766.

Said, E. 1978. *Orientalism*. New York: Pantheon Books.

Schmid, B.V., Büntgen, U., Easterday, W.R., Ginzler, C., Walløe, L., Bramanti, B. and Stenseth, N.C. 2015. Climate-driven introduction of the black death and successive plague reintroductions into Europe. *Proceedings of the National Academy of Sciences*, 112(10), pp. 3020–3025.

Smith, P. 2021. *Impact of the 2019–20 Fires on the Greater Blue Mountains World Heritage Area – Version 2*. Report to Blue Mountains Conservation Society. Blaxland: P & J Smith Ecological Consultants.

Smith, C., Jackson, A. and Pollard, K. 2020. Not all blackened landscapes are bad. We must learn to love the right kind. *The Conversation*, 2 December 2020. Available at: https://theconversation.com/not-all-blackened-landscapes-are-bad-we-must-learn-to-love-the-right-kind-129547 [Accessed 2 February 2022].

Steffensen, V. 2020. *Fire Country: How Indigenous Fire Management Could Help Save Australia*. Melbourne: Hardie Grant.

tebrakunna country and Lee, E. 2019. 'Reset the relationship': Decolonising government to increase indigenous benefit. *Cultural Geographies*, 26(4), pp. 415–434.

UNDRR (United Nations Office for Disaster Risk Reduction). 2021. *Understanding Disaster Risk*. Available at: https://www.preventionweb.net/understanding-disaster-risk/key-concepts/disaster-risk-reduction-disaster-risk-management [Accessed 23 January 2022].

UNESCO. 1972. *Convention Concerning the Protection of the World Cultural and Natural Heritage*. Available from: http://whc.unesco.org/en/conventiontext/ [Accessed 24 January 2022].

UNESCO World Heritage Centre. 1989. *Tasmanian Wilderness*. Available from: https://whc.unesco.org/en/list/181 [Accessed 27 January 2022].

UNISDR (The United Nations Office for Disaster Risk Reduction). 2015. *Sendai Framework for Disaster Risk Reduction 2015–2030*. Available at: https://www.undrr.org/publication/sendai-framework-disaster-risk-reduction-2015–2030 [Accessed 23 January 2022].

United Nations. 2015a. *Transforming Our World: The 2030 Agenda for Sustainable Development*. Department of Economic and Social Affairs. Available at: https://sdgs.un.org/2030agenda [Accessed 2 February 2022].

United Nations. 2015b. *The Paris Agreement*. United Nations Climate Change. Available at: https://unfccc.int/process-and-meetings/the-paris-agreement/the-paris-agreement [Accessed 2 February 2022].

University of East Anglia. 2020. Climate change increases the risk of wildfires confirms new review. *Science Daily*, 14 January 2020. Available from: www.sciencedaily.com/releases/2020/01/200114074046.htm [Accessed 27 January 2022].

van Eeden, L.M., Nimmo, D., Mahony, M., Herman, K., Ehmke, G., Driessen, J., O'Connor, J., Bino, G., Taylor, M. and Dickman, C.R. 2020. *Impacts of the Unprecedented 2019–2020 Bushfires on Australian Animals*. Report prepared for World Wildlife Fund (Australia), Ultimo NSW.

van Oldenborgh, G.J., Krikken, F., Lewis, S., Leach, N.J., Lehner, F., Saunders, K.R., et al. 2021. Attribution of the Australian bushfire risk to anthropogenic climate change. *Natural Hazards and Earth System Sciences Discussions*, 21, pp. 941–960. doi: 10.5194/nhess-21-941-2021.

Vea, J. 2020. *Vox*, 24 January 2020. Available at: https://www.vox.com/2020/1/24/21063638/australian-bushfires-2019-experience [Accessed 27 January 2022].

Woolhouse, M.E.J. and Gowtage-Sequeria, S. 2005. Host range and emerging and reemerging pathogens. *Research*, 11(12), pp. 1842–1847.

Worboys, G. 2003. A brief report on the 2003 Australian Alps bushfires. *Mountain Research and Development*, 23(3), pp. 294–295.

11 The Dangers of Romanticising Local Knowledge in the Context of Disaster Studies and Practice

Demet Intepe, Robert Šakić Trogrlić,
Maria Evangelina Filippi, Thirze Hermans,
Hannah Bailon, and Anuzska Maton

People have always lived in areas exposed to various hazards, including natural hazards (e.g., floods, landslides, droughts, earthquakes), as these areas have often been closely tied with their main sources of livelihoods. For instance, with their fertile soils, floodplains were and continue to be a home and food basket for many farmers across the world. Staying in these hazard-prone areas and banking on livelihood benefits means being ingenious. Through their frequent encounters with hydrological hazards, farmers living alongside the Lower Mekong River in Cambodia developed a specific knowledge for spatial and temporal characterisation of floods and droughts, which directly informs their livelihood actions and options (Pauli *et al.*, 2021). In the volcanically active areas of Central Java's highlands in Indonesia, local communities often have a multifaceted understanding of the causes, consequences, and impacts of volcanic gas eruptions (Griffin and Barney, 2021). Thousands of miles away, in south-east Iceland, knowledge from local communities is proving to be a useful asset even for emerging hazards arising from a fracture in the mountainside of Svínafellsheiði (Matti & Ögmundardóttir, 2021). Communities have always found means of coping with long-term and direct exposure to frequent natural hazards and developed rich knowledges of how to prepare for, adapt to, deal with, and recover from hazards. In scholarship, these rich sets of knowledges have been studied under the rubrics of local, Indigenous, or traditional knowledge. As the latest report by the Intergovernmental Panel on Climate Change (IPCC) lays out, it is not possible to pinpoint a single definition for each of these knowledges (IPCC, 2022), and thus there is an urgent need for contextualisation.

In this chapter, we will focus on local knowledge, which disaster studies and policy are increasingly drawing from, to understand and improve the ways in which communities at risk tackle hazards and disasters. Before we go further, it might be helpful to clarify our position regarding our term of preference, as other alternatives such as traditional knowledge and Indigenous knowledge also exist in scholarship and policy, and these terms are often used interchangeably (Iloka, 2016). A particular preference for any of these terms will depend on the context, academic discipline, and language (Kelman *et al.*, 2012). As authors specialising in different disciplines, we will be bringing a mixture of positionalities and interdisciplinarity into this chapter, mainly drawing from disaster studies and cultural studies. Our goal in this chapter is to highlight local, traditional, and Indigenous knowledges as heterogenous, dynamic forms of knowledge with permeable geographical boundaries (e.g., through migration and displacement). We should also note that none of the authors of this chapter originate from Indigenous communities, and thus, we are not equipped to discuss Indigenous knowledge in particular. Considering the dynamism and flexibility of knowledges, we do not prefer to use the term 'traditional knowledge' either. Therefore, our term of preference in this chapter will be 'local knowledge', which we believe encompasses these various types of knowledges without staking a claim to speaking on behalf of the communities that create, nurture, and pass down these knowledges through generations. However, our use of 'local knowledge' in this chapter should not be taken as a shorthand for a one-size-fits-all definition that erases the complexities of these knowledge types. Instead, we will strive to understand what local knowledge can realistically achieve, especially within the context of disaster risk studies, how it interacts with 'Western' or 'scientific' knowledge, and how this interaction ultimately affects local communities coping with frequent hazards and disasters.

One thing we know about local knowledge is that it was long ignored by mainstream disaster risk reduction policy and practice (Oliver-Smith, 2016). However, this has been changing recently, with many recent studies and global policies highlighting the importance of local knowledge. For instance, the Sendai Framework for Disaster Risk Reduction 2015–2030 (UNISDR, 2015), as a global policy for managing disasters and associated risks, sees local, traditional, and Indigenous knowledges as an important component for informing risk assessments and central to designing locally appropriate plans and policies. Similarly, the latest IPCC report on climate impacts, adaptation, and vulnerability recognises the value of diverse forms of knowledges, including traditional and local knowledge, especially in relation to understanding processes of climate change adaptation and actions to reduce climate risks (IPCC, 2022). Increased interest is also reflected in academic literature, with a growing number of publications over the last two decades (Mercer *et al.*, 2010; Mavhura *et al.*, 2013; Hiwasaki *et al.*, 2014a; Hermans *et al.*, 2022; Vasileiou *et al.*, 2022).

This increased interest might be attributed to, among other reasons, the highly place-specific nature of local knowledge (Fernando, 2003) and its ability to provide contextual information. Local knowledge is developed through the long and close relationships of local communities with a surrounding environment through a process of innovation, trial and error, and experimentation (Shaw *et al.*, 2009a). It is also a highly dynamic and diverse knowledge, constantly shifting with changing circumstances, environments, new layers of understanding, and through interactions with 'scientific' or 'Western' knowledge (Mitchell *et al.*, 2016; Acharya & Prakash, 2019). However, this final feature is also a key point of contention: It is striking that local knowledges are only labelled as 'local' when they are pitted against 'scientific' or 'Western' knowledge, which oftentimes indicates a hierarchy of knowledges (see also Faas and Marino, 2020) and begs the question, 'whose knowledge is "new", whose is "existing", and who decides?' (Bohensky & Maru, 2011).

Another reason behind the growing interest in local knowledge in disaster policy and practice might be attributed to the increasing attention to decolonial approaches to cultural heritage and its preservation in the face of hazards, especially those exacerbated by anthropogenic climate change. This is accompanied by an expanded understanding of cultural heritage that incorporates living heritage such as cultural practices and skills (Jigyasu, 2019). Cultural heritage is now discussed under the two distinct categories of 'tangible' and 'intangible' heritage, the former mostly indicating cultural monuments and objects, and the latter indicating 'traditions or living expressions [...], such as oral traditions, performing arts, social practices, rituals, festive events, knowledge and practices concerning nature and the universe' (Unesco.org). According to UNESCO, intangible cultural heritage is traditional, contemporary, and living at the same time, it is also inclusive, representative, and community-based (Unesco.org). The importance of intangible cultural heritage lies in the wealth of knowledge and skills that is passed down from one generation to the next (ibid). Local knowledge is, therefore, part and parcel of communities' cultural heritage across the world. UNESCO reports that local and Indigenous knowledges and their holders are increasingly included in science and policy fora, particularly on issues such as climate change (IPCC, UNFCCC), disaster preparedness (UNDRR), sustainable development (Rio+20, Future Earth), and biodiversity assessment and management (CBD, IPBES) (Unesco.org). Alongside these, there are other new studies that explore solutions to integrate cultural heritage management into disaster risk reduction (García, 2019).

Yet, less attention has been placed on the legitimacy of local knowledge as part of the cultural heritage that should be recognised in processes towards reducing and managing disaster and climate change risk. In other words, local knowledge not only needs to be 'preserved' as cultural heritage but actively made part of decision-making spaces for reducing risk. This, in turn, has two implications: first, we need to recognise the value and legitimacy of local knowledges on their own, and second, we need to place local knowledges in relation to other types of knowledge.

The Romanticisation of Local Knowledge: Causes and Associated Problems

While the growing interest in local knowledge in scholarship and policy is a positive trend (Tozier de la Poterie & Baudoin, 2015), it is not without dangers. The majority of the current scholarship focuses on cataloguing local knowledges, in other words, documenting knowledges that exists in different settings. While useful in building a body of evidence, this approach does not reveal the processes through which these knowledges are created, perceived, negotiated, and transmitted (Briggs, 2005, 2013; Smith, 2011). From this narrow focus on cataloguing follows a treatment of these knowledges as things to be extracted and decontextualised, thus disregarding the social, political, cultural, and economic realities they indicate (Briggs, 2005, 2013; Klenk et al., 2017). Following this extractive approach, local knowledge is too readily assigned to geographically marginalised contexts. This highlights power dynamics in the use and presentation of local knowledge by other non-marginalised geographical contexts, therefore often resulting in the romanticisation of local knowledge (Briggs, 2005) and risking its co-option (Brosius, 1997). Romanticisation of local knowledge denies agency to the identities and lived experiences of knowledge holders (Bawaka Country et al., 2018). For example, local knowledges, and especially Indigenous knowledge, are often conceptualised as 'environmental', as they might provide knowledge on precipitation patterns, seasonal changes, and soil fertility, among others. This leads to the romanticisation of such knowledges in two ways: focusing on the environmental aspect firstly reinforces Western and colonial categories about how the world is structured and ordered by science, and secondly, erases the plethora of local knowledges that are not environmental but rather social, cultural, and political.

Our aim in this chapter is to call for a critical and balanced engagement with local knowledges. An uncritical engagement with local knowledge, as exemplified above, leads to the further marginalisation of these knowledges and knowledge holders, ultimately resulting in misleading assumptions around what these knowledges are and what they can realistically achieve regarding disaster risk reduction at the local level. We argue that there is an urgent need for a balanced engagement with local knowledge in disaster studies that acknowledges the plurality of knowledges, their strengths as well as limitations, highlighting the power dynamics between local knowledge and other types of knowledge. This balanced engagement should be based on the understanding that neither dismissing nor consecrating local knowledges is helpful. Our engagement should instead focus on the legitimacy and recognition of knowledges in their own right and in relation to other knowledges. A critical analysis of the romanticisation phenomenon is, therefore, an important step towards decolonising disaster studies (Cadag, 2022).

Romanticising local knowledge leads to a perception of local knowledges as utopian and monolithic, and therefore, cannot be used to inform current planning and decision-making, whereas knowledges are diverse even within a given community or region. Romanticisation thus erases the dynamic and changing nature of local knowledge. This should be considered within the expanding definitions of cultural heritage as outlined above; neither cultural heritage at large nor local knowledge as one of its components should be 'preserved' only for the sake of historical value. Instead, we need to understand how local knowledge connects communities' past experiences to the present and future in helping them tackle hazards and preventing disasters. Romanticised views of local knowledge ultimately impose limitations on what local knowledge can achieve. For example, while local knowledge is often conceptualised as relating to the environment, it can also convey knowledge about processes such as colonisation and untold histories, which should not be silenced or erased. It is also crucial to acknowledge the limitations of local knowledge; it does not have to stand for the ultimate good for the people and the environment, but it can present pragmatic solutions to hold survival above environmental benefits. Equally, local knowledge cannot be a sole solution in large scale processes such as climate change or environmental degradation. Turning to local knowledge and involving communities to help themselves prepare for disasters does not mean that decision-making should be entirely led by the holders of local knowledge, as communities are not homogenous and sometimes local solutions can transfer the problem to another location, such as the construction of flood dikes or dams transferring flood risk problems downstream (Kelman *et al*, 2012). Broader knowledge integration and consensus are required to make local knowledge work across various locations and to the benefit of communities without local knowledges becoming over-romanticised and the capabilities of their holders over-estimated.

In the following sections, we will first focus on unpacking (un)realistic expectations around what local knowledge can and cannot achieve. We will centre our discussion on the concepts of vulnerability and resilience, given that disaster risk reduction is ultimately aimed at reducing vulnerability and enhancing the resilience of individuals and groups. Understanding the extent to which local knowledges can contribute to these simultaneous objectives provides a more nuanced perspective that can both recognise the value of local knowledges without further marginalising them. After this, we will make the case for local knowledges as dynamic bodies of knowledge that shift alongside the changing nature of hazards (most notably, of climate change) and socio-demographic changes such as migration from rural to urban areas. While there are many other triggers that lead to a shift in how local knowledges are created and transmitted, for the purposes of this chapter, we will narrow down our focus to climate change and migration. Finally, we will make the case for an urgent need for decolonising disaster studies by analysing how disciplinary-based thinking that is built on the 'local' versus 'scientific' or 'Western' dichotomy reinforces hierarchical and colonial structures that ignore the complexities of local knowledges. Before concluding the chapter, we will present a case study from Malawi through which we will unpack various strands of our arguments.

Towards a Realistic Understanding of Local Knowledge in Disaster Studies

Local Knowledge, Vulnerability, and Resilience

Although involvement of communities and their local knowledge is being increasingly promoted in the literature and policy (Tozier de la Poterie & Baudoin, 2015), it is recognised that one must caution against romanticisation and understand there are limits to what people can do themselves. Often, local people do not have enough power to influence decision-making agendas nor the distribution of resources, meaning that they have limited capacity to influence and tackle underlying vulnerability drivers, such as poverty creation processes, discriminatory policies, and colonial legacies (Maskrey, 2011; Oven *et al.*, 2017). Therefore, it is almost 'unfair' to expect that local knowledge can be a silver bullet and a core component of building disaster resilience. Local knowledge is indeed a component of community resilience, but one needs to be careful in clearly depicting what local knowledge can or cannot do. For instance, climate change and environmental degradation directly influence the nature of flooding at local levels (through changing rainfall patterns and intensities and changes in catchment characteristics) and the use and applicability of local knowledge, yet people in downstream communities will often have no direct influence on deforestation rates in the uplands nor on reversing the impacts (or causes) of climate change.

Vulnerability has been associated with the capacity or ability of a group or household to cope with, and recover easily from, the harmful effects of a hazard (Anderson and Woodrow, 2019; Eade, 1998; Wisner, 2003; IFRC, 2020). Following a line of argument posed by Bauman (2000) and Beck (1992) which posits that due to changing conditions in late modernity through which less developed countries find themselves at increased risk of hazards, Wisner *et al.* (2014) argue that the vulnerability of a group is closely linked to its socio-economic position, race, gender, and age, among other factors. According to this view, the roots of vulnerability lie in the 'rampant consumerism of contemporary rich societies' (Wisner *et al.*, 2014). Other scholars argue that vulnerability can be traced back to 'ecological modernisation' and 'organised irresponsibility'. Ecological modernisation occurs when those who hold technical knowledge of the risk society attempt to solve environmental problems while still overlooking its root causes. Organised irresponsibility, on the other hand, is the blatant denial of environmental problems (Beck quoted in Goldblatt, 1999, p. 379). These approaches indicate that the root causes of disasters lie in a realm outside the reach of local communities. Disasters are 'in the making' over the long term through rampant consumerism in high-income countries combined with environmental inequality, and efforts at addressing disasters remain focused on technological fixes that ignore local knowledges, thereby leaving out those groups that suffer the worst effects of disasters from recovery and decision-making processes. Aldunce *et al.* (2014) found out that community-based programs are not focused on technology and enhancement of government agencies but rather in community self-reliance and participation. Self-recovery efforts highlighted during disasters echoes the trend with neoliberal discourse that communities can manage themselves by means of 'owning' the risks. Furthermore, Aldunce *et al.* (2014) argue that community-based devolution of responsibility must be accompanied with political will and sufficient institutional support to enable more realistic community empowerment.

As individuals become more liberated in society, they are expected to find solutions to socially produced problems (Beck, 1992, in Bauman, 2000). This is what Joseph (2013) argues as a neoliberal idea that promotes free market norms as a specific form of social rules that institutionalises a rationality of competition which, in turn, enterprises individualised responsibility. Citizens are 'free' to take responsibility for their own life choices. Moreover, he emphasizes Foucauldian arguments on laissez-faire governance being legitimated through the liberal concern that one must not 'govern too much' (Joseph, 2013). Thus, if we argue that local communities are resilient because of their local knowledge, we are assuming that the danger of 'helping people' can create a state of dependency. Hence, the state steps back and encourages free conduct of individuals through the idea of governmentality from a distance.

The concept of resilience is often approached as a label to be deconstructed: When can we call someone or a group resilient? What concepts need to be considered and how does politics shape resilience? Or resilience to what? What are these groups or people trying to cope with? Besides this multidimensionality of resilience, it is important to equally recognize its dynamic nature. As knowledges (scientific, local, or hybrid) develop over time and in response to the surrounding context, the resilience of a person or group also develops in conjunction with its environment. Knowledge systems, therefore, develop in correspondence with their surrounding systems in ways that have ecological, political, and socio-economic resonance.

The concept of resilience focuses on the mitigation of the potential impacts of disasters through preventative policymaking. C.S. Holling famously deployed this concept in his 1973 article 'Resilience and Stability of the Ecological Systems'. Here, Holling (1973) defined ecological resilience as 'the persistence of relationships within a system' and 'a measure of ability of these systems to absorb changes […] and still persist' (Holling 1973, p.17). In a 2002 article, Folke *et al.* linked resilience to sustainable development, arguing that sustainable development aims at 'prosperous social, economic, and ecological systems' (Folke *et al.*, 2002). These systems, they argued, are closely linked, as '[h]umanity receives many ecosystem services, such as clean water and air, food production, fuel, and others' (ibid). Therefore, resilience plays and essential role for 'a prosperous development of society' (ibid). If socio-ecological systems are resilient, they can 'cope, adapt, or reorganize without sacrificing the provision of ecosystem services', and function without disruptions in unexpected situations (Folke *et al.*, 2002, p. 438). However, Holling *et al.*'s reading of the Earth's ecosystems through the lens of a 'service' economy, in which human beings are on the receiving end of the goods and services that nature provides, reduces the vast web of relationships between these two actors into a simple transaction.

Recently, other scholars have contested the concept of resilience, pointing out how it enables the neoliberal drive towards limitless growth by propagating a social model in which individuals must assume personal initiative and risk to ensure their survival in the face of disasters. Brad Evans and Julian Reid (2013) argue that the concept of resilience feeds into a neoliberal discourse that deliberately stunts people's capacity for political action and replaces those capacities with adaptive ones. Adaptation, as opposed to resilience, they argue, is predicated on the acceptance of insecurity and precarity, therefore only aiming to reduce the degree of individuals' suffering rather than alleviating the source of suffering through resistant political acts. This reactionary mode of thinking leads to a 'conflation of resistance with resilience', and indicates a fundamental nihilism towards policies that tackle climate change and other catastrophic phenomena (ibid). Jonathan Joseph (2018) argues that neoliberalism, in the process of constructing neoliberal subjects, relates to them as 'citizens or consumers who are "free" to take responsibility for their own life choices, but who are expected to follow competitive rules of conduct' (Joseph, 2018). Neoliberalism thus creates the illusion of freedom by managing the conditions in which one can be free, while resilience reinforces this illusion by stressing 'heightened self-awareness, reflexivity and responsibility', and promoting 'active citizenship' in which individuals take initiatives to ensure their own socio-economic wellbeing instead of relying on the state to provide necessary provisions (ibid).

As seen in Holling *et al.*, the resilience framework is built on the assumption that human life and Earth's ecosystems are separate spheres. According to this, humans exert control over ecosystems in order to receive 'services', which enable not only the continuation of human life but also unchecked economic growth. However, this framework does not consider the uneven relationships among humans across various geographies or the inequalities that cause and exacerbate the occurrence of disasters. Resilience obscures the conditions that lead to social and environmental crises on a global scale, suggesting that the most vulnerable communities impacted by these crises must adapt to post-disaster conditions with their own provisions. This denies vulnerable communities the possibility of even contemplating the existence of a world beyond a procession of disasters (Evans & Giroux, 2015). Thus, rather than uncritically labelling knowledge holders as 'resilient', attention should be focused on solving and addressing the underlying causes of vulnerabilities. Assigning a label of 'resilient' to vulnerable communities simply because they possess local knowledge to deal with hazards, disasters, and their aftermath can shift the attention away from the social, economic, and political inequalities that limit people's capacities to deal with the potential negative impacts of natural hazards.

Local Knowledge as a Dynamic Body of Knowledges

Scholarship on disaster risk reduction has raised a need for understanding the processes that influence the production and use of local knowledge in the light of environmental, socio-economic, political, and cultural changes (e.g., Wisner *et al.*, 1977; Blaikie *et al.*, 1997; Sillitoe, 1998b; Mercer, 2012; Wamsler & Brink, 2014). In addition to climate change, pressures include migration from rural to urban areas, modernisation resulting in lifestyle changes, market economy, urbanisation, deforestation, influence of external knowledge and technologies, population growth, and changes in land use. These have influenced the use and applicability of local knowledge significantly, and threaten local lifestyles, cultures, and values (Agrawal, 1995). Furthermore, the process of modernisation has brought challenges in terms of the intergenerational transmission of local knowledge, whereby younger generations show a lack of interest in local methods and are motivated to live 'modern' and 'technological' lifestyles (Iloka, 2016; Molina & Neef, 2016). Outcomes of approaches in which development and disaster risk reduction were delivered in a top-down manner have led to an increased dependence on technology and people abandoning their local knowledge (Mercer *et al.*, 2009). For instance, Rautela (2005) reported how in the Indian Himalayas, local communities have disregarded local earthquake-resistant construction practice in favour of more 'modern' and 'scientific' approaches. Similarly, Delica-Willison and Gaillard (2012) point out that people neglect their local knowledge because external stakeholders, such as government workers in communities, have treated them as ignorant and not qualified to make decisions.

In reality, local knowledge is extremely dynamic and adaptable as it is able to shift following external 'pressures' on the socio-economic system (e.g., urbanisation or technological change). Its dynamism is especially evident in terms of its continuous interplay with 'scientific' knowledge, resulting in hybrid knowledge (Wisner, 2009). However, it is becoming increasingly apparent that due to human-driven climate change and wider environmental change (e.g., soil degradation, deforestation), expected the pace of change of local knowledge is unprecedented, resulting in some of the loss of this knowledge. For instance, a study by Makondo and Thomas (2018) argues that climate has always been changing and there is nothing new about climate change to local knowledge holders. However, what is new is the rate and variability of this change and associated magnitude of impact. For instance, a recent review of the literature on climate-driven losses to local knowledge (Pearson *et al.*, 2021) indicates that local knowledge which was traditionally used to deal with climate-related risks is becoming less effective due to climate change impacts. Importantly, this study shows that this loss of effectiveness is not only due to climate change but also associated processes, such as population growth, deforestation, and lifestyle changes.

Based on the above, it is apparent that climate change also influences the availability and effectiveness of local knowledge used for disaster risk reduction, as the severity and frequency of climate induced hazards (such as flooding, droughts, landslides, storms, etc.) are on the increase. For instance, Kagunyu *et al.* (2016) report how climate variability has resulted in more frequent and intense droughts in Kenya, which has led to unreliability in the local drought forecasting indicators used by pastoralists. Šakić Trogrlić *et al.* (2019) found similar results in Malawi for flood forecasting based on local knowledge (Box 11.1). In the Philippines, Molina (2016) found that the applicability of local knowledge is challenged by the changing disaster contexts brought

154 Demet Intepe et al.

Box 11.1 Local knowledge for flood risk management in Malawi

Malawi is a country with significant flood risk, especially in its two most southern districts, Chikwawa and Nsanje, forming the Lower Shire Valley. In the Lower Shire Valley, communities experience frequent flooding and have developed rich local knowledge to manage flood risks (Šakić Trogrlić et al., 2019), with multiple dimensions of this knowledge used before, during, and after floods presented in Figure 11.1.

Although previous research with flood prone communities in the Lower Shire argued that due to their local knowledge, communities are confident to face flooding (Chawawa, 2018), a study by Šakić Trogrlić (2020) called for a more critical outlook on local knowledge and found several reasons to caution against its romanticisation, including:

- **Local knowledge is not equally accessible to everyone**: The study found that despite a vast array of local knowledge available, different components are not equally

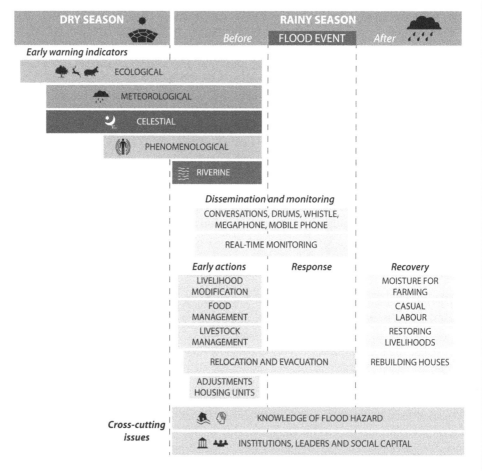

Figure 11.1 Dimensions of local knowledge for flooding in the lower Shire Valley in Malawi.
Source: Šakić Trogrlić et al. (2019).

available to everyone. Rather the access is conditioned by access to resources (e.g., community members who cannot store food in anticipation of floods, lack of financial capacity to build stronger houses) and position within society (e.g., being a member of local disaster management committee thus getting an opportunity to be continuously exposed to dynamic process of local knowledge creation, differences in age, education, and gender).

- **Local knowledge is influenced by local politics, limiting what knowledge is shared with external stakeholders**: Local knowledge within communities is highly heterogenous, with different individuals and groups within a community having different local knowledge, determined by age, gender, livelihood, education, access to technology, among other factors. When external organisations deal with flood risks at local levels (e.g., NGOs, government agencies), they interact primarily with the local disaster management committee (Village Civil Protection Committee – VCPC). The study found that the selection of the members of the VCPC is influenced by local level village politics, meaning that the venues for those most vulnerable to share their local knowledge, including framing of the problems and required solutions, are limited.
- **Local knowledge is heavily influenced by processes that create disaster risks**: Participants in the study shared that local knowledge is heavily influenced by larger environmental and climate change, making some of the knowledge not applicable or reliable. For instance, deforestation in the uplands changed the nature of local flooding by an increased number of flash floods, while climate change modified season, and onset and nature of rainfall, changing local warning signs. Thus, 'solving' climate change or drivers of deforestation upstream is out of the realm of local communities in the Lower Shire.
- **Local knowledge and its applicability for extreme events**: Participants explained that some of the methods to respond to flooding are not effective during extreme floods. For example, dikes constructed from local materials collapse during the floods of higher magnitude, reeds used for measuring the water heights are carried away, canoes are difficult to operate. These results suggest that the applicability of local knowledge is limited to what people already know and have experience with. Some communities reported that in 2015, raised platforms (i.e., '*chete*'), commonly used as a temporary accommodation option during floods, failed. Participants described that these were simply not strong enough to withstand a flood of that magnitude. These results indicate that the use of local knowledge and views of its effectiveness can be negatively influenced by extreme events that might render some local knowledge not applicable.

about by climate change. As argued by Wamsler and Brink (2014), the unprecedented pressures that local communities are exposed to mean that people have a different starting point from which to cope. Hence, people's ability to adapt and incorporate new learnings into their local knowledge can be hindered (ibid). Therefore, it is of extreme importance to critically explore the applicability and effectiveness of local knowledge (Gaillard & Mercer, 2013) without consecrating or entirely dismissing it.

Local Knowledge and Socio-Demographic Changes

Besides the environmentally driven dynamic of knowledge development and adaptation, generational changes may also lead to a loss of local knowledge. Socio-demographic changes over generations are formed through multiple factors. One of the most significant factors among these is rapid urbanisation, which may lead to a loss of local knowledge. Migration of people tends to flow towards urban areas, where rurally based local knowledge is less applicable and access to scientific or Western knowledge is larger. Further to this, the fabric of urban areas is often a mosaic of cultures consisting of different historical and cultural values. Urbanisation accelerates the demand for housing and health and social resources, which places detrimental pressure on already heightened risk and vulnerability to disasters (Warn & Adamo, 2015). Migrants of low socio-economic status are often vulnerable as they are more likely to reside in high-risk areas and are not equipped with robust housing and do not have networks and assets to prepare for environmental hazards. When migrants leave their area of origin, they take their local knowledge with them, thus depriving these areas of such knowledge. However, migrants also likely lack local knowledge of their new surroundings which can cause more pressure on their ability to cope with natural hazards (Warn & Adamo, 2015).

As urbanisation intensifies, families also increasingly disintegrate and globalise, and hence local knowledge continues to become marginalised. This marginalisation may stem from the advancement of technologies and its spill over to private spaces (e.g., intangible social interaction) where local knowledge once thrived (Ocholla, 2007). Scientific knowledge often takes over in urbanised areas and holders of local knowledge become vindicated and condemned for being 'outdated', which is a characteristic that most people find demeaning (Ocholla, 2007). Hence, generations that do not understand, recognise, or value the use of local knowledge are formed and, over time, local knowledge is fully abandoned. Younger generations, with more exposure to nationally programmed education and scientific knowledge, are increasingly placing trust in technocratic-based scientific knowledge (Choudhury et al., 2021). The fact that local knowledge is mainly passed down through generations orally and thus rarely documented makes this generational change more likely (Chang'a et al., 2010; Chanza and Mafongova, 2017).

These two factors are among the few socio-demographic changes that can impact local disaster knowledge. The generational transfer of local knowledge will therefore fall more to the background as a common language other knowledges takes over. Local knowledge is also often regionally bound and the movement of knowledge holders to other areas will decrease the use and further development of this knowledge.

Decolonial Approaches to Local Knowledge: Power, Participation, and the Co-Production of Knowledge in Disaster Studies

In her seminal work on decolonising methodologies, Linda Tuhiwai Smith (1999) shows how different types of knowledges have interacted historically and the present-day implications of this. Fundamentally, 'Western' and 'scientific' knowledge were part and parcel of colonisation: explorers and scientists travelled to lands that they colonised in the name of science. Once there, they researched those lands and peoples using the scientific tools of the time to create knowledge they viewed as legitimate. What they ignored were the cosmologies, epistemologies, and, therefore, knowledges of those they colonised, delegitimising their knowledge (Akena, 2012). Much local (and Indigenous) knowledge viewed the world as inextricably linked, whereas Western knowledge divided parts of the world into disciplines (Watts, 2013). Disciplinary-based thinking is the hegemonic and dominant way Western science is taught. There is a risk that when local knowledge is used in Western science that it is separated from its broader cosmologies and epistemologies, and instead divided up to fit into disciplinary-based thinking, thus losing much of its richness and power and resulting in extractive approaches to local knowledge (Latulippe & Klenk, 2020) that also effectively depoliticises such knowledge systems. Local knowledge is often developed over thousands of years and the survival of people who hold that knowledge is testament to its validity and rigor. Yet, when this knowledge is brought into hierarchical and institutionalized structures of universities (Dei, 2008), it is labelled as 'local' and 'traditional', separating it (arbitrarily) from 'scientific' knowledge.

When discussing the process of engagement, the wording 'integration' is commonly used. Integration, as a term, suggests it concerns two contrasting elements with one merged into another. To this end, Tengö *et al*. (2014) point out 'it is important to differentiate among (a) integration of knowledge, (b) parallel approaches to developing synergies across knowledge systems, and (c) co-production of knowledge' (ibid, p. 43). Parallel approaches focus on knowledge complementariness and synergies instead of one knowledge merging into another. Co-production, on the other hand, refers to the ongoing process of constructing new knowledge in all phases. However, most literature on the integration of local and scientific knowledge for disaster risk reduction uses a point of departure in the scientific knowledge (Hermans *et al*., 2022). This is fitting with the romanticisation of local knowledge as a mystified, static, and less acknowledged process of knowledge construction. For example, Ahmed *et al*. (2019), Grey (2019), Hiwasaki *et al*. (2015), and Lin & Chang (2020) approach integration as the validation of local knowledge observations and indicators. In cases like Hiwasaki *et al*. (2015), the overlap between local and scientific knowledge and the ability to validate local knowledge based on scientific knowledge is mapped. This uses the assumption that only local knowledge indicators validated by scientific approaches are useful for integration. This is illustrative of the lack of attention paid to the presence of power relations on a local level (Šakić Trogrlić *et al*., 2021) and between stakeholders (e.g., scientists and participants). However, these relations highly impact decisions on what knowledge is used and what an integration process looks like.

Considering the risk of romanticisation of local knowledge, what we need to work towards is a process that acknowledges a plurality of knowledges, hence being aware of knowledge boundaries and domains. It is important to reflect on what local and scientific knowledge can or cannot do, e.g., exaggerating the role of local knowledge in 'solving' problems created by larger socio-economic and political forces, such as environmental degradation and climate change intensifying flooding at local levels. A review paper on the integration of knowledges in disaster risk reduction shows that various studies suggest developing knowledges separately to safeguard value systems (Hermans *et al*., 2022). Moreover, Dube *et al*. (2016) point out that integration, co-creation or co-production are not inventions driven by the romanticisation of local knowledge but that knowledge has always been developed in a hybrid manner. The current binary approach to knowledge (local or scientific) (e.g., Ziervogel and Opere, 2010; Masinde, 2015) needs to focus on the hybrid realities (e.g., Lauer, 2012; Mutasa, 2015; Dube *et al*., 2016; Lin and Chang, 2020). The search for which knowledge is the most distinct, risks marginalising the other knowledges, as one knowledge becomes dominant (Mutasa, 2015). Therefore, our understanding of knowledge as binary, hybrid, or plural has fundamental implications on how the interaction process (e.g., cocreation, coproduction, participation) unfolds.

Further to this, local knowledge holders, whom practitioners and researchers often approach to learn and use local knowledge from, may form part of the local elite (Hitomi and Loring, 2019; Mosurska and Ford, 2020). Similar to the vast literature that warns against viewing communities as benign and homogenous entities (Titz *et al*., 2018; Faas & Marino, 2020), there needs to be a critical awareness of who the knowledge holders are and how this may influence the transmission of knowledge. Otherwise, researchers and practitioners risk perpetuating harmful local power structures. This has been a main and validated critique in the participatory methodology, which was popularised for its 'inclusion' of local knowledge (Cooke & Kothari, 2001). In particular, the idea of 'participation' solving issues of 'hearing unheard voices' through 'handing over the stick', has arguably created a new participation tyranny. Here, the attempt of inclusion has focused on moulding local decision-making into dominant local and global paradigms, overriding existing knowledge and decision-making processes (Cooke and Kothari, 2001).

Participation in the design and implementation of policies is becoming increasingly embedded in the discussion of the division of roles for actors in climate change adaptation and disaster risk reduction. Two prevailing yet opposing processes are highlighted through the concept of participation: decentralisation and recentralisation of control. Participation is viewed as a chance to bridge the gap between the local indigenous and national stakeholders; although it can also be exploited as a neoliberal technique to mobilise local actors. Participation as a new leitmotif has a stark resemblance to neoliberalist ideas of governance rationality where responsibility for mitigating and adapting to climate change is transferred from governments to the locale.

Some local knowledges, such as Indigenous knowledges, represent 'speaking back' to the production, categorisation, and position of cultures, identities, and histories by colonizers (Dei, 2008; Akena, 2012). Local knowledge, then, is political, and in certain cases, might serve as a tool for resistance. When romanticised, these political dimensions are obscured, moulding local knowledge into a more palatable form for researchers and/or practitioners from non-marginalised geographical contexts. For example, Aboriginal peoples in Australia have significant knowledge on how to manage land through fire. Many of these practices have been taken up within Australia for wildfire prevention and management. However, it has been used to protect *private* property specifically, which may not align with Indigenous concerns around self-determination. Romanticisation can, therefore, mask peoples' priorities and agendas. This can lead to local knowledge being used for agendas that do not align with the goals of local knowledge holders (Moyo, 2022).

Conclusion: Towards a Realistic View on Local Knowledges for Disaster Risk Reduction Research, Policy, and Practice

In this chapter, we looked at the danger of romanticising local knowledge given its recent popularity in research and policy, especially in discussions around the 'integration' of local and scientific knowledge in disaster studies. Romanticisation emerges as a tendency to 'make use of' local knowledges instead of 'learning across', in other words, processes that indicate collaboration and collegiality. We suggested that this 'learning across' should consist of not an extraction of content but an equitable process of engagement. Equally, there is a need to recognize the hybrid reality of knowledge use, where local knowledge and scientific knowledge are both drawn on by knowledge holders. This means that methods used in analysing local knowledges can be part of different epistemic processes, such as a rigorous process to gather evidence agreed upon among a group of people. This means that knowledge holders active in research, policy, or practice are drawing on both local and scientific knowledge and that there is already an ongoing active role and need for local knowledge as well as local researchers. Therefore, our focus should not fall on preservation but on the already existing and future role of local knowledges in decision-making and knowledge construction processes.

Besides the relation of local knowledge vis-à-vis other knowledges, we have also outlined the importance of acknowledging local knowledge as dynamic and fluid and open to pressures from social or environmental factors. This dynamic adaptation is crucial in the context of resilience and vulnerability. Local knowledge may become outdated or irrelevant due to climate change or social changes; however, supporting its active and ongoing role will also support its dynamic development and, subsequently, its continued relevance and possible synergy with other knowledges. While the increased attention to local knowledge after decades of marginalisation in policy and literature is welcome, there is also a strong need for further critical engagement with it to understand its true limitations, so that communities are not left to fend for themselves in becoming more resilient against hazards and disasters, otherwise, it is inevitable that communities will find themselves further marginalised. The key to ensuring this is going beyond merely documenting instances of local knowledge across the world and instead trying to understand how local knowledges work in real-world scenarios with more attention to and funding for local researchers who are best placed to guide such research projects.

Key Points

- Romanticisation of local knowledge can bring more harm than good for the integration of local knowledge in disaster risk reduction by evoking unrealistic expectations of what local knowledge can achieve.
- A more balanced approach to local knowledge is crucial to understand the hybrid nature of such knowledges and to acknowledge their hybrid and dynamic nature.
- Local knowledges can also not be imagined separately from social, cultural, economic, and environmental factors as well as the relations of power that lead to their emergence and continued existence, meaning that unrealistic assumptions around the 'integration' of local knowledges into scientific knowledge need to be critically questioned in relation to concepts such as 'resilience' and 'vulnerability'.
- A true recognition of local knowledges is only possible through an in-depth rethinking of the conventional knowledge production apparatus and our research methods.

References

Acharya, A. & Prakash, A. (2019). When the river talks to its people: Local knowledge-based flood forecasting in Gandak River basin, India. *Environmental Development*, 31, pp. 55–67. https://www.sciencedirect.com/science/article/pii/S2212420922004745

Agrawal, A. (1995). Dismantling the divide between Indigenous and scientific knowledge. *Development and Change*, 26 (3), pp. 413–439.

Ahmed, B., Sammonds, P., Saville, N.M. *et al*. (2019). Indigenous mountain people's risk perception to environmental hazards in border conflict areas. *International Journal of Disaster Risk Reduction*, 35, 101063. https://doi.org/10.1016/j.ijdrr.2019.01.002

Akena, F.A. (2012). Critical analysis of the production of Western knowledge and its implications for Indigenous knowledge and decolonization. *Journal of Black Studies*, 43(6), pp. 599–619.

Aldunce, P., Beilin, R., Handmer, J. & Howden, M. (2014). Framing disaster resilience: The implications of the diverse conceptualisations of 'bouncing back'. *Disaster Prevention and Management*, 23(3).

Anderson, M.B. & Woodrow, P.J. (2019). *Rising from the Ashes: Development Strategies in Times of Disaster*. NY: Routledge.

Bauman, Z. (2000). *Liquid Modernity*. Cambridge: Polity Press.

Bawaka Country, Burarrwanga, L., Ganambarr, R., Ganambarr-Stubbs, M., Ganambarr, B., Maymuru, D., Lloyd, K., Wright, S., Suchet-Pearson, S., & Hodge, P. (2018). Meeting Across Ontologies: Grappling with an Ethics of Care in our Human-more-than-human Collaborative Work. In J. Haladay, & S. Hicks (Eds.), *Narratives of educating for sustainability in unsustainable environments*. East Lansing: Michigan State University Press, pp. 219–243.

Beck, U. (1992). From industrial society to the risk society: Questions of survival, social structure and ecological enlightenment. *Theory, Culture & Society*, 9(1), pp. 97–123.

Blaikie, P., Brown, K., Stocking, M., Tang, L., Dixon, P. & Sillitoe, P. (1997). Knowledge in action: Local knowledge as a development resource and barriers to its incorporation in natural resource research and development. *Agricultural Systems*, 55 (2), pp. 217–237.

Bohensky, E.L. & Maru, Y. (2011). Indigenous knowledge, science, and resilience: What have we learned from a decade of international literature on 'integration'? *Ecology and Society* 16(4), p. 6. http://dx.doi.org/10.5751/ES-04342-160406

Briggs, J. (2005). The use of indigenous knowledge in development: Problems and challenges. *Progress in Development Studies*, 5(2), pp. 99–114.

Briggs, J. (2013). Indigenous knowledge: A false dawn for development theory and practice? *Progress in Development Studies*, 13(3), pp. 231–243.

Brosius, P.J. (1997). Endangered forests, endangered people: Environmentalist representations of indigenous knowledge. *Human Ecology*, 25(1), pp. 47–69.

Cadag, J.R. (2022). Decolonising disasters. *Disasters*, 46(4), pp. 1121–1126.

Chang'a, L.B., Yanda, P.Z., Ngana, J. (2010). Indigenous knowledge in seasonal rainfall prediction in Tanzania: A case of the south-western highland of Tanzania. *Journal of Geography and Regional Planning*, 3(4), pp. 66–72.

Chanza, N., Mafongova, P.L. (2017). Indigenous-based climate science from the Zimbabwean experience: From impact identification, mitigation and adaptation. In: Mafongoya PL, Ajayi O (eds) Indigenous Knowledge Systems and Climate Change Management in Africa. Wageningen: CTA.

Chawawa, N.E. (2018) *Why do smallholder farmers insist on living in flood prone areas? Understanding self-perceived vulnerability and dynamics of local adaptation in Malawi.* [Doctoral thesis, University of Edinburgh] Edinburgh Research Archive. https://era.ed.ac.uk/handle/1842/31421

Choudhury, M.-U.-I., C. E. Haque, A. Nishat, and S. Byrne. (2021). Social learning for building community resilience to cyclones: role of indigenous and local knowledge, power, and institutions in coastal Bangladesh. *Ecology and Society* 26(1), pp. 5. https://doi.org/10.5751/ES-12107-260105

Cooke, B. & Kothari, U. (2001). *Participation: The New Tyranny?* London: Zed Books.

Dei, G.J.S. (2008). Indigenous knowledge studies and the next generation: Pedagogical possibilities for anti-colonial education. *The Australian Journal of Indigenous Education*, 37(S1), pp. 5–13.

Delica-Willison, Z. & Gaillard, J. (2012). Community action and disaster. In B. Wisner, J.C. Gaillard, & I. Kelman, eds. *The Routledge Handbook of Hazards and Disaster Risk Reduction*, London: Routledge, pp. 711–723.

Dube, T., Moyo. P., Ndlovu, S., Phiri, K. (2016). Towards a framework for the integration of traditional ecological knowledge and meteorological science in seasonal climate forecasting: the case of smallholder farmers in Zimbabwe. *Journal of Human Ecology*, 54, pp. 49–58. https://doi.org/10.1080/09709274.2016.11906986

Eade, D. (1998). *Capacity Building*. Oxford: Oxfam.

Evans, B. & Giroux, H.A. (2015). *Disposable Futures: The Seduction of Violence in the Age of Spectacle*. San Francisco: City Lights Books.

Evans, B. & Reid, J. (2013). Dangerously exposed: The life and death of the resilient subject. *Resilience*, 1(2), pp. 83–98. https://doi.org/10.1080/21693293.2013.770703

Faas, A.J., Marino, E.K. (2020). Mythopolitics of 'community': An unstable but necessary category. *Disaster Prevention and Management: An International Journal*, 29(4), 481–484.

Fernando, J.L. (2003). NGOs and production of indigenous knowledge under the condition of postmodernity. *The ANNALS of the American Academy of Political and Social Science*, 590(1), pp. 54–72.

Folke, C., Carpenter, S., Elmqvist, T., Gunderson, L., Holling, C.S. & Walker, B. (2002). Resilience and sustainable development: Building adaptive capacity in a world of transformations. *Ambio*, 31(5), pp. 437–440. https://doi.org/10.1579/0044-7447-31.5.437

Gaillard, J.C. & Mercer, J. (2013). From knowledge to action: Bridging gaps in disaster risk reduction. *Progress in Human Geography*, 37(1), 93–114.

Goldblatt, D. (1999). Risk society and the environment. *Thinking Through the Environment: A Reader*, London: Routledge, pp. 373–382.

Grey, M.S. (2019). Accessing seasonal weather forecasts and drought prediction information for rural households in Chirumhanzu district, Zimbabwe. *Jàmbá: Journal of Disaster Risk Studies*, 11, pp. 1–9. https://doi.org/10.4102/jamba.v11i1.777

Griffin, C. & Barney, K. (2021). Local disaster knowledge: Towards a plural understanding of volcanic disasters in Central Java's highlands, Indonesia. *The Geographical Journal*, 187, pp. 2–15. https://doi.org/10.1111/geoj.12364

Hermans, T.D.G., Šakić Trogrlić, R., van den Homberg, M.J.C. et al. (2022). Exploring the Integration of Local and Scientific Knowledge in Early Warning Systems for Disaster Risk Reduction: A Review. *Nat Hazards* 114, pp. 1125–1152. https://doi.org/10.1007/s11069-022-05468-8

Hitomi, M.K., Loring, P.A. (2018). Hidden participants and unheard voices? A systematic review of gender, age, and other influences on local and traditional knowledge research in the North. *Facets*, 3(1). https://doi.org/10.1139/facets-2018-0010

Hiwasaki, L., Luna, E., Syamsidik & Shaw, R. (2014a). Process for integrating local and indigenous knowledge with science for hydro-meteorological disaster risk reduction and climate change adaptation in coastal and small island communities. *International Journal of Disaster Risk Reduction*, 10, pp. 15–27.

Hiwasaki, L., Luna, E., Adriano Marçal, J. et al. (2015). Local and indigenous knowledge on climate-related hazards of coastal and small island communities in Southeast Asia. *Climatic Change*, 128, pp. 35–56. https://doi.org/10.1007/s10584-014-1288-8

Holling, C.S. (1973). Resilience and stability of ecological systems. *Annual Review of Ecology, Evolution, and Systematics*, 4, pp. 1–23. http://dx.doi.org/10.1146/annurev.es.04.110173.000245

IFRC. (2020). World Disasters Report, 1998. *International Committee of the Red Cross*. https://www.ifrc.org/document/world-disasters-report-2020

Iloka, N.G. (2016). Indigenous knowledge for disaster risk reduction: An African perspective. *Jàmbá: Journal of Disaster Risk Studies*, 8(1), p. 272.

IPCC. (2022). *Climate Change 2022: Impacts, Adaptation, and Vulnerability*. Contribution of working group II to the sixth assessment report of the intergovernmental panel on climate change [H.-O. Pörtner, D.C. Roberts, M. Tignor, E.S. Poloczanska, K. Mintenbeck, A. Alegría, M. Craig, S. Langsdorf, S. Löschke, V. Möller, A. Okem, B. Rama (eds.)]. Cambridge University Press, Cambridge, UK and New York, NY, 3056 pp, doi:10.1017/9781009325844.

Jigyasu, R. (2019). Managing cultural heritage in the face of climate change. *Journal of International Affairs*, 73(1), pp. 87–100. https://www.jstor.org/stable/26872780

Joseph, J. (2013). Resilience as embedded neoliberalism: A governmentality approach. *Resilience*, 1(1), pp. 38–52.

Joseph, J. (2018). *Varieties of Resilience: Studies in Governmentality*. Cambridge: Cambridge University Press.

Kagunyu, A., Wandibba, S. & Wanjohi, J.G. (2016). The use of indigenous climate forecasting methods by the pastoralists of Northern Kenya. *Pastoralism*, 6(1), p. 7.

Kelman, I., Mercer, J. & Gaillard, J. (2012). Indigenous knowledge and disaster risk reduction. *Geography*, 97(1), pp. 12–21.

Klenk, N., Fiume, A., Meehan, K. & Gibbes, C. (2017). Local knowledge in climate adaptation research: Moving knowledge frameworks from extraction to co-production. Wiley Interdisciplinary Reviews: Climate Change, 8 (5), e475.

Latulippe, N. & Klenk, N., (2020). Making room and moving over: Knowledge co-production, indigenous knowledge sovereignty and the politics of global environmental change decision-making. *Current Opinion in Environmental Sustainability*, 42, pp. 7–14.

Lauer, M. (2012). Oral traditions or situated practices? Understanding how indigenous communities respond to environmental disasters. *Human Organization*, 71(2), pp. 176–187. https://doi.org/10.17730/humo.71.2.j0w0101277ww6084

Lin, P.S.S. & Chang, K.M. (2020). Metamorphosis from local knowledge to involuted disaster knowledge for disaster governance in a landslide-prone tribal community in Taiwan. *International Journal of Disaster Risk Reduction*, 42, 101339. https://doi.org/10.1016/j.ijdrr.2019.101339

Makondo, C.C. & Thomas, D.S.G. (2018). Climate change adaptation: Linking indigenous knowledge with western science for effective adaptation. *Environmental Science & Policy*, 88, pp. 83–91. https://doi.org/10.1016/j.envsci.2018.06.014

Maskrey, A. (2011). Revisiting community-based disaster risk management. *Environmental Hazards*, 10(1), pp. 42–52.

Matti, S. & Ögmundardóttir, H. (2021). Local knowledge of emerging hazards: Instability above an Icelandic glacier. *International Journal of Disaster Risk Reduction*, 58. https://doi.org/10.1016/j.ijdrr.2021.102187

Masinde, M. (2015). An innovative drought early warning system for sub-Saharan Africa: integrating modern and indigenous approaches. *African Journal of Science, Technology, Innovation and Development*, 7(1), pp. 8–25, https://doi.org/10.1080/20421338.2014.971558

Mavhura, E., Manyena, S.B., Collins, A.E. & Manatsa, D. (2013). Indigenous knowledge, coping strategies and resilience to floods in Muzarabani, Zimbabwe. *International Journal of Disaster Risk Reduction*, 5, pp. 38–48.

Mercer, J. (2012). Knowledge and disaster risk reduction. In B. Wisner, J.C. Gaillard, & I. Kelman, eds. *Handbook of Hazards and Disaster Risk Reduction*. London, UK: Routledge, pp. 97–109.

Mercer, J., Kelman, I. & Dekens, J. (2009). Integrating indigenous and scientific knowledge for disaster risk reduction. In R. Shaw, A. Sharma, & Y. Takeuchi, eds. *Indigenous Knowledge and Disaster Risk Reduction: From Practice to Policy*, New York City, UK: Nova Science Publishers, pp. 115–131.

Mercer, J., Kelman, I., Taranis, L. & Suchet-Pearson, S. (2010). Framework for integrating indigenous and scientific knowledge for disaster risk reduction. *Disasters*, 34(1), pp. 214–239.

Mitchell, J.K., O'Neill, K., McDermott, M. & Leckner, M. (2016). Towards a transformative role for local knowledge in post-disaster recovery: Prospects for co-production in the wake of Hurricane Sandy. *Journal of Extreme Events*, 03(01), 1650003. https://doi.org/10.1142/S2345737616500032

Molina, F. (2016). Intergenerational transmission of local knowledge towards river flooding risk reduction and adaptation: The experience of Dagupan City, Philippines. In M.A. Miller, & M. Douglass, eds. *Disaster Governance in Urbanising Asia [Online]*, Singapore: Springer Singapore, pp. 145–176.

Molina, J.G.J. & Neef, A. (2016). Integration of indigenous knowledge into disaster risk reduction and management (DRRM) policies for sustainable development: The case of the Agta in Casiguran, Philippines. In J.I. Uitto, & R. Shaw, eds. *Sustainable Development and Disaster Risk Reduction [Online]*, Tokyo, Japan: Springer, pp. 247–264.

Mosurska, A. & Ford, J.D. (2020). Unpacking Community Participation in Research. *Arctic*, 73(3), pp. 347–367. https://www.jstor.org/stable/26974910

Moyo, I. (2022). Beyond a tokenistic inclusion of indigenous knowledge systems in protected area governance and management in Okhahlamba-Drakensberg. *African Geographical Review*, 43(2), pp. 141–156.

Mutasa, M. (2015). Knowledge apartheid in disaster risk management discourse: Is marrying indigenous and scientific knowledge the missing link? *Jàmbá: Journal of Disaster Risk Studies*, 7(1). https://doi.org/10.4102/jamba.v7i1.150

Ocholla, D. (2007). Marginalized knowledge: An agenda for indigenous knowledge development and integration with other forms of knowledge. *The International Review of Information Ethics* 7, pp. 236–245.

Oliver-Smith, A. (2016). Disaster risk reduction and applied anthropology. *Annals of Anthropological Practice*, 40(1), pp. 73–85. https://doi.org/10.1111/napa.12089

Oven, K.J., Shailendra, S., Shubheksha, R., Wisner, B., Ajoy, D., Jones, S. & Densmore, A. (2017). *Review of the Nine Minimum Characteristics of a Disaster Resilient Community in Nepal: Final Report*. Durham, UK: Durham University.

Pauli, N., Williams, M., Henningsen, S. et al. (2021). 'Listening to the sounds of the water': Bringing together local knowledge and biophysical data to understand climate-related hazard dynamics. *International Journal of Disaster Risk Science*, 12, pp. 326–340. https://doi.org/10.1007/s13753-021-00336-8

Pearson, J., Jackson, G. & McNamara, K.E. (2021). Climate-driven losses to indigenous and local knowledge and cultural heritage. *The Anthropocene Review*, 10(2. https://doi.org/10.1177/20530196211005482

Rautela, P. (2005). Indigenous technical knowledge inputs for effective disaster management in the fragile Himalayan ecosystem. *Disaster Prevention and Management: An International Journal*, 14(2), pp. 233–241.

Šakić Trogrlić, R. (2020). *The role of local knowledge in community-based flood risk management in Malawi*. [Doctoral thesis, Heriott-Watt University] Ros Theses Repository. https://www.ros.hw.ac.uk/handle/10399/4241

Šakić Trogrlić, R., Duncan, M., Wright, G., Van den Homberg, M., Adeloye, A., Mwale, F. & McQuistan, C. (2021). External stakeholders' attitudes towards and engagement with local knowledge in disaster risk reduction: Are we only paying lip service? *International Journal of Disaster Risk Reduction*, 58. Retrieved from https://doi.org/10.1016/j.ijdrr.2021.102196

Šakić Trogrlić, R., Wright, G., Duncan, M., van den Homberg, M., Adeloye, A., Mwale, F. & Mwafulirwa, J. (2019). Characterising local knowledge across the flood risk management cycle: A case study of Southern Malawi. *Sustainability*, 11(6), p. 1681. MDPI AG. Retrieved from http://dx.doi.org/10.3390/su11061681

Shaw, R., Sharma, A. & Takeuchi, Y. (2009a). Introduction: Indigenous knowledge and disaster risk reduction. In R. Shaw, A. Sharma, & Y. Takeuchi, eds. *Indigenous Knowledge and Disaster Risk Reduction: From Practice to Policy*, New York City, USA: Nova Science Publishers, pp. 1–13.

Sillitoe, P. (1998b). The development of indigenous knowledge: A new applied Anthropology. *Current Anthropology*, 39(2), pp. 223–252.

Smith, L.T. (1999). *Decolonizing Methodologies: Research and Indigenous Peoples*. NY: Zed Books.

Smith, T.A. (2011). Local knowledge in development (geography). *Geography Compass*, 5 (8), pp. 595–609.

Tengö, M., Brondizio, E.S., Elmqvist, T. *et al.* (2014). Connecting diverse knowledge systems for enhanced ecosystem governance: The multiple evidence base approach. *Ambio*, 43, pp. 579–591. https://doi.org/10.1007/s13280-014-0501-3

Titz, A., Cannon, T., & Krüger, F. (2018). Uncovering 'Community': Challenging an Elusive Concept in Development and Disaster Related Work. *Societies*, 8(3), 71. http://dx.doi.org/10.3390/soc8030071

Tozier de la Poterie, A. & Baudoin, M.-A. (2015). From Yokohama to Sendai: Approaches to participation in international disaster risk reduction frameworks. *International Journal of Disaster Risk Science*, 6(2), pp. 128–139.

UNESCO. *Indigenous Knowledge and Science Policy*. https://en.unesco.org/links-policy

UNESCO. *What Is Intangible Cultural Heritage?* https://ich.unesco.org/en/what-is-intangible-heritage-00003

UNISDR. (2015). *Sendai Framework for Disaster Risk Reduction 2015–2030*. Geneva. Available at: https://www.unisdr.org/files/43291_sendaiframeworkfordrren.pdf

Vasileiou, K., Barnett, J. & Fraser, D.S. (2022). Integrating local and scientific knowledge in disaster risk reduction: A systematic review of motivations, processes, and outcomes. *International Journal of Disaster Risk Reduction*, 81(15). https://doi.org/10.1016/j.ijdrr.2022.103255

Wamsler, C. & Brink, E. (2014). Moving beyond short-term coping and adaptation. *Environment and Urbanization*, 26(1), pp. 86–111.

Warn, E. & Adamo, S.B. (2015). 'The impact of climate change: migration and cities in South America'. World Meteorological Organization, 63(4).

Watts, V. (2013). Indigenous place-thought and agency amongst humans and non humans (first woman and sky woman go on a European world tour!). *Decolonization: Indigeneity, Education & Society*, 2(1), pp. 20–34.

Wisner, B. (2003). The communities do science! Proactive and contextual assessment of capability and vulnerability in the face of hazards. *Vulnerability: Disasters, Development and People*.

Wisner, B. (2003). "The Communities Do Science! Proactive and Contextual Assessment of Capability and Vulnerability in the Face of Hazards" in G. Bankoff, G. Frerks and T.Hilhorst, (eds.), Vulnerability: Disasters, Development and People. London: Earthscan.

Wisner, B. (2009). Local Knowledge and Disaster Risk Reduction: Keynote during the Side Meeting on Indigenous Knowledge, Global Platform for Disaster Reduction, Geneva, 17 June 2009.

Wisner, B., Blaikie, P., Cannon, T. & Davis, I. (2014). *At Risk: Natural Hazards, People's Vulnerability and Disasters*. London: Routledge.

Wisner, B., O'Keefe, P. & Westgate, K. (1977). Global systems and local disasters: The untapped power of Peoples' Science. *Disasters*, 1(1), pp. 47–57.

Ziervogel, G., Opere, A. (2010). Integrating meteorological and indigenous knowledge-based seasonal climate forecasts for the agricultural sector: Lessons from participatory action research in sub-Saharan Africa. International Development Research, Canada. Climate Change Adaptation in Africa learning paper series. https://idl-bnc-idrc.dspacedirect.org/items/a4b47199-a1ba-4047-a1e4-32ef2bc48c00

12 Challenges with Techno-Centric Approaches in the Implementation of Disaster Risk Management for Cultural Heritage

David A. Torres and Giuseppe Forino

Introduction

Cultural heritage (CH) has different meanings for different individuals and communities; however, it can be generally defined as the physical manifestation of past human activities and interactions with the environment. Article 1 of UNESCO Convention considers CH as monuments (e.g., architectural works, sculpture, painting, inscriptions, and cave dwellings); groups of buildings (groups of separate or connected buildings) and sites (e.g., works of nature and people, and archaeological sites) which are of outstanding universal value from the point of view of history, art, science, aesthetics, or anthropology.[1] However, CH also includes traditions and living expressions such as oral traditions, social practices, rituals, festive events, knowledge, practices, and skills, collectively defined as 'intangible heritage values'.[2] Overall, CH enriches people's lives by providing a meaningful sense of connection across communities and places and to past and lived experiences. In this way, CH has a historic, aesthetic, social, scientific, or spiritual value for past, present, and future generations, and must be preserved from the consequences and marks of development and decay (Forino et al., 2016).

CH is increasingly exposed to various natural and human-induced hazards such as earthquakes, volcanic eruptions, hydrometeorological hazards, as well as terrorism and armed conflicts (Chmutina et al., 2019). Climate and weather-related hazards may increase risks for some CH that can soon become 'last chance to see' items (IPCC, 2018). Besides physical loss, hazards can also lead to the loss of traditional knowledge, practices, skills, and crafts that ensure continuity of living CH as well as means for its maintenance and conservation (Chmutina et al., 2019). Therefore, decisions have to be made about the degree of protection to be assigned to those CH items that, if destroyed or damaged, cannot be regenerated, duplicated, restored, or reintroduced or lose significant value in specific contexts. In this way, protection should be ensured for CH considering not just its economic value but also its social and cultural role in shaping identity values for individuals, communities, and societies (Forino et al., 2016).

National and international organizations have recognized the importance of disaster risk management (henceforth, DRM) as a set of practices aimed to preserve CH by reducing risks from multiple hazards (Ravankhah et al., 2017; Muñiz Trejo, 2018). The Hyogo Framework for Action 2005–2015 (HFA) emphasised 'traditional and indigenous knowledge and culture heritage' (UNISDR, 2005, p. 9). In 2007, UNESCO emphasised the support for DRM at World Heritage sites within relevant global, regional, national, and local institutions.[3] In 2015, SDG 11 aimed at making 'cities and human settlements inclusive, safe, resilient and sustainable' and, for the first time, acknowledged 'heritage' as a target of necessary efforts to protect and safeguard.[4] In the same year, the UN Sendai Framework for Disaster Risk Reduction (SFDRR) 2015–2030 also emphasised the importance of disaster risk for CH, calling for increasing investments in DRM[5] (Ravankhah et al., 2017).

DOI: 10.4324/9781003293019-16

International DRM organisations recognize the importance of considering CH as a target of DRM policies and practice (for more information, see Chapter 3). Accordingly, international CH organisations recognize the need for DRM in CH management, i.e., ICCROM and ICOMOS, along with UNESCO, have developed specific programmes and projects with this goal. In a pioneering guidebook on Risk Preparedness for World Cultural Heritage, Stovel (1998) brought attention to CH conservation from the International Decade for Natural Disaster Reduction. In particular, he suggested that preventive and mitigation measures should be expanded from museums to the wider practice of the built environment. In 2019, ICCROM launched its flagship program, First Aid and Resilience for Cultural Heritage in Times of Crisis (FAR), besides a series of courses and training initiatives that bridge experiences from different sectors (e.g., military sector, humanitarian sector) to the field of CH conservation in a more practical way (Almagro Vidal et al., 2015). Through real hands-on activities, the program has been able to significantly develop local capacities worldwide.

Despite these efforts, and even though DRM has proved to be an important tool that can help prevent loss and damage to CH, questions remain open about the overly technical approach that has prevailed so far. In the light of this perspective, the chapter aims at discussing these points with the following outline. In the next section, the chapter will present the main characteristics of a techno-centric approach (henceforth, TCA) as applied into the DRM realm. Then, it will illustrate the specificities of TCAs within CH conservation, with some examples from CH management in Mexico after different earthquakes. From this, the chapter will discuss the implications of a technocratic approach for CH conservation before concluding with some reflections that might foster a shift from a TCA to a more comprehensive understanding of CH and DRM.

Techno-Centric Approaches to Disaster Risk Management

At the end of the 1970s, governments began to institutionalise DRM as a decision-making process regarding risks and its components. Risk measurement tools, analyses, and management processes and practices were developed and supported by technical resources that were considered the most suitable to better understand the hazard behaviours and their consequences (Scolobig et al., 2015). TCAs, thus, started to emerge as DRM solutions with the application of measuring and monitoring techniques and sophisticated managerial strategies (Alexander, 1991).

TCAs were mostly structured under the form of top-down schemes, with government organizations as solely responsible for DRM. These organizations were in charge of designing and implementing TCA-based solutions by following 'command and control' mechanisms (Alexander, 2002). In this scenario, the civil society was mainly seen as a passive receiver of information, while DRM as a centralised set of solutions proposed by those holding technical skills, capacities, knowledge, and experience (Gaillard et al., 2010). Environmental sciences and engineering were used to understand, monitor, and presumably control natural hazards, relying on ad hoc probabilistic models and technological devices (Scolobig et al., 2015), i.e., protective infrastructure (e.g., sea walls or dams), or hazard-based land use regulations (Gaillard et al., 2010). Furthermore, TCAs assume that individuals and societies have a supposed low risk perception and a limited capacity to adjust to the threat. Risk perception was thus addressed through risk awareness campaigns, frequently insensitive to local cultures and contexts, and through risk evaluation and quantification that inform government policy and planning (Gaillard et al., 2010).

Today, TCAs continue to perceive disasters as a synonym of hazard and investigate it from a phenomenological and technological point of view that do not account for the causes, processes, and effects as consequences of a socially constructed world (Jones et al., 2015). They seem to undermine all the psycho-sociological aspects of DRM, and the nontechnical and rational-analytic factors of individual, community, and collective life (Gaillard et al., 2010). In this way, TCAs do not consider those individual, cultural, and social aspects that have implication on exposure and vulnerability, missing the opportunity to catch the complex and mutual interactions between society and environment (Hilhorst, 2003).

Techno-Centric Approaches within Cultural Heritage Conservation Systems

In recent years, the integration of the fields of disaster studies and of those devoted to the study and conservation of CH has created an interdisciplinary body of knowledge defined as disaster risk management for cultural heritage (henceforth DRM-CH). Its main purpose is to reduce the impact of disasters on CH, to raise awareness about its role in the construction of social resilience, and, therefore, to help improve people's capacity to protect culturally significant assets. However, although important work has been accomplished, integration between both fields has not been fully achieved so far.

An important aspect to achieve integration is the construction of a common vocabulary that enables a dialogue between the CH and the DRM fields (Chmutina et al., 2021). However, this has not yet been reached completely. For instance, the mutually used term 'retrofit' is understood in the CH field as the action of adding extra systems or fittings to make a building more efficient (e.g., in terms of energy consumption). However, in the DRM field, the same term is much more focused on the processes, actions, and decisions aimed at improving and/or reinforcing existing structures so that they are more resistant to the damaging effects of hazards. Far from being just a semantic problem, a common vocabulary is a way to bridge communication between different disciplines. It is an essential component of integration because it can shape action and research with the common goal of reducing disaster risk (Alexander & Davis, 2016). The use of one term or another can even have legal implications in policies and normative documents as they can shape decision-making and planning (Quarantelli, 1999). In that sense, the tendency to apply technological terminology to CH issues brings the problem of overshadowing societal aspects in CH management.

In addition, TCAs applied to CH tend to prioritise the potentially physical and material damage from a hazard as the only foreseeable risk and thus disregard other non-physical consequences. Additionally, when root causes of disaster conditions are explored, TCAs tend to evaluate them as a lack of sufficient technical resources that would allow the reduction of risk (Macías, 1997). Root causes of vulnerability such as financial disadvantage, inadequate policy, lack of institutional governance, or political negligence, among many others, are thus neglected. For CH, this means that only those circumstances that directly endanger its physical and material stability are considered as a threat, while situations that may threaten intangible values are often overlooked. For example, from this perspective, fire extinguishers would be placed in a museum where the incidence of fire has been evaluated as the main threat, but a low capacity to timely respond is disregarded, therefore, no training programs are generated for the personnel who work on the site. In the same way, situations that threaten intangible values in a less evident way, such as over-touristification, might be regarded as not relevant.

While it is true that failing to preserve the physical and material aspects of CH may lead to weakened capacities to cope with the hazards, the prevalence of technocratic solutions disregards the fact that resilience also arises from socially complex and non-structural configurations. There are examples in which grassroot groups responded proactively to a disaster not only to protect CH, but also to catalyse social organisation and coherence. Take the case of Tehuantepec, in Mexico, where in the aftermath of the 2017 earthquakes, community groups carried out emergency actions for the protection and rescue of local heritage against further damage. Culturally significant objects were temporarily evacuated, inventoried, and safeguarded in private houses (Torres, 2021).

Likewise, TCAs do not take into consideration the fact that it is not hazards but the conditions of vulnerability that create disasters. While hazards might be managed through engineering and technical solutions, disasters cannot. While technical solutions may reduce some of the structural vulnerabilities of a building or site, disaster conditions are largely based on unequal power relations, social constructions, and political decisions (Wilches-Chaux, 1993, 2018). Furthermore, even when there are some cases in which pre-disaster strategies are envisioned, TCAs largely denote a reactive standpoint to emergencies. For instance, retrofitting measures are often implemented to improve a building's resistance and performance against a particular threat, say earthquakes, only once the building has been damaged. This is mainly based on the reluctance to invest in preventive measures. However, this means again that funding is only available once a disaster has happened, and thus actions to reduce vulnerabilities are executed only once damage has been observed. In other words, actions aimed at protecting CH are only implemented once cultural assets are lost (see box 12.1).

Box 12.1 Approaches to DRM of CH in Mexico

Mexico has a very long history of earthquakes, largely due to its location at the confluence of five different tectonic plates: Caribbean, Pacific, North America, Rivera, and Cocos (Pérez-Campos et al., 2018). A few cases with important consequences for the country's CH have been selected to explore the prevailing kind of approach.

The 1957 earthquake (Mw 7.8) is known for causing the collapse of the statue of a winged Victory on top of the famous monument named 'Columna de Independencia' (Column of Independence) in Mexico City. Besides the collapse of the gilded bronze statue and its breaking in several sections, the column faced the destabilisation and loss of place of the stonework (Orozco & Reinoso, 2007). After the earthquake, the city's government response was to retrofit the column internally using steel beams, including the replacement of the original stone stairs with metal ones without considering the modification of the original construction system under the idea of reinforcing it against future earthquakes (Martinez Assad, 2005; Fortoul van der Goes, 2019). Additionally, the metal staircase also meant the allowance of touristic visits in a more convenient way.

Similarly, in 1985, the deadliest earthquake in the country's recent history (Mw 8,1) hit Mexico City once again. It has been the most significant disaster in terms of social and cultural impact (Montaño, 2018), and represents a turning point that led to relevant changes and upgrades regarding disaster preparedness. However, these changes only relied on solutions deriving from TCAs. Examples are the creation of the country's Earthquake Early Warning System in 1986, that notifies (with around 45 to 60 seconds in advance) the capital, Mexico City, when a significant earthquake develops in the Pacific coast. Also, stricter construction codes were adopted after the earthquake. Notwithstanding, these solutions did not consider any social strategy for CH in relation to, e.g., mitigation, evacuation plans, risk communication, nor any other non-structural changes within what was called the national civil protection system (SINAPROC, in Spanish). These solutions, instead, focused solely on the technical side to prevent further disasters; for instance, the installation of speakers throughout the city that sounds an alert when the EWS is activated. Even nowadays, after 37 years, those solutions are still insufficient in terms of protecting CH and people from future disasters (Alcántara-Ayala et al., 2019; Velazquez et al., 2020).

Although CH was not included in the government's mitigation strategies in the following years, multiple projects were articulated; however, these notably maintained a TCA perspective. For example, some very important murals, including the work of the artists Diego Rivera and Fermín Revueltas, 'Sueños de una tarde dominical en la Alameda Central' (Dreams of a Sunday afternoon in the Alameda Central) and 'Alegoría de la producción' (Allegory of production), respectively, were detached from their original placement due to building damage. Interestingly, while Rivera's mural was moved from the Hotel El Prado of Mexico City to a purposely built museum just a few metres away from its original site, the work from Revueltas was detached, restored, and sold by the Mexico City's government to the government of Monterrey, a state in northern Mexico, where it currently sits. The decision might have protected its material value but lost the social dimension that might have been associated with the use of the public space, public perception, and its relationship with the original site for which it was created.

Another case was the 1999 earthquake (Mw 7.4), where technical solutions were also privileged during the post-disaster recovery phase as well. After the earthquake, damage assessment pointed out that damages concentrated on the states of Puebla and Oaxaca, recorded mainly in religious buildings of small to medium size, largely from the 17th century. During the recovery stages, heritage authorities and decision-making groups – mainly architects and engineers – privileged the reinforcement and retrofitting of the affected buildings through the addition of external elements such as concrete beams, steel strappings, metal tensioners, and the use of steel mesh jacketing and cement plasters (INAH, CONACULTA, 2000). Reconstruction of lost sections was done in many sites with modern materials, or with construction systems different from the original ones. This also fostered the idea that maintenance was not necessary since modern modifications were robust enough. However, after the two earthquakes (Mw 8.2 and 7.1) of September 2017, it was observed that the interventions made after the 1999 earthquake were largely the cause of greater damage in many historical buildings. This damage was associated with the modification of construction techniques but also with the absence of effective maintenance and preventive conservation measures that would have secured better conditions. These examples show that, throughout recent history in Mexico, a TCA has not only prevailed but, in some cases, even added vulnerability to CH.

This chapter does not claim that technical measures are not necessary. Nevertheless, they are marked as insufficient, as they can become a limited solution that shadows other needs and fundamental elements to consider around CH, i.e., its role as a catalyser of social resilience (Jigyasu, 2015). Furthermore, excessively invasive technical practices produce a significant change in the physical behaviour of a historical building by modifying its construction system. Unfortunately, their effects on CH are often only evident in post-disaster situations, where damage allows us to observe the construction systems insides. Even when a structural retrofitting intervention is needed, an overly TCA shows a poor understanding of the attributes of a historical building since the methodologies used to assess damage cannot successfully take into consideration the values CH represents, and what houses, chapels, and historic places might signify for local communities in terms of identities, landmarks or cultural continuity. The consequence is that historical structures are often categorised as unsafe and worthy of demolition during emergencies (Jigyasu, 2015). For example, in 2017, traditional chapels, small churches, and historical buildings in the state of Puebla were assessed as inadequate against earthquakes and therefore classified as hazardous, without considering that modern materials and elements (i.e., steel strapping and concrete beams) were added in previous years.

In terms of post-disaster recovery, the practice of framing hazards as the only sources of risk, excluding vulnerable conditions as part of the equation, is the foundation of insufficient DRM programs that manage to recover material losses but fail to cover deeper social and cultural needs at the local level. This is closely related to the idea that physical damage can only be recovered once it happens, and since it is the hazard that causes damage, it is not possible to prevent it. Restoration then becomes necessary every time a hazard occurs, perpetuating a continuous cycle of interventions that require an equally continuous availability of resources in order to secure an ideal state of conservation. This is particularly true for cyclical hazards such as typhoons and tropical storms. Similarly, consideration is given primarily to funding strategies that secure the cost of restoring the material consequences of a hazard. Even though technical solutions are often required, omitting other non-structural factors can lead to the failure of mitigation efforts. Moreover, it can even create more risk as root causes of vulnerability are disregarded (Scolobig et al., 2015). Contrastingly, an integrated approach that recognizes that hazards are not the only sources of risk, and CH has both tangible and intangible dimensions, is paramount.

Another characteristic of TCAs is that there is poor or no collaboration between national, international, and subnational organizations regarding DRM and CH. This is another consequence of framing social needs as a lack of technical resources (Roberts & Hernandez, 2017), particularly in post-disaster situations where organisations responsible for managing the emergency tend to assert that all solutions can be covered entirely by a single source, namely the government itself. This issue can set a biased assessment that fails to include different disciplines and points of view, including the local community and CH experts. The result is the delivery of emergency solutions that are often disassociated from the real needs of the affected communities. An inclusive approach, on the contrary, can allow different needs to be considered, like the need to implement temporary spaces for religious activities, or the urgency to recover cultural objects that are key to continue with the community's ways of life.

The case of the city of *Tehuantepec*, in south Mexico, is another example. On 7th September 2017, a Mw 8,2 earthquake hit the region causing extensive damage. It was followed by several days of strong out-of-season rainfall that caused intensive flooding. The overall damage of the compound emergency included a large percentage of households affected to some degree, and more than 50 historic buildings (traditionally, the core of most cultural activities) with severe damage (Prieto Hernández, 2018).

Ten days after the earthquake, personnel from the Secretariat of National Defence (SEDENA) settled in the region, including *Tehuantepec* and nearby towns. The operation lasted for approximately two months, during which SEDENA organized community kitchens, delivered relief products, and offered medical aid. Even though emergency relief proved essential, it turned out problematic as it failed to communicate with other organisations (García Souza, 2018). For instance, the operation included a rapid damage assessment of both historic and modern buildings. Many buildings were marked down as to demolish – due to safety issues without being assessed by a CH expert. Plus, SEDENA did not inform the national CH institution (National Institute of Anthropology and History, INAH) about this assessment, nor included local CH officers that were already working on the ground. When INAH's inspectors arrived a few days later, some historic buildings were already bulldozed and many more were marked for demolition as well. This example shows how the categorization of any impact or loss as a problem of techno-material nature can hinder effective collaboration between agencies and other disciplines and accentuate institutional vulnerabilities.

From a similar perspective, the centralization of emergency response that can derive from TCAs is a practice that favours carrying out acts of corruption as commercial-driven actors may seek to provide relief products without actually ensuring cultural needs are covered, just for the sake of profit. This also comes from the idea that pre-packaged, technical solutions are the answer to risk and CH damage, including social and cultural recovery. For instance, the politically attractive idea to provide canned food – and show it in the media – when the real urge may be to have temporary spaces devoted to religious and cultural practices. Consequently, social and cultural needs are not fulfilled, hindering a prompt recovery. Plus, a clientelism relationship is established between state agencies and the affected communities.

Going back to the example of 2017, during the early recovery phase, the Mexican government designed a 'reconstruction' program that aimed to provide new housing for the affected communities. Again, this was done without carrying out a proper assessment of cultural needs that should have included CH. Therefore, the grassroots necessities of the local groups, specifically based on their traditions, livelihoods, and context, were neglected. The government's program offered funding to build small concrete one-room housing instead of offering means to recover traditional ones. It even paid for the demolition of traditional houses as a requisite to access recovery funds. The outcome is that newly built houses are not used for residence purposes but for storage, to shelter domestic animals, or are de facto abandoned due to their disassociation from the cultural context. A place-based evaluation would have shown that concrete single-room housing is not suitable for a region with an average temperature of 36° Celsius.

From a different perspective, technical solutions most often represent an important investment and need specialized labour for its installation and maintenance, i.e., seismic isolation systems in historic buildings. This might not be affordable for many low-budget organizations or institutions such as local museums. In that sense, it is worth pointing out that not all CH managers might have access to sufficient resources to (a) implement structural measures, (b) have trained staff constantly available or develop their organizational capacity to efficiently use that technology, or (c) to secure continuous funding to maintain the structural measures in the future. These three aspects compromise the sustainability of the proposals and, therefore, of the CH site itself. Additionally, it is also important to mention that technology, at some point, might become obsolete or inadequate to an ever-changing context and require additional resources to be removed, disassembled, or upgraded.

It is frequent to see that technical solutions are applied to single historical buildings or sites, while the unavailability of resources makes it almost impossible to implement on a large scale. However, implementing highly technological solutions to only some buildings means that those assets will have a different behaviour from the rest of their context, especially when in urban and peri-urban areas. In that sense, those heritage sites are not promoting cohesion, collaboration, and thus social resilience among the community. Instead, they establish communication barriers based on each one's access to resources and funding. The disparity might promote disadvantage for other less fortunate cultural assets such as non-world heritage sites, and therefore inequality, which is one of the root causes of vulnerability and disasters.

Similarly, many cultural sites are immersed in such spatial interactions that the vulnerability of any single building can (and probably will) influence the vulnerability of the entire area (Bosher et al., 2019). In this sense, the proportionality of any one intervention must take into consideration the situation of the surrounding buildings.

Implications of a Technocratic Approach for Cultural Heritage Conservation

The unbalanced appreciation of CH as a purely physical and material entity is also the foundation of the idea that every heritage asset is recoverable. From this notion, cultural institutions prioritise the creation of mechanisms that guarantee access to recovery funds. These might include two different options. The first one is to buy insurance policies based on the single idea of securing access to restoration funds in case of a disaster. Alternatively, they might choose to set aside a part of their own funds to meet expected losses when needed through an arrangement known as 'self-insurance' (Hartwig & Wilkinson, 2007).

Although both strategies are useful mechanisms to finance the recovery process including restoration of CH, they assume that every loss resulting from disasters is recoverable or open to reconstruction (see Chapters 15 and 16). Therefore, insurance has become the one-fits-all 'risk management' solution chosen by organizations or institutions to manage risk by transferring the cost of recovery to an insurance company. Nonetheless, cultural values cannot be always replaceable in the same way as the fabric is as their loss has deeper social consequences than the destruction of a building, object, or site. Additionally, considering just the physical or material characteristics of CH denies its other dimensions that cannot be valued in terms of revenue. The monetisation of CH, in terms of social value, is then only partial. Therefore, DRM is often fictitious, as it neither facilitates the development of protective plans nor prevents damage, and because CH managers can only access funds once a disaster has occurred and CH is damaged or lost.

In that same line of thought, in emergency or disaster situations, it is taken for granted that cultural values lie in the invariable continuity of the material remains, therefore, they should remain physically unchanged (Bosher et al., 2019). Instead of promoting change during post-disaster recovery, framing CH as a fixed asset prevents adaptation. In that sense, the overly materialistic TCAs pose a false dilemma in which cultural values and attributes are outlined as something static that cannot evolve, becoming a passive element that only demands attention and care (Smith, 2006), particularly during emergencies. Also, the idea that authenticity is based merely on material stability has resulted in a much better funded search for technological solutions and the underestimation of the dynamic nature of CH and the chance to reduce its socially constructed vulnerability through adaptation. What is more, the idea that heritage must remained unchanged has contributed to the conceptualization of disasters as 'inevitable' and undervalue the beneficial impacts of pre-disaster risk reduction activities that are often associated with mitigation (i.e., structural and non-structural risk reduction measures) and preparedness (i.e., emergency planning and capacity building) activities (Bosher et al., 2021).

On top of that, CH can also be incorporated into narratives that are used to exercise power relations. Frequently, these are top-down models that reflect a poor or absent channel for communication with society and its needs and ways of understanding CH. Examples exist of built heritage that has been recovered and restored after several disasters but exclusively by following the criteria of institutional 'specialists' and of their 'approved' knowledge, without listening (and even less, addressing) the needs and demands of local actors and communities (and their related knowledge). Even when these are formally recognized and incorporated into the decision-making process. On the opposite side, a broader and more inclusive understanding of CH should try to protect the values and meanings that are assigned to it by society in the same way as a building is retrofitted to prevent damage before an earthquake. This idea aims at shifting the vision in which CH is seen as a passive and victimised asset in need of assistance towards one that acknowledges the multiple ways in which heritage actively contributes to the construction of social resilience (Jigyasu, 2015; Torres, 2021).

On a different side, TCAs can bring a biased understanding of risk as an 'intrinsic factor', an unavoidable condition of the fabric. This means that some construction systems or techniques might be weighed as incapable or not inclined to resist hazards (e.g., masonry as inherently weak against earthquakes). There is no doubt that some systems and materials behave better than others to specific hazards; nonetheless, other non-structural factors that also play a role in creating vulnerable conditions are disregarded (e.g., lack of maintenance). This produces a preconceived notion that subtracts responsibility from the people who make decisions (e.g., lack of maintenance funding) and establishes the idea of facing damage as something somehow acceptable, normalizing its occurrence.

Towards More Integrated Approaches to DRM-CH

From 1970 where UN's state members were asked to submit recommendations on pre-disaster planning by means of technology, to the current SFDRR and SDGs, the field of DRM has shifted towards a notion of disaster risk that acknowledges the influence of systemic factors (i.e., policy and power distributions) in the creation of vulnerability. This has paved the way to the idea that non-technocentric approaches, i.e., robust governance and people-centred systems, have the capacity to effectively act as coping mechanisms (Bosher & Chmutina, 2017). This approach puts people and their needs, including CH, at the centre of policy and practices. This has also created room for recognizing CH beyond its physical and material characteristics as part of a cultural system where different components (social, cultural, political, economic) play an essential role. This is achieved, for example, by including intangible dimensions of CH in terms of its psychological and social values (individually and collectively speaking), such as traditional risk governance and management systems, and their role as fundamental coping and self-recovery mechanisms, in the same degree as the physical attributes (Jigyasu, 2021).

However, despite its recognition as an essential component of social structures, and away from a passive asset, CH, in both its tangible and intangible dimensions, is still often not adequately integrated in wider DRM plans. Urban planning and risk management processes still lack understanding of the complexity and ubiquity of intangible CH values and their direct relationship with the physical elements that create CH. Thereupon, CH is still seen as a separate component (Jigyasu, 2015). Thus, while DRM policies and strategies remain focused on the physical and material aspects of CH, its intangible values in all their forms (social, cultural, individual, collective) face greater challenges for their preservation.

In that sense, cultural diversity is still hindered by institutional structures that homogenise CH in its attempt to standardise risk. Communities cannot find a way to express their cultural needs and the risks they consider a priority, while centralised structures do not include them in the decision-making processes. Alternative approaches to TCAs are those based on the premise that involving people in DRM regarding their own heritage and culture empowers them, encourages ownership, co-responsibility, and social cohesion while fostering diversity. These may result in more effective DRM. Community participation must be integrated in DRM for CH as a way to strengthen local capacities to respond and protect CH in emergencies (Marchezini et al., 2019; Torres, 2021).

Furthermore, risk governance and management models need to acknowledge CH as an evolving and multi-dimensional system, depending on the values that communities assign to it. The fact that CH can become controversial when different values clash can be taken as an opportunity to assess its multi-dimensional nature that may not have been evident before.

Conclusions

Throughout the chapter, it has been explored the way multiple understandings of CH play an important role in CH conservation and management. Each actor frames CH depending on the values and functions it admits, and designs related strategies accordingly. However, it has been also asserted that there are not yet sufficient tools, policies, and stakeholders converging and working together towards a more comprehensive understanding of CH and the risks it faces.

We also unpacked the way solutions based on TCAs tend to mask root causes that are deeply responsible for creating disaster conditions. Nonetheless, excessively unbalanced and invasive interventions based on technical solutions not only can prevent risks from being identified and reduced but can even create other factors of vulnerability.

In that sense, it is urged to move beyond TCA towards truly integrated approaches able to catch the complexity of CH. This approach certainly must include collaborative work by academics, researchers, and specialists from multiple disciplines, but also practitioners and local communities, in the construction of risk reduction strategies.

Summary of Key Points

- Techno-centric approaches undermine the psycho-sociological aspects of disaster risk management.
- Techno-centric approaches for cultural heritage prioritise hazard over non-physical and collective aspects.
- After multiple earthquakes in Mexico, techno-centric solutions for recovering physical aspects of cultural heritage did not consider social values and collective meanings.
- Techno-centric approaches to cultural heritage in Mexico also hindered cross-agencies collaboration while favouring corruption.
- A comprehensive risk management of cultural heritage needs to address and include also social, intangible, and cultural values.

Notes

1 https://whc.unesco.org/en/conventiontext/
2 https://ich.unesco.org/en/what-is-intangible-heritage-00003
3 https://whc.unesco.org/archive/2007/whc07-31com-72e.pdf
4 https://sdgs.un.org/goals/goal11
5 https://www.preventionweb.net/files/resolutions/N1516716.pdf

References

Alcántara-Ayala, I., Salinas, M.G., García, A.L., Rueda, V.M., Orozco, O.O., Aguilar, S.P. & Rangel, G.V. (2019). Integrated disaster risk management in Mexico: Reflections, challenges, and proposals from the academic community seeking a transformation on policy making. *Investigaciones Geográficas*, 98. https://doi.org/10.14350/rig.59784

Alexander, D. (1991). Natural disasters: A framework for research and teaching. *Disasters*, 15(3), pp. 209–226.

Alexander, D. (2002). From civil defence to civil protection–and back again. *Disaster Prevention and Management: An International Journal*, 11(3), pp. 209–213.

Alexander, D. & Davis, I. (2016). *Recovery from Disasters*. London & New York: Routledge.

Almagro Vidal, A., Tandon, A. & Eppich, R. (2015). First AID to cultural heritage. Training initiatives on rapid documentation. *International Archives of the Photogrammetry, Remote Sensing and Spatial Information Sciences – ISPRS Archives*, 40(5W7), pp. 13–19.

Bosher, L. & Chmutina, K. (2017). *Disaster Risk Reduction for the Built Environment*, Chichester: John Wiley & Sons, pp. 21–44.

Bosher, L., Chmutina, K. & van Niekerk, D. (2021). Stop going around in circles: Towards a reconceptualisation of disaster risk management phases. *Disaster Prevention and Management: An International Journal*, *30*(4/5), pp. 525–537.

Bosher, L., Kim, D., Okubo, T., Chmutina, K. & Jigyasu, R. (2019). Dealing with multiple hazards and threats on cultural heritage sites: An assessment of 80 case studies. *Disaster Prevention and Management: An International Journal*, *29*(1), pp. 109–128.

Chmutina, K., Jigyasu, R. & Okubo, T. (2019). Editorial for the special issue on 'securing future of heritage by reducing risks and building resilience'. *Disaster Prevention and Management: An International Journal*, 29(1), pp. 1–9.

Chmutina, K., Sadler, N., von Meding, J. & Abukhalaf, A.H.I. (2021). Lost (and found?) In translation: Key terminology in disaster studies. *Disaster Prevention and Management: An International Journal*, *30*(2), pp. 149–162.

Forino, G., MacKee, J. & von Meding, J. (2016). A proposed assessment index for climate change-related risk for cultural heritage protection in Newcastle (Australia). *International Journal of Disaster Risk Reduction*, *19*, pp. 235–248.

Fortoul van der Goes, T. (2019). La columna de la Independencia. *Revista de La Facultad de Medicina de La UNAM*, Mexico.

Gaillard, J.C., Wisner, B., Benouar, D., Cannon, T., Creton-Cazanave, L., Dekens, J. & Vallette, C. (2010). Alternatives for sustained disaster risk reduction. *Human Geography*, *3*(1), pp. 66–88.

Hartwig, R.P. & Wilkinson, C. (2007). An overview of the alternative risk transfer market. In J. D. Cummins & B. Venard, eds. *Handbook of International Insurance: Between Global Dynamics and Local Contingencies*, vol. 26, Boston: Springer, pp. 925–952.

Hilhorst, D. (2003). Responding to disasters. Diversity of bureaucrats, technocrats and local people. *International Journal of Mass Emergencies and Disasters*, *21*(1), pp. 37–56.

Instituto Nacional de Antropología e Historia (INAH) & Consejo Nacional para la Cultura y las Artes (CONACULTA). (2000). *Memoria Fonden 2000. Rehabilitación de inmuebles históricos dañados por los sismos de junio y septiembre de 1999 en los estados de Guerrero, Estado de México, Morelos, Oaxaca, Puebla, Tlaxcala y Veracruz*. INAH. Ciudad de México.

Intergovernmental Panel on Climate Change. (2018). Global Warming of 1.5°C. An IPCC Special Report on the impacts of global warming of 1.5°C above pre-industrial levels and related global greenhouse gas emission pathways, in the context of strengthening the global response to the threat of climate change, sustainable development, and efforts to eradicate poverty [Masson-Delmotte, V., P. Zhai, H.-O. Pörtner, D. Roberts, J. Skea, P.R. Shukla, A. Pirani, W. Moufouma-Okia, C. Péan, R. Pidcock, S. Connors, J.B.R. Matthews, Y. Chen, X. Zhou, M.I. Gomis, E. Lonnoy, T. Maycock, M. Tignor, and T. Waterfield (eds.)]. Cambridge University Press, Cambridge, UK and New York, NY, USA, 616 pp., doi:10.1017/9781009157940.

Cannon, T., & Schipper, L. (eds.). (2014). World Disasters Report 2014: Focus on Culture and Risk. *International Federation of Red Cross and Red Crescent Societies*.

García Souza, P. (2018). Los efectos de los terremotos y las inundaciones de septiembre de 2017 en San Mateo del Mar. *Rutas de Campo*, *3*, pp. 52–68.

Jigyasu, R. (2015). The intangible dimension of urban heritage. In F. Bandarin & R. van Oers, eds. *Reconnecting the City: The Historic Urban Landscape Approach and the Future of Urban Heritage* (1st Ed). Hoboken, NJ: John Wiley & Sons, Ltd.

Jigyasu, R. (2021). Mainstreaming cultural heritage in disaster risk governance. In *Strengthening Disaster Risk Governance to Manage Disaster Risk*. Amsterdam: Elsevier pp. 21–26.

Jones, S., Manyena, B. & Walsh, S. (2015). Disaster risk governance: Evolution and influences. In J. F., Shroder A.E. Collins, S. Jones, B. Manyena & J. Jayawickrama, eds. *Hazards, Risks, and Disasters in Society*, Elsevier, pp. 45–63.

Macías, J.M. (1997). La sociedad y los riesgos naturales. Estudio de algunos efectos de los sismos recientes en Ciudad Guzmán, Jalisco. In A. Lavell, ed. *Viviendo en riesgo. Comunidades vulnerables y prevención de desastres en América Latina*, Red de Estudios Sociales en Prevenciòn de Desastres en Amèrica Latina, pp. 236–266.

Marchezini, V., Iwama, A.Y., Pereira, D.C., da Conceição, R.S., Trajber, R. & Olivato, D. (2019). Designing a cultural heritage articulated warning system (CHAWS) strategy to improve disaster risk preparedness in Brazil. *Disaster Prevention and Management: An International Journal, 29*(1), pp. 65–85.

Martinez Assad, C. (2005). *la Patria en el Paseo de la Reforma. Universidad Nacional Autónoma De México*. Mexico: Fondo de Cultura Económica.

Montaño, E.A. (2018). Memorias imbricadas: Terremotos en México, 1985 y 2017. *Revista Mexicana de Sociologia, 80*, pp. 9–40.

Muñiz Trejo, E. (2018). Esfuerzos globales: del DIRDN al marco de Sendai. In H. Castaños Rodriguez & E. Muñiz Trejo, eds. *La ciudad bajo amenaza. Reducción de riesgos de desastre en México*, Ciudad de México: Universidad Nacional Autónoma de México; Instituto de Investigaciones Económicas.

Orozco, V. & Reinoso, E. (2007). Revisión a 50 años de los daños ocasionados en la Ciudad de México por el sismo del 28 de julio de 1957 con ayuda de investigaciones recientes y sistemas de información geográfica. *Revista de Ingeniería Sísmica, 87*(76), pp. 61–87.

Pérez-Campos, X., Espíndola-Castro, V.H. (2018). La realidad geológica, una amenaza al patrimonio cultural de méxico (los sismos de 2017). In B. Cottom, ed. *Sismos y patrimonio cultural. Testimonios, enseñanzas y desafíos, 2017 y 2018*. Dirección General de Publicaciones, México: Secretaría de Cultura.

Prieto Hernández, D. (2018). Sismos y patrimonio cultural. Destrucciòn y restauración. In *Sismos y patirmonio cultural. Testimonios, enseñanzas y desafíos, 2017 y 2018*. INAH. Mèxico.

Quarantelli, E.L. (1999). *The Disaster Recovery Process: What We Know and Do Not Know from Research*. University of Delaware Disaster Research Center.

Ravankhah, M., Chmutina, K., Schmidt, M. & Bosher, L. (2017). Integration of cultural heritage into disaster risk management: Challenges and opportunities for increased disaster resilience. In Albert, M., Bandarin, F. & Pereira Roders, A. *Going Beyond: Perceptions of Sustainability on Heritage Studies No. 2*, Berlin: Springer, pp. 307–321.

Roberts, T., & Hernandez, K. (2017). The techno-centric gaze: Incorporating citizen participation technologies into participatory governance processes in the Philippines, Making All Voices Count Research Report, Brighton, Institute of Development Studies.

Rodriguez, E. & Muñiz Trejo, eds. (2018). *La ciudad bajo amenaza. Reducción de riesgos de desastre en México*. Ciudad de México: Universidad Nacional Autónoma de México; Instituto de Investigaciones Económicas.

Scolobig, A., Prior, T., Schröter, D., Jörin, J. & Patt, A. (2015). Towards people-centred approaches for effective disaster risk management: Balancing rhetoric with reality. *International Journal of Disaster Risk Reduction, 12*, pp. 202–212.

Smith, L. (2006). *Uses of Heritage*. London & New York: Routledge.

Stovel, H. (1998). *Risk Preparedness: A Management Manual for World Cultural Heritage*.

Torres, D.A. (2021). Community organization for the protection of cultural heritage in the aftermath of disasters. *International Journal of Disaster Risk Reduction, 60*, 102321.

UNISDR (2005). Hyogo Framework for Action 2005-2015: ISDR International Strategy for Disaster Reduction International Strategy for Disaster Reduction. https://www.unisdr.org/2005/wcdr/intergover/official-doc/L-docs/Hyogo-framework-for-action-english.pdf

Velazquez, O., Pescaroli, G., Cremen, G. & Galasso, C. (2020). A review of the technical and socio-organizational components of earthquake early warning systems. *Frontiers in Earth Science, 8*, pp. 1–19.

Wilches-Chaux, G. (1993). La vulnerabilidad global. In A. Maskrey, ed. *Los Desastres No Son Naturales*. Red de Estudios sociales de Prevención de Desastres en América Latina, *1144*.

Further Reading

Cruz Cervantes, F.A. (2015). La democracia participativa, instrumento de vinculación para la protección del patrimonio cultural. *Cuicuilco*, *22*(63), pp. 63–88.

Mason, R. (2002). Assessing values in conservation planning: Methodological issues and choices. *Assessing the Values of Cultural Heritage*, *1*, pp. 5–30.

Minguez Garcia, B. (2019). Resilient cultural heritage: From global to national levels – the case of Bhutan. *Disaster Prevention and Management: An International Journal*, *29*(1), pp. 36–46.

O'Brien, G., O'Keefe, P., Jayawickrama, J. & Jigyasu, R. (2015). Developing a model for building resilience to climate risks for cultural heritage. *Journal of Cultural Heritage Management and Sustainable Development*, *5*(2), pp. 99–114.

13 Development and Cultural Heritage in the Disaster Capitalism Era

Victor Marchezini, Andrea Lampis, Danilo Celso Pereira, and Adriano Mota Ferreira

Introduction

The concept of 'sustainable development' was defined in 1987 by the World Commission on Environment and Development (WCED) as the 'development that meets the needs of the present without compromising the ability of future generations to meet their own needs' (WCED, 1987, p. 15). Since the WCED, national and international policy frameworks have contributed to shaping and framing 'sustainable development'. The most recent is the Sustainable Development Goals (SDGs) agenda which was published in 2015 and is becoming more popular in public and private sector debates, and scholarly debates. However, while in international fora, such as those promoted by mainstream international cooperation, concepts such as 'sustainable development', the SDGs, climate change, and the energy transition are the object of an ontological transformation, there also is a teleological dimension embedded in the very term 'development'.

These terms can be seen as hegemonic notions because they are performative concepts; this is to say they play a function that serves the purposes of dominant developmental culture and elites. Hegemonic notions are first converted into ontological objects, that is, a sort of entities that have their own being. They are therefore placed above any meaningful discussion regarding their historical construction, they are not any longer the result of negotiations among developmental actors or the outcome of complex institutional frameworks clashing for the control and domination of economic and political power, but they just exist and used unquestioned. Second, these notions are converted into teleological goals; they become the endpoint towards which all actors from nation-states to local government, down (in that logic) to the communities and to households and individuals, have to conform and aspire to.

From classic authors such as Fanon (1963), analysing the dehumanising effects of colonialism, to Said (1978) criticising how the cultural representation the Western culture projects regarding what non-western cultural configurations end up determining those very configurations, scholars have systematised the debates on development. Martinussen (1997), for instance, stressed that the goal-oriented nature of the debate on development is defined as a clash between tradition and modernity that is reconfigured again and again over time. The ontological and teleological features of any developmental model, and the genealogical analysis of concepts and paradigm are relevant to understand the debate on cultural heritage.

Within the wider debate on the nature, goals, and priorities of development, others have analysed several complementary dimensions: for instance, South African scholar Pieterse (1998) analysed how critical scholarship had taken distance from a notion of development as economic growth, pointed out the complexity involved in any attempt to reframe development as 'alternative development' presenting an argument that illustrates well our sceptical position on development at large. First, the fact that in development, the concern is with policy frameworks rather than explanatory frameworks; a point that in spite of having been made some 20 years before the SDGs applies very well to them as much as to all the 'family' of development approaches that do not discuss the roots of unequal power relations but rather the desirable outcomes. Second, because of the proliferation of perspectives on what an alternative development would look like, with contrasting and coexisting positions emphasising either its scale – 'national', 'regional', 'local' – their zones – such as 'urban', 'rural' – and/or types – 'human', 'territorial'. Third, beyond the sometimes confusing plethora of new or supposedly innovative approaches, Pieterse places a question that still resonates with its relevance and cogency in the 2020s: 'whether alternative development is an alternative way of achieving development', broadly sharing the same *goals* as mainstream development, but using different *means,* participatory and people-centred (Pieterse, 1998: 345, italics in the original). Why is this question so relevant when discussing cultural heritage within an era of disaster capitalism?

Critical scholarship has pointed out that the SDG agenda has prioritised economic growth, but it avoids challenging the capitalist status quo, leading people to fend for themselves (Hickel, 2019; Chmutina et al., 2021); it comprises a whole set of contradictions, tensions, and trade-offs when comparing one SDG's goals with those of others (Nerini, 2018). Such interpretation of SDGs also shows the clear links with the concept of disaster capitalism (Klein, 2007) by which Naomi Klein, who coined the term, meant that disasters are functional to capitalist reproduction and make it thrive in areas where previous policies, laws, regulations, and norms protected people's rights and land in rural as much as in urban areas. Our perspective, hence, aims to move the debate a step forward highlighting the contradictions of contemporary narratives and policy practices on cultural heritage, whereby the notion is given high political priority and traditional institutions, such as UNESCO, decisively act in its name, meanwhile, several processes of cultural and territorial destruction are tolerated, with an emphasis in the Global South, in the name – once again – of development and progress.

This chapter discusses the notion of development in its relation to cultural heritage, shedding light on four categories that stem out when we consider this topic in the era of disaster capitalism: economic growth drivers; power; cognitive capitalism, and territory. This discussion is based on a bibliographic review and data analysis of the heritage sector and disaster risk management in Brazil. First, a brief introduction to disaster risk debates within development studies is provided, using mainly the *Pressure and Release Framework* (Wisner et al., 2004, 2012) to point out some root causes and dynamic pressures which influence economic growth and impact cultural heritage. Then, we discuss the power imbalances to classify things, including natural, cultural, and mixed heritage. Finally, we discuss the concept of territory as a category that is the object of development models and where intangible and tangible heritage is usually expressed. Although the focus of this chapter is not on the impacts of disasters in the cultural heritage sector, we understood that the current convergence of the Covid-19 syndemic, climate change, and the potential Second Cold War is an expression of a disaster capitalism era, where social agents, especially nations and corporations, can use these recurrent situations of crisis to catalyse their economic growth (Klein, 2007), impacting the cultural heritage.

Economic Growth or Development? Heritage Always Comes Last!

Thinking outside the development paradigm box is not an easy task. However, when discussing cultural heritage, it is important to remember that the notion of *development* is still a hegemonic one; even when it is declined as 'human' it tends to reproduce exploitation in the relationship between capital and nature (Escobar, 2010; Azevedo-Ramos & Moutinho, 2018). Indeed, development as a project of modernity is still elaborated by a limited group of elite institutions and research centres, the access to which is often severely restricted if not barred to minority groups in terms of the cultural legitimation of their cultural expressions, practices, and traditions, themselves the object of capitalist new forms of cognitive appropriation (Vercellone & Cardoso, 2017).

The two main approaches to class in the social sciences are the Marxian and the Weberian concepts of class (Wright, 2005, cited in Fuchs, 2011). Whereas the Marxian class concept stresses exploitation, the Weberian concept takes class as a group of people who have certain life chances in the market in common. None of them really fits an analysis of cultural heritage and territory. In fact, local cultures, mostly indigenous groups, have successfully pointed out that capitalism operates predatory forms of exploitation that go beyond the notion of class insofar colonial mercantilism domination first and neocolonial hegemonies later operate on a predatory basis with two key targets: identity and territory (Haesbaert, 2004; Lampis et al., 2022).

Despite the aspirations for civil protection and cultural heritage sectors of the 2030 agenda for Sustainable Development Goals and Disaster Risk Reduction (Hosagrahar, 2017; UNDP, 2019), 'development' and 'economic growth' are often considered synonymous, as reported by previous studies (Escobar, 2010). While economic growth is a variation in the Gross Domestic Product (GDP), development is a broader concept that includes several dimensions such as the equal distribution of resources (water, land, etc.) between people, the exercise of freedom of speech and expression, dignity, justice, peace, respect for cultures, identities and differences based on age, gender, race, ethnicity, mobility status, impairments, and forms of life on Earth and so on (Carvalho, 2002; Vieira & Santos, 2012).

Nonetheless, the concepts of economic growth and development also have similarities. The indigenous thinker Ailton Krenak (2019) considers that Western civilizations tend to separate 'nature' and 'society' when planning their settlements in the territories. The nature-culture distinction (discussed in more detail in Chapter 11) has also been applied when thinking about the modes of intervention in the territory (van Riet, 2021). Scientists from Earth System Science and/or Sustainability Science will probably argue they have adopted a holistic/system approach, while sociologists working in these multidisciplinary research teams will criticise the power imbalances about who formulated the social science research questions in research projects, pointing out examples of approaches which don't challenge the status quo (Swyngedouw, 2011; Brulle & Dunlap, 2015).

Despite the similarities and differences in the concepts of economic growth and development – and the diverse epistemologies and political perspectives implicit in their uses, we can also reflect on their interdependency and complementariness.

In his seminal book '*Open veins of Latin America*', the Uruguayan thinker Eduardo Galeano (1973) describes and analyses ways of colonial exploitation, providing examples of atrocities that began at the end of the 15th century and have been perpetuated throughout the following centuries, by intruders such as Spaniards, Portuguese, French, Dutch, British, North Americans, etc. Gold, silver, diamonds, gypsum, hematite, magnetite, tantalite, titanium, thorium, bauxite, zinc, chrome, manganese, and other plentiful minerals were and still are the strategic commodities that nations and global corporations wish to capitalise causing disasters in the territories of local people. The ambition for these strategic commodities has increased the economic growth of some nations, and today enriched a short list of corporations. But it also led to the deterritorialisation (Haesbaert, 2004) – the eviction of peoples from their lands – and extermination of the intangible and tangible heritage of indigenous/native peoples who had their own ways of 'development', according to their epistemologies, their views about how to live, many of them not separating 'nature' and 'society' (Krenak, 2019). In other words, intruders killed, enslaved, and/or colonised the locals in the name of the economic growth of their nations. These injustices are still present. The SDGs agenda set recommendations about the future, by 2030, without questioning the injustices of the past. And those injustices are often embedded in intangible and tangible heritage (as also discussed in Chapter 2).

Extermination of the heritage of some people led to the creation of the heritage of others. Plentiful minerals were drained to the European countries and were used to build monuments and other tangible manifestations of these conquests. In a museum in Paris, France, for instance, there is a big rock from Minas Gerais State, Brazil. Visitors need to pay in Euros to see an artifact from Brazil which, in reality, is a reminder of the numbers of people from tribes in the African continent who were enslaved and survived the transport overseas by the Portuguese exploitation project, to be killed in the gold, silver and diamond mines in Minas Gerais State. This is just one example of how cultural heritage is embedded with colonialism, racism, patriarchy, inequality, impunity, and other root causes of vulnerability, a reason for (paraphrasing Galeano) the veins still being open in the Latin America and Caribbean (LAC).

Disasters and Development

As highlighted in the 1970s (e.g., O'Keefe et al., 1976), it is important to understand disasters as unsolved problems of development that require us to think of the social production of vulnerabilities in the territory. Rather than focusing on episodic 'natural' events – such as earthquakes – that could harm some groups of people in their precarious settlements, O'Keefe et al. (1976) and other scholars (Davis, 1987; Oliver-Smith, 1991; Wisner et al., 2004, 2012) have for decades been questioning the processes and conjunction of reasons that make some groups of people more vulnerable to 'natural hazards' such as hurricanes, floods, droughts. This seems to be strategic since the discussions about development are influenced by the climate change debate (Brulle & Dunlap, 2015; Kelman, 2015; Kelman et al., 2020) and it is not sufficient to blame 'nature' for the problems manufactured in the territories.

Within the development/disasters debate, a contra-hegemonic view of the development-disasters nexus was supported, among others writing on the issue at that time (Wijkman & Timberlake, 1984; Davis, 1987), by Hewitt (1983), who analysed the ways governments and scholars emphasised the biophysical dimension of disasters (for instance, hazards and impacts), frame natural phenomena. A focus on 'hazards' or 'disasters' reflects a technocratic approach to disaster risk management with the not-so-hidden goal to promote the implementation of monitoring practices, engineering, and a whole set of top-town technocratic 'solutions' that ultimately contribute to keeping disasters and development as two separated domains. Technocracy may also be understood as a field of power. As such, one may think of it as a field shaped by values, beliefs, practices, and ideologies that are internalised as a *habitus* (Bourdieu, 1991), separating 'nature' and 'society', unquestioning the differences between economic growth and development, inequalities, political regimes, etc. Technocracy is present in organisations and institutions, including those which deal with development, cultural heritage, natural heritage, or disaster management. Social agents who occupy positions in those organisations and institutions are all shaped by interests and *habitus*. For instance, the LAC national elites and governments – sometimes in partnership with military personnel elites – were co-opted by international corporations to serve their interests of exploring commodities on indigenous lands, as exemplified by Galeano (1973) during dictatorships in Brazil, Chile, Mexico, and usually engaged in corruption practices, using the State apparatuses to serve their private interests.

This inconvenient 'heritage' of militarism – echoing Al Gore's movie title, An Inconvenient Truth – is represented in ideologies but also in monuments and can be considered an important root cause that drives the decisions about the projects of development and the cultural heritage policies. Root causes are defined as those lifelong and remote – geographically and temporally – drivers that have shaped societies over centuries, being the core of the social problems, the DNA of vulnerability in the past and current eras. The root causes of vulnerability – including the institutional vulnerability (Wilches-Chaux, 1993) represented by obsolescence, personal decisions, etc. – involve social and economic structures, such as the characteristics of power, wealth, and resources distribution, as well as ideologies and historical heritage (i.e., war and post-war fragility, militarism) (Wisner et al., 2004, 2012). Authoritarianism, for instance, is rooted in several societies, from democracies to autocracies. Militarism is another source that nurtures the social structure of wealth and power distribution, influencing political regimes in different ways and scales (Galeano, 1973). Inequalities have assumed a central role in explaining the visible and invisible catastrophes of human history. These kinds of root causes shape the economic growth, the access to and utilisation of resources in the territory, the politics, discourses, and practices. Some of them were analysed by the sociologist Florestan Fernandes (1969), who researched the incomplete development and modernisation process of racist Brazilian society, shedding light on the cultural resistance of Brazilian elites to the democratisation process, who usually support militarism, class privilege, and autocratic regimes.

Such root causes may change, albeit rather slowly. They are updated by *dynamic pressures*, a set of processes that changes the order of magnitude of root causes, such as population change, technological change, deforestation, housing market boom, commodity price fluctuations, etc. Dynamic pressures are '*normally decadal-scale trends involving business cycles, population dynamics, land use, and governance. They translate or transmit root causes to local scale and present moment, where they produce unsafe conditions and fragile livelihoods*' (Wisner, 2016, p. 13), such as people living in urban precarious settlements without basic sanitation and water. Such situations happened in cultural heritage sites such as Salvador de Bahia, Recife, and Rio de Janeiro, Brazil, where homeless people used to occupy abandoned buildings listed by the heritage sector in flood- and landslide-prone areas and also endangered by the interests of real-estate speculators who wish to demolish them (Rolnik, 2015).

The consequences of these economic growth drivers and human intervention in the territories have led UNESCO (2022a) to point out the need of identifying threats or factors affecting the outstanding universal value of World Heritage properties (see more on this in Chapter 3). These threats/factors were grouped into 13 categories and some of them are similar to the *root causes* and *dynamic pressures* proposed by the PAR framework (Wisner et al., 2004, 2012). Seven categories proposed by UNESCO (2022a) remind us of some hazards mentioned in the PAR framework (Wisner et al., 2004, 2012). However, UNESCO (2022a) did not nominate them as 'hazards'. All are grouped as threats/factors. The 13 categories proposed by UNESCO (2022a) were:

i human activities (deliberate destruction of heritage, terrorism, etc.);
ii social/cultural uses of heritage (society's valuing of heritage, impacts of tourism, etc.);
iii management and institutional factors (financial and human resources in the heritage sector, governance);
iv buildings and development (housing, industrial areas);
v transport infrastructure;
vi services infrastructure (water infrastructure, for example).
vii local conditions affecting physical fabric (dust, pest, radiation, wind, etc.);
viii biological resource use/modification (commercial wild plan collection, land conversion);
ix physical resource extraction (mining, water, oil, and gas);
x pollution (air pollution, groundwater pollution, etc.);
xi climate change and weather events; xii) sudden ecological or geological events (tsunami, volcanic eruption, earthquake); and,
xii invasive/alien species or hyper-abundant species.

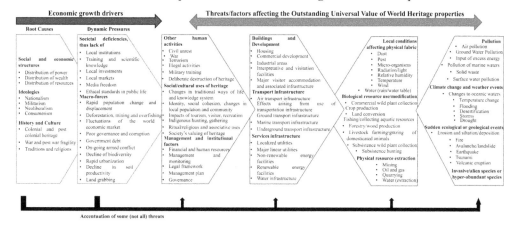

Figure 13.1 Economic growth drivers and threats/factors affecting the outstanding universal value of world heritage properties.

Source: Elaborated by the authors based on the literature review of Wisner et al. (2012) and UNESCO (2022a).

While the Pressure and Release (PAR) framework (Wisner et al., 2004, 2012) is related to the progression of vulnerability to disasters, the threats/factors highlighted by UNESCO (2022a) are not represented as a process. However, there are similarities in these categories (Figure 13.1) which can be an important analytical lens to examine the consequences of economic growth or maldevelopment to the cultural heritage, especially in the disaster capitalism engendered by the neoliberal policies. By a similar token, Ribot's (2010) classical paper has illustrated how vulnerability is politically produced out of processes that 'do not fall from the sky', as in the title of his paper, but are the outcome of choices as also reiterated by Kelman (2020) in his recent book on Disasters by Choice.

The neoliberal policies and market-driven approaches have flooded the State's apparatuses. The State should not be conceived as a monolithic entity (Lund, 2006), as it usually presents different *habitus* across its sectors, some of them with competing and conflicting interests, such as economic department, agriculture, mines and energy, environmental agencies, and disaster management sector. Usually, the State sector which will be responsible for managing disasters, especially in LAC, is civil defence or civil protection. Civil defence services emerged during the Second World War to protect citizens from bombing aerial attacks. Those services were led by military personnel who are still occupying and/or influencing the current management apparatuses, updating their discourses to the Sendai Framework for Disaster Risk Reduction (SFDRR) and SDGs. In practice, civil protection still persists in the reactive approach to responding to and controlling public security when hazards happen. In Brazil, for instance, there isn't any disaster prevention initiative that bridges development, civil protection, and cultural heritage, although this type of recommendation has been emphasised in the SDGs, the SFDRR, and the Urban Agenda Habitat III. The further development of the DRR paradigm with the inclusion in the 1980s of the vulnerability perspectives (Oliver-Smith, 1991; Lampis, 2017) seems not to have managed to significantly alter the above-described feature as one of the central tenets in DRR in the global south. It seems necessary to look at the power dynamics of those social agents that make up the State (Lund, 2006) in these three sectors. The next section offers some findings and insights to develop research on this topic.

Power to Declare 'It is a Cultural Heritage'

People are social agents who have different forms and volumes of economic, political, cultural, and educational capital which define their position in the social structure (Bourdieu, 1991) in various fields of power such as the academic, political sphere, private market, the culture industry, or even personal lives. The access to these forms and volumes of capital/power is unequal and dynamic, varying according to class, gender, race, ethnicity, age, mobility status, impairment, and to the intersection of these dimensions in time and space.

The inequities of access to the political, economic, cultural, and symbolic capital/power will reflect in the positions of agents in the field of power, in their social struggles to classify events, facts, agents, and things in the social fabric (Bourdieu, 1991). This is also reflected in the power to create classifications, to declare something as 'developed', 'underdeveloped', 'natural heritage', 'cultural heritage', etc., to include some topics in the agenda-setting of governments. The heritage sector, for instance, has its own classifications which tend to separate 'nature', 'society', and 'culture', although there is a category entitled 'mixed', which is a further declination of what we described as the processes of creation of ontological and teleological objects within the broader hegemonic, modernist and positivist developmental paradigm (see more on this in Chapters 2 and 11). By the same token, as illustrated in this case borrowing from Bourdieu, definitions 'nature' and 'society' are a sort of categories of 'second rank order'. That way one has 'developed-nature' vs 'underdeveloped-nature' or 'societies with cultural vs societies with natural heritage', which allows us to put forward a set of interesting considerations and insights about how disaster capitalism needs taken advantage of the mechanism of cognitive capitalism in order to define, tame, classify, let rot and die, or rescue and protect. In the remains of this section, we characterise the historical formation of the notion of 'cultural heritage' through the critical lens on development we have been adopting since the beginning of the chapter.

UNESCO (1972) considered that natural features, geological and physiographical formations, and delineated areas that constitute the habitat of threatened species of animals and plants, as well as the natural sites of value from the point of view of science, conservation of natural beauty, can be classified as natural heritage. It includes private and public protected natural areas, zoos, aquaria, botanical gardens, natural habitats, marine ecosystems, etc.

Cultural heritage is defined as a broader array of places such as living cultural landscapes, gardens or sacred forests and mountains, historic cities, technological or industrial achievements in the recent past, and even sites associated with painful memories of massacres and war. It includes material/tangible elements such as collections of movable and immovable items within sites, museums, historic properties, and archives, as well as intangible heritage such as lifestyles, knowledge, beliefs, and value systems which have a powerful influence on people's daily choices and behaviours (Jigyasu et al., 2013, p. 13).

The way societies and people classify their priorities in terms of tangible and intangible heritages are influenced not only by their cultures but also by the power imbalances that triggered the distinction mechanisms to classify heritage in terms of categories (natural, cultural, etc.), evolutionary perspectives of civilisations ('modern', 'advanced' societies), and also racist perspectives. The unequal development around the world is also represented in the number of inscribed properties on the World Heritage List. Fifty States Parties of Europe and North America account for 47.2% (545) of the 1,154 World Heritage Properties. 18 Arab States and 35 countries in Africa have less than 10% of these inscribed properties (Table 13.1). 167 State Parties have inscribed properties in the World Heritage List, being Italy (58), China (56), Germany (51), France (49), and Spain (49) those with more properties listed. 27 State Parties have no properties inscribed on the World Heritage List, most of them in Africa. This unequal distribution of

Table 13.1 Number of world heritage properties by region

Regions	Cultural	Natural	Mixed	Total	%	States parties with inscribed properties
Latin America and the Caribbean	100	38	8	146[a]	12.6	28
Europe and North America	468	66	11	545[a,b]	47.2	50
Asia and the Pacific	195	70	12	277[a,b]	24.0	36
Arab States	80	5	3	88	7.8	18
Africa	54	39	5	98	8.5	35
Total	897	218	39	1154	100.1	167

Source: Elaborated by authors, based on data from UNESCO (2022a).

[a] The property 'The Architectural Work of Le Corbusier, an Outstanding Contribution to the Modern Movement' (Argentina, Belgium, France, Germany, India, Japan, Switzerland) is a trans-regional property located in Europe, Asia, and the Pacific and Latin America and the Caribbean region. It is counted here in Europe and North America.
[b] The properties 'Uvs Nuur Basin' and 'Landscapes of Dauria' (Mongolia, Russian Federation) are trans-regional properties located in Europe and Asia and the Pacific region. They are counted here in the Asia and the Pacific region.

inscribed properties persisted between 1978 and 2021, but it was greater from 1992 to 2000 in terms of the amount inscribed each year by region (Figure 13.2). This inequality in the World Heritage List reflects the power imbalance between those countries that can or cannot shape and drive development in the global capitalist system. But this inequality is also manifested domestically. The next section will discuss the Brazilian case.

A Brief Overview of the Cultural Heritage Sector in Brazil

It was not until 1933 that a federal law was published to nominate the mining town of Ouro Preto, Minas Gerais State (Southeast region), as a National Monument because it was the site

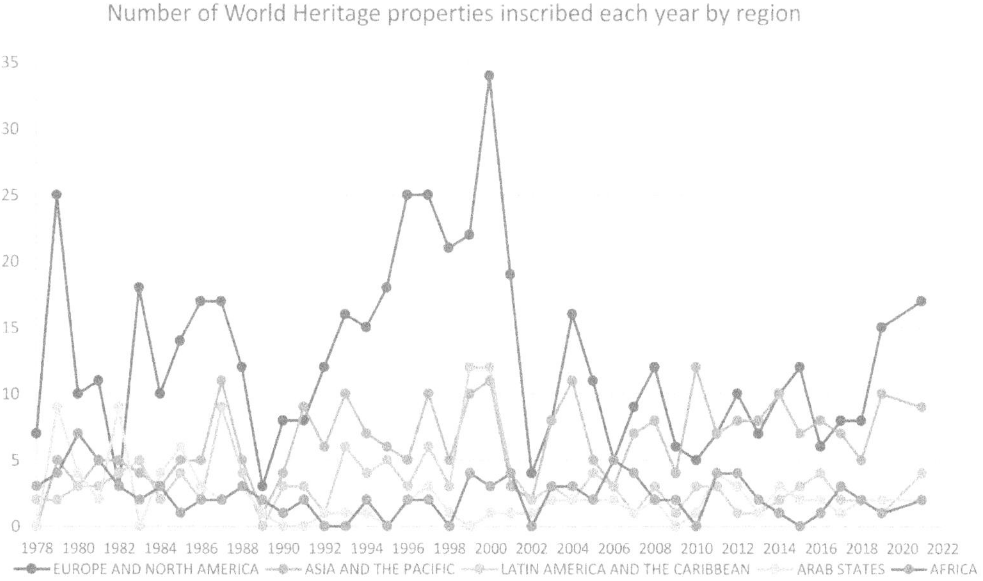

Figure 13.2 Number of world heritage properties inscribed each year by region.

Source: Elaborated by authors based on data of UNESCO (2022a).

of extremely important works of the Brazilian Baroque period. One year later, the National Constitution mentioned the need for protecting the cultural heritage and assigned that the National and State governments should hold the responsibility to protect areas of natural beauty and monuments of historical and artistic value, pointing out the need for preventing the theft of artworks (Brasil, 1934). Later, in 1937, the National Institute of Historic and Artistic Heritage (Iphan) was created with the responsibility of selection and protection of cultural heritage assets through the regulatory protection of places or objects of special cultural or physical significance.

In that context, Iphan defined property as being either movable and immovable and included objects linked to memorable events in Brazilian history or of exceptional archaeological, ethnographic, bibliographic, or artistic value included in one of the four Books of 'Livros do Tombo' National Archive – an inventory of the Brazilian heritage (Brasil, 1937). Between 1937 and 1942, Iphan created a federal preservation policy that set out the official position of the State with regard to safeguarding the memory of the country. The cultural heritage assets selected in this period – inspired by the case of Ouro Preto, Minas Gerais (Southeast region) – lay a great emphasis on aesthetic features while disregarding their social and economic aspects and purposes.

After the 1960s, the cultural heritage policy recorded some significant changes. The responsibility for safeguarding the heritage started to be shared between the Federal, State, and Municipal governments. Another important event was the publication of the new Federal Constitution in 1988, the most important legislative act in a country that was restoring democracy after the military dictatorship (1964–1985). The new Constitution introduced notions such as 'cultural rights' and provided a broad concept of what is meant by the word 'heritage':

> The Brazilian cultural heritage includes assets of a material and immaterial nature, when taken individually or together, and includes a reference to the identity, activities, and memory of different formative groups in Brazilian society. These include the following: I – forms of expression; II – ways of creating, doing, and living; III – scientific, artistic, and technological creations; IV – works, objects, documents, buildings, and other spaces destined for artistic-cultural expression; V – urban complexes and sites of historical, picturesque, artistic, archaeological, paleontological, ecological and scientific value.
> (BRASIL, 1988, Article 216)

The New Constitution of 1988 triggered a further transformation of the heritage policy with the introduction of new institutional practices. Concerning regulatory protection, the main shift referred to the significance of aesthetic criteria, which was no longer necessarily given priority because the emphasis was now laid on the historical values of the assets. According to Sant'Anna (1995), the concept of heritage changed from valuing a monument as a traditional relic to regarding it as an object that could provide a wealth of information about the life and social organisation of the Brazilian people in the various phases of its history. However, the main change occurred in 2000, when a federal decree was enacted to institutionalise the preservation of intangible assets. This offered a formal recognition of the responsibility of the States to use public resources for safeguarding these assets. By December 2021, Iphan recognised 1,358 assets as national cultural heritage, of which 1,309 (96.4%) were classified as protected tangible assets and 49 (3.6%) as intangible assets (Table 13.2).

Even within the same country, the indigenous nations, ethnicities, races, religions, forms of knowledge, and regions can be unequally represented in terms of cultural, natural, tangible, and intangible heritage. The cultural heritage is not homogeneous and could consider the diversity and inclusiveness instead of reproducing the hierarchies of colonialism, patriarchal and racist

Table 13.2 Number of assets listed by Iphan, by type, until December 2021

Type of asset	Type/category	Total assets listed	%
Tangible asset	Buildings	447	32.9%
	Buildings and collections	395	29.1%
	Urban complexes	87	6.4%
	Architectonic complexes	100	7.4%
	Movable asset or integrated	63	4.6%
	Urban infrastructure	47	3.5%
	Rural complexes	33	2.4%
	Collections	34	2.5%
	Ruins	33	2.4%
	Natural heritage	28	2.1%
	Historical garden	19	1.4%
	Terreiros – African religious site	12	0.9%
	Archaeological site	8	0.6%
	Paleontological assets	2	0.1%
	Quilombos – historical sites of fugitive African slaves	1	0.1%
Intangible asset	Mode of expression	18	1.3%
	Knowledge	13	1.0%
	Celebrations	14	1.0%
	Places	4	0.3%
	Total	**1358**	**100%**

Source: The authors, based on data from Iphan (2022).

societies through white supremacy in tangible heritage such as monuments and paintings. In Brazil, about 75% of the cultural heritage assets are from the colonial period – 54% of it from the 18th century (Rubino, 1996) – and this situation has not changed in recent years (Marins, 2016). The country also has 23 properties inscribed on the World Heritage List. The Iphan also recognised natural, cultural, and mixed heritage in the country. However, regional inequities in terms of development are also manifested in the number of tangible cultural heritage per region (Figure 13.3). Municipalities in the Southeast and Northeast regions of Brazil concentrate more than half of the assets listed by Iphan, while municipalities in the North and Midwest regions have about five percent of the national heritage list. Cities such as Rio de Janeiro, Salvador, and Recife concentrate a significant part of the tangible heritage listed by Iphan (Figure 13.3).

This inequality measured by the number of heritage assets per region can also be represented by the number of public institutions and plans related to the culture sector in the Brazilian municipalities. In 2014, the Brazilian Institute of Geography and Statistics (IBGE) launched a survey that analysed the profile of the cultural sector institutions in municipalities. The survey identified that only 19.3% of the 5,570 Brazilian municipalities have an exclusive municipal secretariat to deal with cultural issues. Usually, the municipal secretaries deal with more than one sectoral issue (54.1%) – such as sports, culture, and education. But there are some municipalities (5.5%) that don't have any organisation to deal with cultural issues, most concentrated in the Midwest, North, and Northeast regions.

Of the 5,260 municipalities which have a secretariat or department to deal with cultural issues, 11.8% had a website, most of them in the Southeast and South regions. A significant portion (23.4%) did not have a telephone and 11.3% worked without a computer. 14% (732) of the Brazilian municipalities with a secretariat or department for cultural issues (5,260) had a system of information that permitted them to carry out several activities. However, only 253 (34.5%) performed cultural heritage management activities – most (41.1%) in the Southeast

CULTURAL HERITAGE PROTECTED AT FEDERAL LEVEL, IN BRAZILIAN MUNICIPALITIES

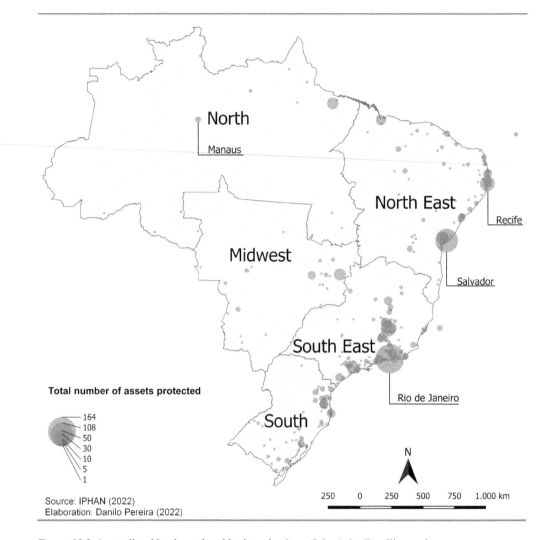

Figure 13.3 Assets listed by the national heritage institute (Iphan), by Brazilian regions.

Source: The authors, based on data from Iphan (2022).

region. Another interesting aspect is related to the head of these secretariats or departments of cultural issues: 40% had an undergraduate degree.

More than half (54.6%) of the Brazilian municipalities (5,570) have a municipal policy for culture. Among those cities which have cultural policies (3,042), the proportion is higher in bigger municipalities (Table 13.3). However, the existence of municipal cultural policies is not a guarantee that they will include the preservation of cultural heritage on it. On average, 64.9% of cities, which have cultural policies, also incorporate cultural heritage preservation. Another challenge is the formulation of plans which guide the implementation of policies: on average,

Table 13.3 Brazilian cities' profile in the sector of culture

Size of municipalities, considering the number of inhabitants	Total of cities	% of cities with cultural policies	% of cities with cultural policies which include the preservation of cultural heritage	% of cities with municipal plans for cultural sector	% of cities which promote capacity building for cultural heritage preservation and restoration	% of cities which have Municipal Council for Heritage Preservation	% of cities which have Municipal Funding for Heritage Preservation
Brazil	5,570	54.6	64.9	8.4	7.5	18.7	11.9
By 5,000	1,243	43.1	59.7	4.2	3.9	14	8.7
5,001–10,000	1,216	44.8	62.4	4.8	3.9	17.9	10.1
10,001–20,000	1,383	52.4	64.6	7.1	6.7	15	10.1
20,001–50,000	1,080	65.9	65.2	11.7	9.8	19.2	13.7
50,001–100,000	348	76.1	73.6	16.7	15.2	27.9	19.5
100,001–500,000	261	84.7	71.5	24.1	21.8	44.1	23
More than 500,000	39	97.4	76.3	33.3	23.1	64.1	35.9

Source: Elaborated by authors, based on data from IBGE (2014).

only 8.4% of the Brazilian municipalities (5,570) have municipal plans for the cultural sector. The lack of capacity building strategies and mechanisms for social participation can be an additional component of this institutional vulnerability in terms of lack of plans and financing mechanisms. Only 7.5% of cities promote capacity building for cultural heritage preservation and restoration, while 18.7% informed they have a municipal council for heritage preservation, and 11.9% have municipal funding mechanisms (Table 13.3).

It is important to remember that these institutional capacities to preserve and protect the cultural heritage are those 'management and institutional factors' mentioned by UNESCO (2022a) when referring to the 13 categories which may affect the outstanding value of the World Heritage properties. Although the 1,358 assets listed by Iphan are not included in the World Heritage list, these 13 threats and factors proposed by UNESCO (2022a) can be an important mechanism to improve the management and institutional capacities of agencies dealing with cultural heritage and protect it from maldevelopment practices and disaster capitalism. However, there are threats and factors which need to be included when thinking about the connections between development, heritage, and disasters. Many of these threats and factors happened in the territory, as a unit of analysis which can be useful to our discussion, to think about the inclusion of disaster risk management practices in the development projects which also influence cultural heritage.

The cultural heritage in Brazil reflects the 'development' of the colonial exploitation project which drained the resources from indigenous lands and rural areas to domestic elites in Brazilian urban centres – such as São Paulo and the coastal cities of Salvador, Rio de Janeiro – and also to the European commercial partners of these Brazilian elites. Most surnames of these domestic elites are still the same: their generations are usually landowners of large farms and are still in politics occupying or influencing key positions in the Brazilian Senate and Chamber of Deputies, keeping their lobbies to deregulate environmental protection and advance deforestation, cattle ranching, mining activities, and soybean production close to indigenous lands in the Brazilian Amazon, and other biomes such as Pantanal and Cerrado. The territories of indigenous people and traditional communities – which have unlisted forms of natural, cultural, and mixed heritage – also have the commodities necessary to keep the neoliberal capitalist system. The power to classify these territories as 'heritage' or 'commodities' is dependent on who is in the key positions to decide the pathways to 'development'.

Territory, Disasters, and Climate Change

The concept of 'territory' has many definitions, but for the purposes of this chapter, in line with the conceptual framework of interpretation presented in the introduction, we use it as an area where the power is exercised in order to dominate and use resources, as well as to construct symbolic meanings to its control (Haesbaert, 2004). Paramount for the comprehension of the notion of territory in Brazil and Latin America is the fact that, unlikely the notion of landscape, that of territory is culturally and politically owned by local populations. As such, the territory is subject to processes such as deterritorialisation and reterritorialisation. Deterritorialisation means a denial or a misrecognition by (more) powerful actors of the cultural and often economic sovereignty of the people who inhabit a territory on its material resources and immaterial heritage. In many respects, the latter implies the misrecognition of the right to have a place and, as a consequence, the satisfaction of basic needs, the right to have a 'home' and housing tenure, or the exercise of 'cultural rights', as a fundamental part of what constitutes their territoriality for several local communities in Brazil get jeopardised.

Governments and/or corporations – and even paramilitary groups – tend to displace people from their territories to carry out activities such as deforestation, mining, livestock farming, crop production, building resorts, as well as to implement megaprojects such as dams, highways, soccer stadiums, etc. Reterritorialisation is the process of a new intervention in some area to control it, which usually involves conflicts. The connection between territory and the intervention through projects of development or maldevelopment is clear, but cultural heritage is sometimes invisible in those relations or is represented as a barrier to the de- and reterritorialisation process. The Ministry of Culture, for instance, was created in 1985 and ceased in January 2019, affecting the preservation, restoration, and safeguard policies for cultural heritage. Moreover, the budget for disaster risk reduction activities by the National Secretary of Civil Protection was reduced between 2019 and 2022, as well as the budget of the Ministry of Environmental Affairs, and the Ministry of Science, Technology, and Innovation. These shocks doctrine – to paraphrase Naomi Klein (2007) – broke the monitoring apparatuses of these ministries and agencies which deal with culture, environment, and disaster risk management, paving the way for the maldevelopment projects and disaster risk creation, such as exemplified in the previous section when discussing the situation in the Brazilian Amazon. Coherently with our conceptual framework, in this section, we illustrate how the notion of cultural heritage and the attached practices of domination deployed by disaster capitalism are operationalised through the territories from their physical to the economic and up to the symbolic and representational dimensions.

There are additional threats that can accentuate the de- and re-territorialisation processes. UNESCO (2022a) nominated several of them, but in the context of this chapter, climate change and weather events are of particular salience. The intention is not to include here the climate change scenarios published in the Sixth Assessment Report of the IPCC, but to provide an overview of some governance mechanisms to cope with disaster risks, since development projects create them and increase greenhouse emissions, especially when megaprojects such as cattle ranching, and large scale soybean production are implemented. This situation can create additional risks to the precarious situation that cultural heritage suffers in Brazil – lack of preservation and maintenance is a permanent problem. Another risk driver refers to institutional vulnerability to manage these disaster risks. Bringing some data and information about the institutional capacities of the disaster risk management sector, based on a recent survey published by the IBGE (2021), can be important to understand the challenges posed to cultural heritage protection against disaster risks and climate change.

Table 13.4 Brazilian cities' profile for disaster risk management and climate change adaptation

Size of municipalities, considering the number of inhabitants	Total of cities	% of cities with DRR plans	% of cities with laws or instruments related to climate change adaptation or mitigation	% of cities with contingency planning	% of cities with early warning system	% of cities with civil defence units	% of cities with DRR campaigns
Brazil	5,570	13.1	7.0	25.3	7.8	76.0	15.0
By 5,000	1,249	5.6	3.7	21.9	2.5	70.5	6.9
5,001–10,000	1,200	8.1	4.2	17.2	2.5	70.7	9.7
10,001–20,000	1,334	10.8	6.7	22.1	6.7	74.7	11.7
20,001–50,000	1,110	15.1	9.5	26.8	10.0	80.3	19.2
50,001–100,000	351	29.3	10.3	39.0	16.2	87.2	29.1
100,001–500,000	277	42.2	15.9	58.1	33.2	95.7	47.6
More than 500,000	49	61.2	36.7	73.5	53.1	95.9	65.3

Source: Elaborated by authors, based on data from IBGE (2021).

In the Brazilian DRM sector, the situation of institutional vulnerability is almost self-evident. In 2021, 13.1% of the Brazilian municipalities had plans for disaster risk reduction (DRR) (IBGE, 2021). The percentage of those who have municipal laws or management mechanisms related to climate change adaptation and mitigation was also alarming: 7%. Despite the existence of a civil defence unit in 76% of Brazilian municipalities, the challenges of implementation are visible in the lack of basic policy instruments: 25.3% have contingency planning, 15% promoted DRR campaigns, and 7.8% have warning systems (Table 13.4). Based on these findings about cultural heritage and DRM sectors, it seems that parts of the State are being imploded to pave the way for neoliberal policies of far-right ideologies.

Disaster risk management (DRM) plans are important to define interventions to reduce and mitigate the current and future disaster risks in the territory in order to address the hazards (floods, landslides, etc.), the vulnerabilities (lack of sanitation and urban drainage, people in precarious settlements in slopes, lack of maintenance of cultural heritage buildings, etc.) and capacities (contingency plan, drills, educational campaigns, protocols to safeguard cultural heritage, training to rescue the cultural heritage assets, etc.). DRM plans are important to development planning in the territory, including at the municipal level. However, the governance mechanisms to formulate and implement these DRM plans are fragile, which adds new layers of institutional vulnerability to safeguard cultural heritage.

As seen in Figure 13.3, the cultural heritage assets listed are concentrated in the Southeast and Northeast regions, especially in big cities – except for Ouro Preto (Box 13.1) which has about 75,000 inhabitants – like Rio de Janeiro, Salvador, Recife, and São Paulo, which accounted for 42% of those assets listed. However, there are regional differences (Figures 13.3 and 13.4). Regarding cities with cultural heritage listed by Iphan and DRR plans, those cities in the Midwest region have the lower percentages (14%), while, on average, 27% of the cities from other regions in Brazil had both. The percentage of cities with laws or instruments related to climate change adaptation (CCA) or mitigation is 11.6% in the Southeast and 18.1% in the Northeast region. The North region has only 14 municipalities with cultural heritage, and 28.6% of them have CCA plans. It is important to highlight the lack of DRR and CCA plans in about 61% (190) of municipalities that have cultural heritage assets listed by Iphan (309), which are equivalent to 36% of the 1,309 tangible cultural heritage assets.

> **Box 13.1 Ouro Preto, a legacy of cultural heritage and disaster risk creation**
>
> The Ouro Preto settlement has started in the XVII century, after the discovery of abundant gold mines. The city was settled in a valley close to rivers and surrounded by steep slopes containing fragmented rocks. Rapid urbanisation led Ouro Preto to be the second most populated urban centre in LAC in that period.
>
> However, intensive gold mining drained out this mineral at the end of the same century. When the capital of Minas Gerais was shifted in 1897 to Belo Horizonte, Ouro Preto suffered an economic recession and depopulation which happened until the 1950s, when the Canadian mining company Alcan moved to the region to explore aluminium, triggering a new wave of migration and urban sprawl. The urbanisation process was intensified by the creation of the Federal University of Ouro Preto in the 1970s and the increase in touristic activities after the 1980s.
>
> The urban sprawl was intense. In the 1950s, the urban area has 115 ha. 28 years after, it reached 382 ha. In 2004, the urban area was 687 ha. (Oliveira & Sobreira, 2015). In five decades, the Ouro Preto urban area increased almost six times, ignoring land use and building codes that forbid settlements on steep slopes, even in urban zones close to cultural heritage assets. This process led to disaster risk creation and many disasters associated with landslides and floods were reported. In January 2022, a mountain landslide in Morro da Forca collapsed two historic buildings – one of them was the city's first neo-colonial building called Solar Baeta Neves (UNESCO, 2022b).

The percentage of Civil Defense units is 72% of the 309 municipalities that have cultural heritage listed by Iphan. The Southeast region, which has 38% (116) of the Brazilian municipalities with assets – the highest of the regions, holds the higher percentage of cultural heritage cities with civil defence units (96%) (Figure 13.4). However, this is not synonymous with having DRR and CCA plans. While the Southeast region has a higher percentage of cities with cultural assets and DRR plans (about 40%), the North region has a higher percentage of CCA plans in cities with cultural heritage assets (about 30%). These findings are not exclusively dependent on the levels of Gross Domestic Product or development indexes. For instance, the Midwest region has a similar Human Development Index (HDI) to the Southeast and South regions, but the percentage of cities with cultural assets and civil defence units, DRR and CCA plans are the lowest among the five regions (Figure 13.5). Further studies shall investigate the reasons for these findings since the Midwest and North regions are prone to megaprojects and dynamic pressures – such as deforestation, mining exploitation, cattle ranching, etc. – and extreme events related to climate change.

Conclusion

This chapter has discussed three interrelated aspects bridging development, cultural heritage, and disasters, namely: (i) economic drivers; (ii) power to declare 'it is a heritage'; and (iii) territory.

The economic drivers are represented in the PAR framework – root causes, dynamic pressures, unsafe livelihoods, and hazards – and in the threats to the Outstanding Value of World Heritage Properties (Figure 13.1). The agents who occupy positions in the fields of power – political, economic, cultural, scientific, educational, etc. – have different forms and quantities of capital to define their positionality, and their *habitus* reflect cultural beliefs, ideologies, degrees

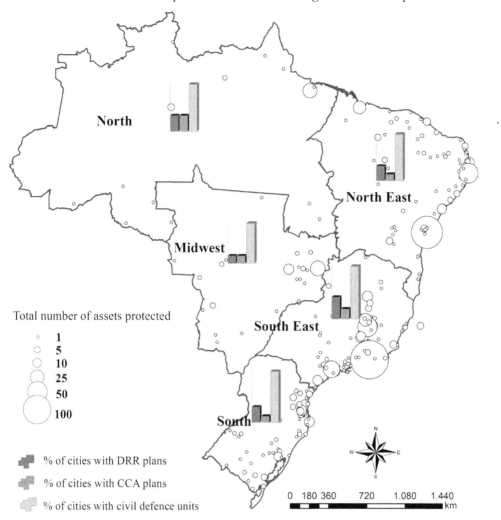

Figure 13.4 Culture heritage assets listed by the national heritage institute (Iphan) and the presence of civil defence units, disaster risk reduction (DRR), and climate change adaptation (CCA) plans, according to Brazilian regions.

Source: Elaborated by authors, based on data from Iphan (2022) and IBGE (2021).

of authoritarianism, militarism, racism, and other root causes which shape the discourses and practices of their organisations. The State is not a monolithic entity, and the agencies that compose it have different *habitus* and interests. In Brazil, the Ministry of Culture, created in 1985, was extinguished in 2019, affecting the preservation, restoration, and safeguarding of cultural heritage. The Ministry of Cities was also banned. The National Civil Defense, which is responsible for DRR activities in Brazil, also suffered a cut in its budget, as well as the Ministry of Environmental Affairs. Those institutions 'are/were' responsible for managing the culture, cities, environmental issues, and disaster risk issues in the territory to create a governance mechanism.

The governance situation was worrying before these aforementioned examples of 'shock doctrine' (Klein, 2007). In 2014, only 8.4% of the Brazilian municipalities (5,570) had municipal

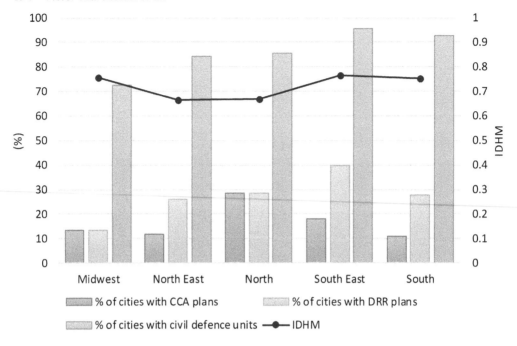

Figure 13.5 Percentage of civil defence units, disaster risk reduction (DRR), and climate change adaptation (CCA) plans in cities with assets listed by the national heritage institute (Iphan), according to the Brazilian regions and their human development index (HDI).

Source: Elaborated by authors, based on data from Iphan (2022) and IBGE (2021).

plans for the cultural sector. Only 7.5% of cities promoted capacity building for cultural heritage preservation and restoration, while 18.7% informed they have a municipal council for heritage preservation, and 11.9% have municipal funding mechanisms. Unfortunately, there is no data available to analyse the cultural sector profile in Brazilian municipalities after the shutdown of the Ministry of Culture in 2019.

Two overall reflections stem from the elements presented throughout the text as an invitation for further research:

a Disaster capitalism drives cultural heritage destruction through the dismantling of those institutions that had been created to protect the territory in several of its dimensions (Klein, 2007; Matthewman, 2015).
b Cultural heritage, colonialism, and the financialisation of space and life, analysed in great depth by Latin American scholarship with notions such as the 'commodity consensus' (Svampa, 2013), can hardly be considered as separate domains.

Key Points

- SDGs agenda sets recommendations about the future, by 2030, without questioning the injustices of the past. It prioritises economic growth and avoids challenging the status quo, tolerating several processes of cultural and territorial destruction, with an emphasis on the Global South, in the name – once again – of development and progress;

- The process approach proposed by the *Pressure and Release Framework*, which is used to understand the progression of vulnerability to disasters, can also be useful to analyse the threats or factors affecting the outstanding universal value of World Heritage properties highlighted by UNESCO (2022a);
- The State is not a monolithic entity, but it presents different *habitus* and technocracies across its sectors, some of them with competing and conflicting interests, that are not stopping disaster capitalism and disaster risk creation;
- While some of these technocracies have the power to drive the development projects, the cultural heritage sector and civil protection have been ignored and have faced institutional vulnerabilities, especially at municipal level, where there is a lack of funding mechanisms and plans for the cultural sector, for disaster risk reduction and climate change adaptation.

Acknowledgements

Victor Marchezini acknowledges the São Paulo Research Foundation – Fapesp (Grant Number 2018/06093-4). Danilo Celso Pereira and Adriano Mota Ferreira acknowledge their PhD scholarships awarded by the Coordenação de Aperfeiçoamento de Pessoal de Nível Superior (CAPES). Andrea Lampis acknowledges the São Paulo Research Foundation – FAPESP (Grant no. 2018/17626-3).

References

Azevedo-Ramos, C. & Moutinho, P. (2018). No man's land in the Brazilian Amazon: Could undesignated public forests slow Amazon deforestation? *Land Use Policy*, 73: https://doi.org/10.1016/j.landusepol.2018.01.005

Bourdieu, P. (1991). *Language and Symbolic Power*. Cambridge: Polity Press.

Brasil (1934). 'Constituição da República Federativa do Brasil (1934)', *Diário Oficial da União*, Rio de Janeiro, available at: http://www2.camara.leg.br/legin/fed/consti/1930-1939/constituicao-1934-16-julho-1934-365196-publicacaooriginal-1-pl.html (accessed July 1, 2018).

Brasil (1937). 'Decreto-Lei n° 25 de 30 de novembro de 1937. Organiza a proteção do patrimônio histórico e artístico nacional', *Diário Oficial da União*, Rio de Janeiro, available at: http://portal.iphan.gov.br/uploads/legislacao/Decreto_no_25_de_30_de_novembro_de_1937.pdf (accessed July 1, 2018).

Brasil (1988). 'Constituição da República Federativa do Brasil (1988)', *Diário Oficial da União*, Brasília.

Brulle, R.J. & Dunlap, R.E. (2015). Sociology and global climate change. In R.E. Dunlap, & R.J. Brulle, eds. *Climate Change and Society: Sociological Perspectives*, New York: Oxford University Press, pp. 1–31.

Carvalho, J.M. (2002). *Cidadania no Brasil: o longo caminho*. Rio de Janeiro: Civilização Brasileira.

Chmutina, K., von Meding, J., Sandoval, V. et al. (2021). What we measure matters: The case of the missing development data in Sendai framework for disaster risk reduction monitoring. *Int J Disaster Risk Sci*, 12: 779–789. https://doi.org/10.1007/s13753-021-00382-2

Davis, I. (1987). Safe shelter within unsafe cities: Disaster vulnerability and rapid urbanisation. *Open House International*, 12(3): 5–15

Escobar, A. (2010). *Territorios de diferencia: Lugar, movimientos, vida, redes* (2a ed.). Envión. https://doi.org/10.1017/CBO9781107415324.004

Fernandes, F. (1969). Beyond poverty: The negro and the mulatto in Brazil. *Journal de la Société des Américanistes. Tome*, 58: 121–137.

Fanon, F. (1963). *The Wretched of the Earth*. New York: Grove Press.

Fuchs, C. (2011). Cognitive Capitalism or Informational Capitalism? The role of class in the information economy. In M. Peters, E. Bulut, eds. *Cognitive Capitalism, Education and Digital Labor*, New York: Peter Lang, pp. 75–119.

Galeano, E. (1973). *Open Veins of Latin America: Five Centuries of the Pillage of a Continent*. New York: Monthly Review Press.

Haesbaert, R. (2004). *O Mito da Desterritorialização: do 'fim dos territórios' à Multiterritorialidade 1ªed*. Rio de Janeiro: Bertrand Brasil.

Hewitt, K. (1983). *Interpretations of Calamity*. Boston: Allen and Unwin.

Hickel, J. (2019). The contradiction of the sustainable development goals: Growth versus ecology on a finite planet. *Sustainable Development*, 27(5): 873–884.

Hosagrahar, J. (2017). Culture: At the heart of SDGs. *The Unesco Courier*, 1: 12–14. Available at: https://unesdoc.unesco.org/in/documentViewer.xhtml?v=2.1.196&id=p::usmarcdef_0000248106&file=/in/rest/annotationSVC/DownloadWatermarkedAttachment/attach_import_81c3cb12-523c-4cec-80f8-edc98d210673%3F_%3D248106eng.pdf&locale=en&multi=true&ark=/ark:/48223/pf0000248106/PDF/248106eng.pdf#%5B%7B%22num%22%3A55%2C%22gen%22%3A0%7D%2C%7B%22name%22%3A%22XYZ%22%7D%2C0%2C842%2Cnull%5D (accessed Sept 25, 2022).

IBGE – Brazilian Institute of Geography and Statistics. (2014). Available at: https://biblioteca.ibge.gov.br/visualizacao/livros/liv95013.pdf (accessed Feb 01, 2022).

IBGE – Brazilian Institute of Geography and Statistics. (2021). Available at: https://www.ibge.gov.br/estatisticas/sociais/saude/10586-pesquisa-de-informacoes-basicas-municipais.html?=&t=destaques (accessed Jan 01, 2022).

Iphan – National Heritage Institute (2022). Bens Tombados. Available at: http://portal.iphan.gov.br/pagina/detalhes/126 (accessed March 01, 2022).

Jigyasu, R., Murthy, M., Boccardi, G., Marrion, C., Douglas, D., King, J., O'Brien, G., Dolcemascolo, G., Yongkyun, K. & Albrito, P. (2013). Heritage and Resilience: Issues and Opportunities for Reducing Disaster Risks, United Nations International Strategy for Disaster Reduction (UNISDR); United Nations Educational, Scientific and Cultural Organization (UNESCO); Marsh United Kingdom; International Council on Monuments and Sites (ICOMOS); International Committee on Risk Preparedness (ICORP); International Centre for the Study of the Preservation and Restoration of Cultural Property (ICCROM), Mumbai, available at: http://lib.riskreductionafrica.org/handle/123456789/1187 (accessed July 01, 2018).

Kelman, I. (2015). Climate change and the Sendai framework for disaster risk reduction. *International Journal of Disaster Risk Science*, 6(2): 117–127.

Kelman, I. (2020). *Disaster by Choice: How Our Actions Turn Natural Hazards into Catastrophes*. New York: Oxford University Press.

Kelman, I., Mercer, J. & Gaillard, J.C. (2020). *The Routledge Handbook of Disaster Risk Reduction Including Climate Change Adaptation*. Abingdon: Routledge.

Klein, N. (2007). *The Shock Doctrine: The Rise of Disaster Capitalism*. New York: Metropolitan Books.

Krenak, A. (2019). *Ideias para adiar o fim do mundo*. São Paulo: Companhia das Letras.

Lampis, A. (2017). Concepts, Connections and Disruptions: Disaster Risk Reduction and Climate Change Adaptation. The Routledge Handbook of Disaster Risk Reduction Including Climate Change Adaptation.

Lampis, A., Brink, E., Santos, A.H. & Solórzano, E. (2022). Reparation ecologies and climate risk in Latin-America: Experiences from four countries. *Frontiers in Climate*, https://doi.org/10.3389/fclim.2022.897424.

Lund, C. (2006). Twilight institutions: Public authority and local politics in Africa. *Development and Change*, 37(4): 685–705.

Marins, P.C.G. (2016). Novos patrimônios, um novo Brasil? Um balanço das políticas patrimoniais federais após a década de 1980. *Revista Estudos Históricos*, 29(57): 09–28.

Martinussen, J. (1997). *From development to Post-Development: Society, State and Market*. London: Zed Books.

Matthewman, S. (2015). *Disasters, Risks and Revelation. Making Sense of Our Times*. London: Palgrave Macmillan.

Nerini, F.N., Tomei, J., To, L.S., Bisaga, I., Parikh, P., Black, M., Borrion, A., Spataru, C., Broto, V.C., Anandarajah, G., Milligan, B. & Mulugetta, Y. (2018). Mapping synergies and trade-offs between energy and the Sustainable Development Goals. *Nature*, 260(5552): 566–567. https://www.nature.com/articles/260566a0.

O'Keefe, P., Westgate, K. & Wisner, B. (1976). Taking the 'Naturalness' out of 'Natural' disasters. *Nature Energy*, 3(1): 10–15. https://doi.org/10.1038/s41560-017-0036-5.

Oliveira, L.D. & Sobreira, F.G. (2015). Crescimento urbano de Ouro Preto-MG entre 1950 e 2004 e atuais tendências. *Revista Brasileira de Cartografia*, 67(4): 867–876.

Oliver-Smith, A. (1991). Successes and failures in post-disaster resettlement. *Disasters*, 15(1), 12–23.

Pieterse, J.N. (1998). My paradigm or yours? Alternative development, post-development, reflexive development. *Development and Change*, 29(1998), 343–373. https://doi.org/10.1111/1467-7660.00081.

Ribot, J. (2010). Vulnerability does not fall from the sky: Towards multi-scale, pro-poor climate policy. *Social Dimensions of Climate Change: Equity and Vulnerability in a Warming World*, 319. https://doi.org/10.1088/1755-1307/6/34/342040.

Rolnik, R. (2015). *Guerra dos Lugares: a colonização da terra e da moradia na era das finanças*. São Paulo: Boitempo.

Rubino, S. (1996). O mapa do Brasil passado. *Revista do Patrimônio Histórico e Artístico Nacional*, 24: 97–105.

Said, E. (1978). *Orientalism*. New York: Vintage Books.

Sant'anna, M. (1995). *Da cidade-monumento à cidade-documento: a trajetória da norma de preservação de áreas urbanas no Brasil (1937–1990)*, Dissertação de Mestrado em Arquitetura e Urbanismo, Universidade Federal da Bahia, Salvador.

Svampa, M. (2013). «Consenso de los commodities» y lenguajes de valoración en américa Latina. *Nueva Sociedad*, 244: 30–46. https://doi.org/ ISSN: 0251-3552.

Swyngedouw, E. (2011). Depoliticized environments: The end of nature, climate change and the post-political condition. *Royal Institute of Philosophy Supplement*, 69: 253–274. http://doi.org/10.1017/s1358246111000300.

UNDP. (2019). Fortalecimento da conexão entre agentes e multiplicação de conhecimento são focos de projeto da Defesa Civil com o PNUD. Available at: https://www.undp.org/pt/brazil/news/fortalecimento-da-conex%C3%A3o-entre-agentes-e-multiplica%C3%A7%C3%A3o-de-conhecimento-s%C3%A3o-focos-de-projeto-da-defesa-civil-com-o-pnud (accessed Sept. 25, 2022).

UNESCO. (2022a). World Heritage Statistics. Available at: https://whc.unesco.org/en/list/stat?msclkid=695e9c2eaeda11ec9b539efcb0e14e96 (accessed Mar. 17, 2022).

UNESCO. (2022b). Rain and Damage to the World Heritage sites in Minas Gerais (Brazil). Available at: https://www.unesco.org/en/articles/rain-and-damage-world-heritage-sites-minas-gerais-brazil (accessed Sept. 09, 2022).

UNESCO (1972). Convention Concerning the Protection of the World Cultural and Natural Heritage. Available at: https://whc.unesco.org/archive/convention-en.pdf (accessed Mar. 17, 2022).

van Riet, G. (2021). The Nature–Culture distinction in disaster studies: The recent petition for reform as an opportunity for new thinking?. *Int J Disaster Risk Sci*, 12, 240–249. https://doi.org/10.1007/s13753-021-00329-7.

Vercellone, C. & Cardoso, P. (2017). Nueva división internacional del trabajo, capitalismo cognitivo y desarrollo en América Latina. *Chasqui. Revista Latinoamericana de Comunicación*, 2017(133): 37.

Vieira, E.T. & Santos, M.J. (2012). Desenvolvimento econômico regional – Uma revisão histórica e teórica. *G&DR – Revista Brasileira de Gestão e Desenvolvimento Regional*, 8(2): 344–369, mai-ago, Taubaté, SP, Brasil.

WCED (World Commission on Environment and Development). (1987). *Our Common Future*. Brutland Report. https://sustainabledevelopment.un.org/content/documents/5987our-common-future.pdf (accessed Nov 8, 2021).

Wijkman, A. & Timberlake, L. (1984). *Natural Disasters: Acts of God or Acts of Man?* Nottingham: Earthscan.

Wilches-Chaux, G. (1993). La vulnerabilidad global. In A. Maskery, ed. *Los desastres no son naturales*, Ciudad de Panamá: República de Panamá, Red de Estudios Sociales en Prevención de Desastres en América Latina, pp. 11–41.

Wisner, B. (2016). Vulnerability as concept, model, metric, and tool. *Oxford Research Encyclopedia of Natural Hazard Science*. http://naturalhazardscience.oxfordre.com/view/10.1093/acrefore/9780199389407.001.0001/acrefore-9780199389407-e-25.

Wisner, B., Blaikie, P., Cannon, T. & Davis, I. (2004). *At Risk: Natural Hazards, People's Vulnerability and Disasters*. London: Routledge.

Wisner, B., Gaillard, J. C. & Kelman, I. (2012). Framing disaster: Theories and stories seeking to understand hazards, vulnerability and risk. In B. Wisner, J. C. Gaillard, & I. Kelman, eds. *The Routledge Handbook of Hazards and Disaster Risk Reduction*, London: Routledge, pp. 18–34.

Wright, E. O. (2015). Social class. In G. Ritzer, ed. *Encyclopedia of Social Theory*, Thousand Oaks: Sage, pp. 717–724.

14 Cultural Heritage and Post-Disaster Recovery

Wesley Cheek

Introduction

New Orleans

On 27 June 2005, Allison 'Tootie' Montana, the Big Chief of Yellow Pocahontas tribe of Mardi Gras Indians, stood in front of the New Orleans City Council to plead for a change in the decades of police harassment and hostility towards Black citizens. The previous March, on St. Joseph's Night – a night when Mardi Gras Indians prowl the streets of the city engaged in performative battles with rival tribes – the New Orleans Police Department blocked the streets and violently dispersed the crowds. As Big Chief Montana framed this violence in the context of his life at the heart of New Orleans's cultural heritage for over 80 years, his heart gave out. He collapsed and died.

Big Chief Montana's suits were striking. White and gold awash in a sea of cloudy feathers, strikingly pink with smaller headdresses protruding from his costume, bright yellow with a smaller version of himself in the same suit extending down from his chest. He had been awarded a National Heritage Fellowship in 1987 through the National Endowment of the Arts. This was both in recognition intricately decorated hand sewn suits and his role in bringing the Mardi Gras Indian tribes from their rougher past into becoming a cultural force in neighbourhoods across New Orleans. Like Many Mardi Gras Indians, Big Chief Montana worked in the traditional building trades. He was a lather, hand finishing walls and ceilings in buildings across the city.

Two months after Big Chief Montana's death, Hurricane Katrina made landfall on the Mississippi coastline just east of New Orleans. These two months span a lifetime in the history of a city. The tensions between authority and culture. The death of the traditional building trades along with the deaths of the people who practiced them. The destruction of an old city, the sorrow over the loss of what that city meant, paired with bizarre by the indifference by many outside the city to its survival.

In the aftermath of the devastating flooding brought about by the failure of the Federal levee system, the city of New Orleans struggled to recover. This struggle was economic, physical, infrastructural, and cultural. The citizens of New Orleans would be asked to play a vital, contested, and generally unpaid role in the recovery of their city. See more on this in Box 14.2.

Kessenuma

Six years after Hurricane Katrina, on the other side of the world from New Orleans, a massive ship came to rest across a major road running south out of the fishing city of Kessenuma. Kessenuma lies at the back of a deep bay in Miyagi Prefecture on Japan's rugged northeast coastline. When a colossal tsunami struck this region on 11th March 2011, this ship – the Kyotoku-Maru – floated

DOI: 10.4324/9781003293019-18

in on the waves and failed to be dragged back to sea. For some, including those who had travelled from outside the area, the ship was a symbol of a time and a place. A disaster that had fractured a region and challenged our perceptions of the physical world. For others, the ship represented the horrible deaths of neighbours and family members, as well as the destruction of their homes and neighbourhoods. See more on this in Box 14.1. There were calls to keep the ship, to maintain it as a symbol of the tsunami, a warning about the power of the ocean, and as a curiosity that drew people

---

Box 14.1   The Former Disaster Management Center – Minamisanriku, Japan

The Former Disaster Management Center located in the Shizugawa area of Minamisanriku, Japan, has become something of a shrine to the 3.11 disaster (Fig. 14.1). While the events that took place at the Disaster Management Center as the successive waves of the tsunami hit the town were in many ways intensely local, the story itself has become symbolic of the disaster as a whole. Bus tours from around Japan, and with tourists from outside the country as well, bring scores of people to lay flowers, pray, and pay their respects at the steel frames that remain standing amidst the destruction and reconstruction of Shizugawa.

Looking at the ruins of the Disaster Management Center, it is possible to think that it has remained in this state since the disaster over a decade ago. This is not the case, however. This site has been at the centre of an intense – and intensely local – debate that touches on many issues in the broader world of cultural heritage and disasters.

What most people who visit this site know about it is the well-told story of the immense bravery of a young public worker named Endo Miki, who remained behind at her post broadcasting warnings of the arrival of the tsunami over the community loudspeakers. Local residents recall her voice urging them to higher ground as reports of the size

Figure 14.1 Looking at the SunSun shopping area from Kaminoyama Hachiman Shrine. Minamisanriku, Japan. 9th March 2021.

Source: Photo by Wesley Cheek.

of the tsunami came in. Endo's heroism has become a parable of heroism and sacrifice. There is another story entwined with this one. That is the story of the town's mayor, Sato Jin, who evacuated to the roof of the build and managed to be one of the ten survivors out of the one-hundred and thirty people who were in the building. For many in the town, the Disaster Management Center was not a symbol of Endo Miki's heroism, it was a monument to the death of a large section of public workers, family members, fathers, mothers, sons, and daughters. Many in the community did not feel inspiration from looking at the remains of the location where so many people that they knew died tragic deaths.

A debate arose as to what to do with these ruins. Were they a symbol to heroism, a memorial for the dead, or a divisive reminder of personal suffering? The complicated truth is that these red steel frames rising from the now barren earth were all of three of these things at once and much more. This is the complicated position of cultural heritage because it is rich with meaning, it is also fraught with meaning. This is especially true after a disaster.

One of the first issues that had to be addressed was, if the structure was to be preserved, how was it to be maintained. It might seem that the stripped down remains of a steel frame building would be good as is. This is not the case. A destroyed building left exposed to the elements presents complicated conservation challenges. Conservation of a heavily damaged structure is expensive. Miyagi Prefecture included the maintenance of the Former Disaster Management Center in its reconstruction budget. However, this only covers a ten-year period from 2021 to 2031. After that, the building will be Minamisanriku's responsibility.

Currently (as of 2023), a local group called Thinking about What to Do with the Former Disaster Management Center has formed. Their mission is just as their name states. Members of the group have all of their individual connections to the site. Some of them lost family members in it. They are facing issues of what to do with the site when the funding runs out in a decade, but also more pressing issues such as how to program activities for a park that is also the place where many people they cared about died. This has led to debates about the ethics of having traditional *hanami* (flower viewing parties) under the cherry blossom trees in the park. Is it okay to drink alcohol and carry on respectful to those that died and their families? Is it enough to maintain the building as a site for tourism? Tourism is important to the recovery economy, but if it is just a way to make money, then how does this affect the town? Is it okay to eventually take the building down and move on? Does this mean that those who died here have been forgotten? These are difficult issues and ones that are not being taken lightly. Ultimately the people of Minamisanriku will decide for themselves. Currently, the group is running hide-and-seek tournaments for local elementary school students, which is another way to continue the culture of the town for future generations.

Box 14.2 Parades of New Orleans

To give a further example, some festivals serve to mark important places that no longer exist outside of the collective memory of the local population. In New Orleans, Louisiana, on most Sunday afternoons, outside of the hottest days of August, one can find what is known as a second line parade winding its way through the streets of a relatively normal neighbourhood. The second line tradition comes from a period after the American Civil War when newly freed enslaved people had to form their own associations for social insurance. These associations provided funerals for members as well as a parade once a year. A second line parade consists of a main line of marchers who are members of the club and a brass band. The second line is largely people from the neighbourhood who come to dance and cheer on the club members and brass bands as well as people from outside the neighbourhood who want to join in the revelry.

A week or so before the parade, a route sheet is published in the local media and also distributed through local bars, restaurants, and social networks, alerting people as to the course that the moving festival will take. The choice of routes is an interesting one; it trickles through neighbourhoods that often an outsider would have little cause to visit and makes stops at bars, stores, and clubhouses whose importance to the community cannot be inferred from a distance. While a second line parade will make stops at famous locations such as Commander's Palace, an internationally renowned restaurant in the tony Garden District, it is just as likely to stop at a shuttered business that now only exists in the memory of the community.

Many of the contemporary parades congregate under a raised section of Claiborne Avenue. To look at the location now, it would seem ill chosen. On most days, it is a forlorn stretch of concrete and asphalt with little pedestrian activity. During a second line parade, a Mardi Gras Indian festival day, or Mardi Gras day itself, it is a centre of the African-American community. Before the late nineteen-sixties, this area was not an elevated concrete bypass but an oak tree lined boulevard that acted as the hub of daily life in the area. The expansive boulevard was razed to make way for the new, elevated expressway. If that actual space has died in contemporary life, it still hangs on in the collective memory of the community. Henri Lefebvre contends that 'If space is a product, our knowledge of it must be expected to reproduce and expound the process of production'. In neighbourhoods and communities around the world, local residents harbour this knowledge and continue to reproduce the significance of a space even when the structure itself is, either temporarily or permanently, non-existent.

---

to come and gape at the massive, unsettling spectacle of an ocean-going vessel, prow protruding, and screw airing out in front of the world.

As the Kyotoku-Maru loomed over the newly desolate landscape, 'Safety First' written in Japanese across its wheelhouse, a debate formed around its future. While the city of Kessenuma pondered what to do with the trawler, people from all over Japan travelled to lay flowers at its base, to pray, and to take their photos with this incongruous landmark. They mayor of Kessenuma said that he would hate to see the ship go as it was a 'visible symbol of what happened here' (BBC News 2013). Other residents said that it was heartbreaking to see the ship every day. These residents too viewed the ship as a symbol. Other residents such as a local firefighter saw the Kyotoku-Maru in more practical terms, such as the local firefighter who commented, 'I have no idea how they are going to get that thing out of here' (Herzkovits 2011).

In the summer of 2013, around 70% of local residents voted to scrap the trawler. A ship recycler from nearby Fukushima Prefecture was enlisted to undertake the immense dismantling job. The work of dismantling the ship began with a few workers in reflective safety vests moving amongst the flower bouquets laid out in the summer heat, stepping over the concrete foundations of former homes. By the following year all signs of the Kyotoku-Maru's existence were gone.

La Seu

In 1851, a large earthquake shook the Mediterranean island of Mallorca. Damaged in the disaster was the massive Catedral de Santa María de Palma de Mallorca. Constructed upon an existing Muslim Mosque beginning in 1229, the cathedral – also known as La Seu – would become one of the largest cathedrals in the world and an exemplar of Gothic architecture. Architectural historians have separated the construction of the cathedral into five phases with the two phases following the earthquake of 1851 as phases of reconstruction, stabilisation, and constant intervention (Elyamani & Roca 2018).

One of these interventions would begin in 1901 when legendary Catalan architect Antoni Gaudi would be called in to work on the La Seu. Even though this work began fifty years after the earthquake, it is frequently attributed in literature promoting the city as a direct response to the earthquake damage. In reality, the cathedral had been undergoing various structural failures for decades leading up to the earthquake, in particular to its western façade. The Palma Earthquake facilitated a continuation of the evolution of the Catedral de Santa María de Palma de Mallorca from a Muslim holy site, to a Catholic church, an exemplar of the Gothic style, to a showcase for Catalan Gothic (Beddall 1975).

The Great Fire of London

After the Great Fire destroyed large swaths of London in 1666 – launched in the middle of the night from a bakery in Pudding Lane – a massive rebuilding effort was needed. Astronomer, mathematician, and Renaissance man Cristopher Wren was one of the experts called on to take charge of ambitious plans to put the city back to rights. While his ambitious plans for a new city plan were rejected by Charles II, Wren was tasked with rebuilding parish churches as well as St. Paul's Cathedral. Through this, Wren would both preserve and shift the architectural heritage of England by bringing a new style of Baroque construction. This style – which would later be turned Wren Baroque or English Baroque – became the symbol of a new and rebuilt London through the dominance of St. Paul's on the city's skyline. It also became a larger symbol of a new England, being recapitulated in places such as Liverpool (Sharples 2004), Oxford (Historic England 2004), and Derbyshire (UNESCO 1987).

Why Cultural Heritage Matters

I have started this chapter with these disparate examples to illustrate the broadness and depth of the intersection of cultural heritage and recovery. These vignettes convey to us very different stories about remarkably different places across a large stretch of time. These are only a few such possible examples. What each of these vignettes shares is a twofold connection of cultural heritage and disaster. In this chapter, I will discuss why cultural heritage is critical to both understanding and facilitating post-disaster recovery.

I will not spend too much time sorting through the debates about what culture and heritage mean – these have already been covered in other chapters included in this volume. Tylor's definition of culture as 'that complex whole which includes knowledge, belief, art, morals, law, custom and any other capabilities and habits acquired by man as a member of society' (2017, 1). Tylor originally wrote this in 1881 as an anthropologist, so his definition should be taken with all of the usual caveats, but it has been serviceable over the past century.

Culture engages in a complex interplay with habitus. Defined by Pierre Bourdieu, an influential 20th-century social theorist, habitus is all of the things that humans learn to be interested in, to prefer, or to reject, in some ways their tastes, but more solid and spread amongst a population in ways that are mutually intelligible and enduring (Edgerton & Roberts 2014).

To put that more succinctly, culture is a means by which groups of people can understand each other as parts of an overlapping group. This is facilitated by their personal tastes and behaviours that confirm their memberships of this group. This does not mean that every person within a group is strikingly similar. It also does not mean that one person can only belong to one group. What it does mean is that individual people do not have to spend their daily lives explaining everything that they do and are. It also means that they share activities, objects, and beliefs that work to reaffirm their belonging.

When a disaster occurs, these bonds can be broken. A mutual understanding amongst people can begin to fade. The structures, literal and metaphorical, that have held the society together might have been destroyed. Part of the task of post-disaster recovery is to maintain these bonds, again, both physical and as they exist in how people perceive and relate to the world. In a post-disaster environment cultural heritage can an asset to the recovery process, or a stumbling block.

Examples of cultural heritage and disaster recovery occur in a broad range of contexts. People throughout human history have experienced disasters and reacted to them. Likewise, culture and heritage are not contemporary features of society. We know that the people of Kyoto – far back when it was still known as Heian Kyo – created and maintained the Gion Festival to recover from a massive epidemic as well as to ward off future outbreaks. This cultural practice became not just a ritual of recovery and prevention, but the heart of the annual cultural calendar in the thriving capitol city (Dougill 2006). In a similar fashion, we know that a vibrant theatre scene emerged in the late 1960s from the uprisings in the Watts area of Los Angeles (Davis & Wiener 2021). The legendary Wattstax concert was held as a direct commemoration of the 1965 uprising and as a reconfirmation of local cultural heritage. Individual members of the community who are the subject of this reconstruction process are the bearers of their culture. They stand firmly inside everyday life as parties to, and actors on their surroundings.

Cultural heritage sites are the product of human beings making the built environment and through that process assigning meaning to it. What is important to understand is that both the process of building and the assigning of meaning – as well as how that grows and changes – does not exist outside of the society. In other words, the way that the people who build and occupy the site understand themselves, and their culture, and their place in the world is all wrapped up in the site itself. If we don't make an effort at understanding that, we don't completely understand what we are valuing and what we are working to recover.

Cultural Heritage and Disasters

> But it is wrong to assume, as many people did, that the pattern would disappear as soon as the circumstances changed, like something released from a constraining mould. It had a force of its own.
> – Kai T. Erikson *A New Species of Trouble* (1994, 41)

If we look at cases of recovery from around the world, we can often find examples that failed to understand the role of cultural heritage. The majority of worldwide construction involves neither architects nor planners, much less seeks the advice of historical cultural heritage professionals. These professions are often not a luxury that poverty permits. It is no wonder then, under these circumstances, that aid groups and affected governments opt for results based on more apparent measurable goals, such as speed of construction and numbers housed over complex thinking about the use of space. The pressure placed on the people and institutions tasked with undertaking post-disaster recovery is immense and unenviable. Nevertheless, if the labour of reconstruction is ineffective in the long term, then the process's overall success should be called into question. In the post-disaster settlement of Nueva Choluteca in Honduras, 50% of post-disaster homes that were built by non-governmental organisations were abandoned by the people they were constructed for (similar situation is described in Chapter 14). This happened because existing communities were broken up and removed from their normal cultural setting (Lizarralde & Boucher 2004; Barrios 2014).

The literature on post-disaster reconstruction is unfortunately rife with cases like Nueva Choluteca. After the 1995 Kobe earthquake struck, victims were routed into temporary housing that was assigned by lottery instead of allocated by neighbourhood. Many elderly people died what are termed 'solitary deaths' as no one was around to check up on them and their bodies were discovered long after death had occurred (Edgington 2010). After the 2004 Indian Ocean Earthquake and Tsunami, the Indonesian Development Planning Agency informed the residents of Banda Aceh that they were constructing a new city away from the sea and that all new construction within one and a half miles of the coast was banned. The new, safer housing that had been constructed at a tsunami proof distance from the ocean was quickly rejected and the coastal communities relocated themselves to seaside areas where they could continue their traditional fishing lifestyles. The very same disaster brought to light the lesson of Sri Lankan fishing communities who abandoned their new post-tsunami homes for seaside shacks where they could care for their nets, maintaining their new housing, which lacked the deep porches to shade them from the summer sun, to use as dowries or to rent to other families (Aquilino 2011). These are but a few examples amongst many. These situations must serve as a reminder that even recovery efforts involving trained professionals with well thought out goals can miss the mark.

While each new disaster brings more experience and shared knowledge, problems faced in the past continue to carry themselves into the future. Of course, each situation needs to be viewed as existing in part of a process, not as taking place only within a moment. It could very well be that these disaster-affected communities are in a state of transformation and are discovering new ways to deal with a changed setting. It might be that outside observers are misinterpreting a use previously not contemplated as a misuse or an ineffective use. However, without making an effort to understand how the local built environment works in conjunction with the local culture, it will be difficult to judge whether outcomes are positive and efficient expenditures of resources or are negative elements that can be avoided in future recovery efforts.

Scholars of disaster have also shown us that disasters can act as windows-exposing how our societies function (Erikson 1976, 1994; Tierney 2019). Everyday life often functions without us questioning how our society works, or without us thinking about or questioning why things are the way they are and why we do the things we do. But often – during and after – a disaster we have the opportunity to examine and question the societal process: Why do we do things this way? Is this the best way to do it? Is it fair? Did we even notice this is how things work? The occasion of a disaster allows us to see this exposed society. While none of these concepts are strictly speaking new as such, we have moved towards codifying these terms in evolving iterations of best practices.

In this same vein, it must be considered that the field of cultural heritage is not without its own pitfalls and weak points. To avoid addressing them head-on would render cultural heritage professionals as less than useful participants in disaster recovery and rebuilding. Antoinette J. Lee noted in *A Richer Heritage* that the field 'carries an air of elitism' (Stipe 2003, 392). This presumption might be a carryover from an earlier era when cultural heritage battles formed around monumental architecture or other sites treasured by those with enough free time to rally in support. As Dolores Hayden describes it, 'the focus on great buildings dies hard' (Hayden

1995, 53). In some ways, we are still trapped within the dynamic of sociologist Herbert J. Gans and architectural critic Ada Louise Huxtable talking past each other in their 1975 series of debates in *The New York Times* established a paradigm that sticks with us today (Giovannini 1995). A framework in which Gans believed that architecture encompassed the totality of the built environment and Huxtable thought of architecture as the construction of noble repositories of our culture echoes in present-day dialogues on the subject (Hayden 1995). Culture, in Huxtable's formulation, meant high culture and architecture meant well-financed, well-planned buildings. In some respects, the field is still in the process of moving out of that rather large shadow.

These concerns are, however, an essential element of the working process of those involved in the field of cultural heritage. In particular, they are the part of a way of thinking within the larger cultural heritage movement that argues for casting a wider net around what are important elements of the public memory. This section of the cultural heritage movement seeks to expand the scope of cultural memory and include elements outside of the traditional areas of architectural cultural heritage professionals, such as working class neighbourhoods and vernacular structures. This work has been described as 'politically conscious' as well as 'time-consuming, interdisciplinary, and politically controversial but not particularly expensive' (Hayden 1995, 11, 246). The first four factors here could just as easily describe the task of disaster reconstruction. Regrettably, post-disaster recovery can never be characterised as anything akin to a bargain. It is the inherently interdisciplinary nature of both the forward-looking factions within cultural heritage and disaster recovery that calls for collaboration between the fields.

Cultural heritage exists beyond human lifespans. It is a way of transmitting culture to future generations, of forming individual habitus. Across diverse cultures, many buildings are themselves narratives of public memory, the Mayan pyramids, for example, or the monuments in Augustan Rome. The Buddhist caves of the Silk Road or the doctrinal scriptural programs for Gothic cathedrals are a part of this process, as are the monuments to George Washington that began cropping up across America during the Great Depression. These forms of monumental architecture tell stories that enshrine public memory. In 'Patrimony and Cultural Identity' de Lourdes Luz and Viera dos Santos (2004, 99) frame the dynamic well, 'Architecture represents the history, tradition, and culture of a specific community. By protecting the cultural patrimony, we are contributing to the rescue and consolidation of the community's social identity in its historical evolution'.

Miller and Rivera (2008, 2) bring us back to the linkage of cultural heritage and disasters in the aftermath of Hurricane Katrina in New Orleans stating that 'the constitution of the physical, cultural, economic and political landscapes serve as the basis for understanding the context in which people connect physically, socially, and emotionally to their ecological surroundings'. Architectural historian Carel Bertram says that 'memory is a malleable carrier, one that allows objects to be remembered differently according to the needs of specific groups in specific situations or times; thus it is memory that allows places or objects to take on emblematic value or to become the centerpiece of a group's collective identity' (2004, 165)While the object or structure has the special ability of being able to move through time in a relatively recognisable state, unlike human beings, the memory and the meaning that it represents is subject to change and redefinition.

In *The Same Ax, Twice,* Howard Mansfield states, 'A tool has a double life. It exists in the physical sense, all metal and wood, and it lives in the heart and the mind. Without these two lives, the tool dies' (2001, 4). His point is well taken, however, the reaffirmation of missing places expressed through second line parades points to an afterlife for space, a shelf in the memory of a community reserved for preserving something meaningful. It is worth recalling that within the African-American community, as well as other marginalised communities around the world, preserving their own public memory without official sanction is a long-standing tradition. Storytelling and the naming of prominent buildings after important leaders in their community were means of perpetuating cultural memory. Cedar Hill, the home of abolitionist Fredrick Douglass, didn't leap directly from his ownership into the care of the National Park Service; members of the community looked after the home for years before receiving official recognition of its importance.

Thoughtful consideration of processions such as second line parades and shrine festivals can assist people from outside the community in understanding the larger social fabric of a neighbourhood or community. *The Power of Place,* written in an effort to tell a broader social history of the City of Los Angeles, tells us that 'While a single, preserved historic place may trigger potent memories, networks of such places begin to reconnect social memory on an urban scale' (Hayden 1995, 78). Embodying one's cultural surround with meaning and using it to preserve memory is not indigenous to one culture, it is a worldwide, human experience. In their work documenting the plantation systems of Brazil, Maria de Lourdes Luz and Ana Lucia Viera Dos Santos observe that 'Without memory, change would create alienation and dissociation...the built environment provides a firm, supporting ground at times of rapid cultural change' (2004, 99) Lewis Mumford, writing in *The City in History,* contends that, 'Soon after one picks up man's trail in the earliest campfire or chipped-stone tool one finds evidence of interests and anxieties that have no animal counterpart ...' (1968, 6).

The residue of humanity is, whether calculated or not, a communication across generations through broad expanses of otherwise impassable time. It is the stability of the built environment that allows for this kind of collective memory. Even in its malleable state, human architecture is far less fragile than human existence, as are the rituals that take place within these structures' confines. The Gion Matsuri, a festival that transpires amidst the sweltering summer of downtown Kyoto, has been taking place every year for over a thousand years with very few interruptions. Originally a purification ritual pleading to the gods to end a devastating plague, the festival persists in modern Kyoto even though it would seem that this city, with its abundance of teaching hospitals, private clinics, and the support of the Japanese national health care system would be relatively plague free.

One means by which we can look at disasters and society is through the lens of culture. Writing of the devastating flood in Buffalo Creek, yet speaking to disasters in general, Kai Erikson wrote that 'To speak of culture is to speak of elements that help shape human behavior – the inhibitions that govern it from the inside, the rules that control it from the outside, the languages and philosophies that serve to edit a people's experience of life, the customs, and rituals that help define how one person should relate to another. To speak of culture is to speak of those forces that promote uniformity of thought and action' (Erikson 1976, 81).

Stuart Hall tells us that culture, or cultural identity can be the product of multiple, overlapping, collectively held ideas of a shared culture (1990). Both Hall and Erikson assert that culture is not just a confluence of historical circumstances and shared experiences and mores but that it also is the force that generates possible futures. Hall puts it this way: 'Cultural identity ... is a matter of "becoming" as well as "being". It belongs to the future as much as the past' (1990, 223). Erikson states that '... there are other forces at work in a culture too. Cultural forms help determine how people will think and act and feel, but they also help determine what people will *imagine*' (1976, 81).

If we are to think about these characterisations in terms of disasters and cultural heritage, we can see that a specific heritage of an affected area is not simply a historical anomaly, a compelling context in which a disaster unfolds, it is also a hinge point that the future leans upon. What is possible, what can be imagined, what is feasible, as well as what is desirable, is rooted in cultural heritage. When understood in these terms, culture heritage becomes more than an artifact or an attraction, it becomes a crucial element in both understanding a disaster and facilitating recovery.

Anthropologist David Graeber cites Clyde Kluckhohn's assertation 'that what makes cultures different is not simply what they believe the world to be like, but what they feel one can justifiably demand from it' (2001, 6). This viewpoint helps to synthesise Hall and Erikson in trying to understand the role of cultural heritage and disasters. In terms of what we actually look at as being cultural heritage, we tend to gravitate towards unique folkways, or remarkable festivals, remarkable architecture, or revealing vernacular practices. These are, of course, critically important and crucial to maintain. However, if we are to contemplate cultural heritage and disasters from a theoretical perspective, it is helpful to dig into what is actually meant by these terms. As Graeber shows us in Kluckhorn's formulation culture is the medium in which overlapping understandings and hopes are suspended. Culture is also the filter through which desires and beliefs are sifted. Because of this, how a hazard, natural or otherwise, affects a specific area is largely contingent of the specific culture or cultures which are prevalent in the region.

International Frameworks for Cultural Heritage and Disasters

Coming out of the Second World War, there was an international recognition that the vast scale of the damage done throughout multiple countries could not be simply reconstructed. Formulated at Dumbarton Oaks in 1944 and formalised in San Francisco in 1945, the United Nations was born out of a desire to bring a certain form of stability to the world. In 1946 in London, a committee generated by the formation of the United Nations ratified the constitution of the United Nations Educational, Scientific and Cultural Organization (UNESCO).

In the decade following the 3.11 Triple Disaster, there has been international recognition that 'non-structural measure' should be a part of disaster recovery and mitigation. In *Learning From Megadisasters,* produced by the World Bank following 3.11, what these non-structural measures are is not spelled out fully, but their use is called for (Ranghieri 2014). We can infer, however, that rituals, festivals, vernacular practices, and folkways would fall into this category. Many organisations beside the World Bank have come to understand that strictly relying on civil engineering and large-scale infrastructure projects is not the totality of disaster mitigation or recovery. Bringing back damaged areas and protecting them against future calamities necessarily involves taking into account the cultural heritage present in the area. The Sendai Framework for Disaster Risk Reduction, enshrined in 2015 in the region struck by the 3.11 Triple Disaster, specifies that cultural resilience is a necessary component of DRR (see Chapter 3). In fact, the protection of cultural assets is called for in the Guiding Principles of the document.

In concrete terms, the Sendai Framework calls for participating states and local actors to assess and inventory their cultural assets. This is helpful after a disaster in understanding the scale of the disaster itself. It is also likely that through the assessment and inventory process a better understanding of the existing cultural heritage can be obtained. This is an important step forward in international frameworks for DRR. This understanding of the importance of cultural heritage has come out of decades of study and activism over the last several decades. It is no small victory to have these ideas made official. That the success of mitigation and rebuilding efforts is tied up in the material objects, buildings, building practices, rituals, festivals, folkways, and vernacular lives of people is an important realisation. It can also be that making culture 'official' or placing it in an inventory brings along with it issues of power and exclusion that can hamper or snuff out vibrant, yet deviant, or transitory cultures.

Conclusion

Cultural heritage is a complex, malleable, and evolving aspect of our societies. It can be difficult to pin down, and a challenge to maintain. And yet it is cultural heritage that drives our societies forward and provides them with meaning. Cultural heritage is anchored in the past but not anchored to it. Generations in the past shaped the heritage we come into contact with today, and generations in the future will work from this and transform their worlds. In the present, our mission is to recognise cultural heritage, to accept its complexities and contradictions, and to ensure that the diverse streams of human society in all its different iterations and aspects flow ever forward. Disasters present an opportunity – an opportunity for cultural practices to cease, an opportunity to continue, or an opportunity to transform. We cannot predict how cultural heritage will grow, change, and transform, but we can grasp that it is important to society, and that it is possible – with some effort – to play a role in its continuation.

We have seen numerous examples of it changing to suit human needs and existing conditions. Cultural heritage is also fragile. It can be disrupted, broken, and fractured. Disasters are a process. It isn't that disasters jump in from nowhere and unexpectedly wreak havoc on existing cultural practices. Oftentimes the seeds of disaster and the practices of heritage work alongside one another. That being said, whether the disaster expresses itself as through a gradual onset or an abrupt interruption, the existing heritage can be disrupted, put under pressure, or pushed into non-existence. On the other hand, cultural heritage can be utilised in post-disaster recovery. Heritage sites, folkways, and festivals can work to resecure the bonds of community. Both tangible and intangible cultural heritage can operate as a sign-post that an area is back, that it is functioning, that it has survived. Researchers, practitioners, and anyone else engaging with post-disaster recovery would be well served in understanding the local contexts of cultural heritage before a disaster occurs, and be able to utilise this knowledge to expedite a robust recovery.

Key Learning Points

- The disruption of cultural heritage presents a challenge for recovery.
- Cultural heritage can facilitate post-disaster recovery.
- Cultural heritage offers a malleable tool for reengaging people in their community.
- Knowledge of local cultural heritage prior to a disaster is critical to using cultural heritage for effective recovery.

References

Aquilino, M.J. 2011. *Beyond Shelter: Architecture and Human Dignity*. New York: Metropolis Books : Available through D.A.P./Distributed Art Publishers.

Barrios, R.E. 2014. "'Here, I'm not at Ease': Anthropological perspectives on community resilience." *Disasters* 38(2): pp. 329–50.

BBC News. 2013. "Japanese Town to Scrap Marooned 'Tsunami Boat.'" *BBC News*, August 13.

Beddall, T.G. 1975. "Gaudí and the Catalan Gothic." *Journal of the Society of Architectural Historians* 34(1): pp. 48–59. doi: 10.2307/988956.

Bertram, C. 2004. "Imagining the Turkish house." Pp. 165–90 in *Memory and Architecture*, edited by E. Bastéa, Albuquerque: University of New Mexico Press.

Davis, M., and J. Wiener 2021. *Set the Night on Fire: L.A. in the Sixties*. Verso Books.

Dougill, J. 2006. *Kyoto: A Cultural History*. 1st edition. Oxford University Press.

Edgerton, J.D., and L.W. Roberts. 2014. "Cultural capital or habitus? Bourdieu and beyond in the explanation of enduring educational inequality." *Theory and Research in Education* 12(2): pp. 193–220. doi: 10.1177/1477878514530231.

Edgington, D.W. 2010. *Reconstructing Kobe: The Geography of Crisis and Opportunity*. Vancouver: UBC Press.

Elyamani, A., and P. Roca. 2018. "One Century of Studies for the Preservation of One of the Largest Cathedrals Worldwide: A Review." doi: 10.5281/ZENODO.1214557.

Erikson, K. 1976. *Everything in Its Path: Destruction of Community in the Buffalo Creek Flood*. New York: Simon and Schuster.

Erikson, K. 1994. *A New Species of Trouble: The Human Experience of Modern Disasters*. 1st edition. New York: W.W. Norton & Co.

Giovannini, J. 1995. "No preservation without representation!" *The New York Times*, September 3.

Graeber, D. 2001. *Toward An Anthropological Theory of Value: The False Coin of Our Own Dreams*. 2001st edition. New York: Palgrave Macmillan.

Hall, S. 1990. "Cultural identity and diaspora." Pp. 222–237 in *Identity: Community, Culture, Difference*, edited by Jonathan Rutherford, London: Lawrence & Wishart.

Hayden, D. 1995. *The Power of Place: Urban Landscapes as Public History*. Cambridge, Mass: MIT Press.

Herzkovits, J. 2011. "Ghost Ship Haunts Tsunami-Hit Japanese City." *Reuters*, March 22.

Historic England. 2004. "CHATSWORTH HOUSE, Chatsworth – 1373871 | Historic England." Retrieved 27 April 2022, https://historicengland.org.uk/listing/the-list/list-entry/1373871.

Lizarralde, G., and M. Boucher. 2004. 'Learning from post-disaster reconstruction for predisaster planning." in *Second International Conference on Post-Disaster Reconstruction: Planning for Reconstruction*.

Mansfield, H. 2001. *Same Ax, twice, the: Restoration and renewal in a throwaway age*. First Printing Edition. Hanover, NH; Wantage: University Press of New England.

Maria de, L.L., and A.L.V.D. Santos 2004. "Patrimony and cultural identity: The coffee plantation system – Paraíba Valley, Rio de Janeiro, Brazil." Pp. 99–115 in *Memory and Architecture*, edited by E. Bastéa, Albuquerque: University of New Mexico Press.

Miller, D.M.S., and J.D. Rivera 2008. *Hurricane Katrina and the Redefinition of Landscape*. Lanham: Lexington Books.

Mumford, L. 1968. *The City in History: Its Origins, Its Transformations, and Its Prospects*. Mariner Books.

Ranghieri, F. 2014. "Learning from megadisasters: Lessons learnt from the great East Japan earthquake and tsunami." *Economics and Policy of Energy and the Environment*.

Sharples, J. 2004. *By Joseph Sharples Liverpool: Pevsner City Guide*. Yale University Press.

Stipe, R.E. 2003. *A Richer Heritage Historic Preservation in the Twenty-First Century*. Chapel Hill: University of North Carolina Press.

Tierney, K. 2019. *Disasters: A Sociological Approach*. 1 edition. Cambridge, UK: Polity.

Tylor, E.B. 2017. *Primitive Culture, Vol. 1 of 2: Researches Into the Development of Mythology, Philosophy, Religion, Art, and Custom*. London: Forgotten Books.

UNESCO. 1987. "Blenheim Palace – UNESCO World Heritage Centre." Retrieved 27 April 2022, https://whc.unesco.org/en/list/425.

15 The Politics of Post-Disaster Reconstruction of Heritage

Planning and Funding Mechanisms in Bhaktapur, Nepal, in the Aftermath of the 2015 Gorkha Earthquake

Vanicka Arora

Introduction: Heritage in the Aftermath of the 2015 Gorkha Earthquake

The Gorkha Earthquake that devastated Nepal in 2015 comprised a sequence of seismic events (Gautam, 2017), the first of which was the catastrophic 7.8 M_w (8.1 M_s) earthquake that struck on 25th April. This was followed by thousands of aftershocks, but most significantly, the two aftershocks of 26th April (6.9 M_w) and 12th May (7.3 M_w), which compounded the damage of the first quake and considerably impeded ongoing relief operations. Officially, the death toll recorded from the earthquake stands at 8,790 with over 22,300 seriously injured. However, the estimated impact on lives, livelihoods, and habitation extended to more than 8 million people, or approximately one-third of the nation's population (National Planning Commission, Government of Nepal, 2015). Even these figures are underestimates, given the difficulties in recording the precise scope and nature of the damage, some of which unfolded over several months.

In the days that followed the earthquake, thousands of images, and videos of collapsed and damaged built heritage sites in Nepal, predominantly from the Kathmandu Valley, circulated globally, along with initial reports from the field. One of the key media narratives describing the impact of the 2015 Gorkha Earthquake focused on its catastrophic impact on the country's heritage, with several hundred newspaper articles dedicated to mourning the loss not only of lives and livelihoods but of multiple internationally recognised and beloved heritage destinations that contribute to Nepal's national identity.[1] Several international and national heritage networks coalesced to crowdsource data recording the locations and extent of damage to heritage places (Allain et al., 2015). One of the first responses was initiated by the UNESCO Kathmandu office, which focused on the seven monument zones of the UNESCO World Heritage Site of the Kathmandu Valley (Weise et al., 2017), namely the Durbar Squares in the historic capital cities of Kathmandu, Patan (Lalitpur) and Bhaktapur, as well as the religious ensembles of Swayambhu, Bauddhanath, Pashupati, and Changu Narayan (UNESCO World Heritage Centre, 1979). UNESCO, with support from professional heritage networks and organisations, such as ICCROM and ICOMOS, sent in an independent post-disaster assessment team of experts, which focused on assessing the extent of damage to the World Heritage Sites in Kathmandu Valley, including the three Durbar Squares, and followed up the initial assessment with archaeological explorations carried out by several international institutions and universities (Coningham et al., 2016; DoA Nepal, 2016; UNESCO World Heritage Centre, 2016). It quickly became evident that heritage would emerge as a critical area for intervention in the national plan for post-disaster recovery.

However, the initial frenzy of post-disaster documentation and assessment not only for heritage but across multiple sectors, soon gave way to an extended period of stasis in recovery processes across the country. In part, this could be attributed to an increasingly fraught

DOI: 10.4324/9781003293019-19

political situation within Nepal following the adoption of a new and controversial Constitution in September 2015, as well as the lack of pre-existing national legislation and bureaucratic infrastructure for managing post-disaster recovery at the scale required by an earthquake of this magnitude. Finally, the Nepal Reconstruction Authority (NRA) was constituted in December 2015, as a national body for post-disaster recovery and reconstruction, under the *Act Related to the Reconstruction of Earthquake Affected Structures, 2015*. This was accompanied by a multi-sectoral Post-Disaster Needs Assessment (PDNA) undertaken by the Nepal Government with assistance from the United Nations Development Programme (UNDP), the Japan International Cooperation Agency (JICA), World Bank, Asian Development Bank (ADB), and the European Union, and followed up by a *Post Disaster Recovery Framework* (PDRF) issued by the Nepal Government in 2016.

According to the PDNA, the damage to 'tangible heritage' amounted to NPR 16.9 billion (USD 169 million) out of total estimated losses amounting to NPR 706 billion (USD 7 billion) (National Planning Commission Government of Nepal, 2015). Notably, heritage, specifically built heritage, was identified as an individual sector for reconstruction and recovery in both the PDNA as well as in the PDRF, and specific mechanisms were constituted to oversee the assessment and reconstruction of built heritage sites across the country.

An official list of destroyed and damaged heritage sites was compiled by the Nepal Ministry of Culture, Tourism and Civil Aviation, the Pashupati Area Development Trust, and the Buddhist Philosophy Promotion and Monastery Development Committee, initially identifying over 2,900 buildings that needed reconstruction or significant repairs. Of these structures, 745 were nationally listed monuments, out of which 193 had collapsed completely, while 95 had partially collapsed. The majority of the remainder were monasteries, historic homes and settlements, and religious and community buildings, all of which were included in the official list of destroyed or damaged structures, even though they were not necessarily listed as national monuments. Since then, overall reconstruction progress following the earthquake and its aftershocks has been slow and chaotic, complicated by recurring disasters, including seasonal floods (each year since 2015) and forest fires (2017, 2019, 2020), and the Covid-19 pandemic (2020 onwards). Nevertheless, by the end of 2021, the NRA's tenure was officially completed. Any remaining reconstruction work for heritage sites was handed to the Ministry of Culture, Tourism and Civil Aviation, which oversees the Department of Archaeology (DoA) in Nepal, the national body responsible for official sites of heritage. According to the NRA, by 2021, 586 sites had been reconstructed, while 195 were currently under reconstruction out of a final target of 920 (Nepal Reconstruction Authority, Government of Nepal, 2021).

The trajectory of recovery in Nepal following the 2015 Gorkha Earthquake, particularly regarding heritage reconstruction, reflects some critical shifts in international policy that preceded or followed the earthquake (Arora, 2022). The *Sendai Framework for Disaster Risk Reduction, 2015–2030* (referred to as the *Sendai Framework*) (UNISDR, 2015) had only recently been endorsed by the United Nations General Assembly when the earthquake struck Nepal and subsequently became one of the main policy frameworks that fed into Nepal's PDRF (Nepal Reconstruction Authority, Government of Nepal, 2016). The *Sendai Framework* has come to represent a paradigm shift in disaster risk reduction, with an explicit reference to culture made at several junctures in the document. Heritage – tangible and intangible, cultural and natural – became the subject of prominent sections of Nepal's recovery plans, particularly in relation to build heritage. This reflects not only an international shift in the way that heritage and culture are framed in Nepal's development policy but the critical role played by heritage in Nepal's tourism economy. However, even though international agendas and interests in heritage and culture within post-disaster reconstruction seem to be increasingly aligned (Chmutina et al., 2019), local aspirations, practices and political ecologies may often be at odds with international, or even national interests (Brosius & Michaels, 2020; Arora, 2022).

In the sections that follow, I will discuss cases in Bhaktapur along with a brief overview of reconstruction trajectories in Kathmandu and Patan, to illustrate the ways in which post-disaster reconstruction of heritage acts as a venue for the negotiation of political aspirations and agendas. At seven square kilometres, Bhaktapur is considerably smaller than Kathmandu, which is almost 50 square kilometres. The majority of Bhaktapur's inhabitants are Newar Hindus, who make up almost 80 percent of the city's population. This demographic has continually inhabited the city for several centuries and is credited with maintaining a great degree of continuity in its socio-cultural practices, community associations, festivals, and rituals. Bhaktapur was the centre of political power in the region between the twelfth and the eighteenth centuries during the reign of the Mallas (Whelpton, 2005) and thrived due to its location on the trading route that led to Tibet. Bhaktapur embodies complex interactions between different communities, ethnicities, and caste hierarchies within the city today, and the persisting tensions between city authorities, the central government, and international agencies.

I will describe the trajectories of reconstruction of heritage in Bhaktapur which stands in contrast to the rest of the Kathmandu Valley. By 2022, 124 heritage projects were completed in the city, most of which were funded and administered by the Bhaktapur Municipality.[2] Reconstruction of heritage in Bhaktapur has been repeatedly referenced in local and national media as an exemplar of local leadership, as independent from 'external influence', and for actively engaging the local community (Prajapati, 2018; Pant, 2020; Suji, 2020). However, it has also been critiqued for ignoring international standards and norms, and for effectively writing over the material built past in the city through reconstruction (UNESCO World Heritage Centre, 2017, 2021).

The sustained focus on post-disaster reconstruction of built heritage in Bhaktapur, as well as across Nepal, is illustrative of the convergence of disaster management frameworks and heritage conservation frameworks, as well as highlighting a series of unresolved tensions between the two. For instance, within the broad strategies for built heritage reconstruction, the *Sendai Framework's* emphasis on 'Build Back Better' as a central recovery principle has inevitably led to conflicts because reconstruction of heritage has simultaneously needed to respond to international conservation charters that endorse 'rebuilding the original' as the goal of reconstruction (Hutt, 2015; Shneiderman et al., 2020; Suji et al., 2020). Aspects of material authenticity, loss of value, historicity, and integrity continue to feature prominently in debates surrounding post-disaster reconstruction of built heritage. However, attempts to resolve or negotiate this difference can also be seen in recent policy. For instance, the ICOMOS Guidance on Post Trauma Recovery and Reconstruction of World Heritage Sites (ICOMOS, 2017) acknowledges and describes multiple forms of reconstruction, indicating a response not only to disaster policies but also to a broader context and understanding of the scope of reconstruction practices. These include 'modified reconstruction, partial reconstruction, reconstruction as a recurring process, reconstruction of newly revealed underlying historic layers, reconstruction as an opportunity to improve building or urban conditions, reconstruction as a critical element to maintenance of customary knowledge, practices, beliefs, or as an opportunity to sustain these or other intangible attributes'. The reconstruction of heritage in Nepal has directly influenced this shift in approach, though substantial points of divergence remain in the way reconstruction is framed by disaster risk reduction frameworks and heritage conservation frameworks. Despite this apparent conflict between Build Back Better and reconstructing 'as before', both heritage and disaster management policies in Nepal have enacted mechanisms of operation that reinforce each other, including the disbursement of international aid and expertise and the reliance on technocentric approaches to reconstruction.

The Politics of International Aid and Expertise

Most of the international focus of funding and expertise for built heritage following the 2015 Gorkha Earthquake has been on individual buildings within the seven monument zones of the UNESCO World Heritage Site of Kathmandu Valley (Weise et al., 2017). Nepal's history and politics have often been conflated with the history and politics of the Kathmandu Valley, given the economic significance of the latter, its strategic geo-political location, and its development as an urban centre (Whelpton, 2005). This has meant far greater support for the Kathmandu Valley in terms of both aid and expertise than for the rest of the country, particularly its rural settlements and remote towns. Within the Valley, heritage reconstruction has taken on distinct trajectories across its three major urban centres; Kathmandu, Patan, and Bhaktapur.

Kathmandu

Reconstruction in the political capital of Kathmandu has been critiqued repeatedly for delays due to tangled bureaucracies (UNESCO World Heritage Centre, 2017; Apil et al., 2019; Sharma et al., 2022), interference by foreign funding agencies (Weise et al., 2017), possible corruption, and a lack of sustained local participation (Bajracharya & Michaels, 2017; Weise, 2018). Several reconstruction projects undertaken in the Durbar Square in Kathmandu were stalled at multiple junctures due to disagreements over funding, lack of local consultation, and protracted debates over the appropriate technologies and materials to be used for funding. Others still (Shrestha, 2019; Rimal & Rajbhandari, 2021) pointed out that having multiple stakeholders across multiple sites had caused significant delays in reconstruction, and on multiple occasions, forced individual implementing agencies to delay their projects to accommodate other ongoing reconstruction projects in the vicinity, due to a lack of coordination amongst agencies.

The trajectories of how projects have been planned, managed, and executed in Kathmandu have been fractious, following diverse modalities of funding and implementation. Multiple other countries came forward to fund individual reconstruction projects in Hanumandhoka (Kathmandu's Durbar Square), including Japan, China, India, and the United States, acting out a soft geopolitics of international aid (Figure 15.1). Several media reports described the co-opting and adoption of monuments across the Durbar Square, as political acts rendered visible through display boards and public information material, erected at sites of reconstruction.

Multiple projects have been handled almost exclusively by international agencies such as the reconstruction of Basantpur Palace (Nau Tale), an iconic nine-storeyed palace structure in the square which has funded by the Chinese government and executed by the Chinese Academy of Cultural Heritage. The extensive repairs and reconstruction of two temples in the vicinity, namely Agamchchen and Shiva Temple, were funded by Japan International Cooperation Agency (JICA) and the projects overseen by a consultant appointed by the agency. Still, other projects were funded through UNESCO's World Heritage Programme. On the other hand, the reconstruction of some structures in Hanumandhoka has become exemplars of locally led heritage activism such as the Kasthamandap. The different mechanisms of funding and execution between these projects represent not only the entanglements of political interests but also how different sites and building can elicit different degrees of ownership and association with global institutions or local communities, respectively. As a gathering space built for the community and a building which is considered eponymous with the foundation of Kathmandu city, Kasthamandap represents the people of the capital, which led to local groups coalescing around its reconstruction, who fought hard to prioritise its reconstruction according to locally emplaced practices (Joshi et al., 2021). On the other hand, Basantpur Palace, a nine-storey royal palace valued for its ceremonial and historic value, attracted international attention due to its landmark

Politics of Post-Disaster Reconstruction of Heritage 215

Figure 15.1 (Top) Signage prominently displaying the heritage agencies erected across reconstruction sites in Hanumandhoka (Kathmandu's Durbar Square). (Bottom) Basantpur Durbar funded by Chinese government and right reconstruction of Agamchen funded by JICA.

Source: Author, taken in April 2019.

status and as a visible sign of recovery. As the largest urban ensemble of the three, as well as being the political centre of Nepal, Kathmandu's heritage sites received the most international visibility and attention but simultaneously involved the highest number of stakeholder conflicts and associated delays. Heritage reconstruction became a proxy for developing international relations in the wake of the 2015 Gorkha Earthquake and a symbol for international cooperation. However, the long-term impact of these internationally funded initiatives on national politics is uncertain and replete with unresolved tensions.

Patan

In Patan (Lalitpur), heritage reconstruction has been far less contentious than in Kathmandu, with a greater degree of local participation and more transparent and streamlined processes in place. In part this can be attributed to the presence and engagement of the Kathmandu Valley Preservation Trust (KVPT), particularly in the case of Patan's Durbar Square, where most projects have been managed by the KVPT, at times with Japanese funding (Brosius & Michaels, 2020; Shrestha, 2021). KVPT was founded in 1991 and described itself as 'the only international private non-profit dedicated to safeguarding Nepal's architectural heritage' (KVPT, 2017). It has built a long-standing relationship with local heritage practitioners and activists even prior to the 2015 Gorkha Earthquake, and has been involved in extensive documentation, assessment, and physical conservation initiatives across the Valley, but most prominently in Patan. The organisation's involvement in post-disaster reconstruction was therefore built on decades of active involvement in heritage practice within the city and with its people, which meant a more cohesive and emplaced approach to identifying, prioritising, and executing individual reconstruction work, even though many of its experts are internationally trained. Here too however, the focus on international reconstruction efforts was, as Brosius and Michaels note, on buildings within the boundaries of the UNESCO World Heritage Site, which received the bulk of foreign funding and expertise, while more locally significant places were slower to receive funding and were in general managed through city level or neighbourhood level initiatives. In the years that followed the 2015 Gorkha Earthquake, both Patan and Kathmandu have been supported extensively, through international aid and expertise, for the reconstruction of their built heritage, particularly for buildings within both their Durbar Squares which lie inside the boundaries of the UNESCO World Heritage Site.

Bhaktapur

Reconstruction of heritage in Bhaktapur has repeatedly been referenced in local and national media as an exemplar of local leadership (Prajapati, 2018; Suji, 2020), as independent from 'external influence' (Suji et al., 2020). Unlike its neighbours Kathmandu and Patan, Bhaktapur's reconstruction following the 2015 Gorkha Earthquake has been funded predominantly via the municipality, and through local donations in cash and kind. In fact, in 2018, the Bhaktapur municipality became famous for 'refusing' a third out of a total outlay of 30 million Euros in the form of financial assistance pledged at a donor conference by the German Development Bank (KfW) in 2015. According to Mayor Sunil Prajapati, 'there were many points of the agreement that we [Bhaktapur municipality] did not agree with' (Bhattarai, 2018), including stipulations regarding a tendering process that would be open to global competition, oversight by international consultants, specifically approved by KfW, as well as the potential use of modern materials and technologies (Shrestha, 2019). Local newspapers described the lack of agreement between as a 'clash of cultures' and characterised the municipality as 'fiercely independent' (Bhattarai, 2018; Prajapati, 2018). Within Bhaktapur's residents, this action has come to be referenced with a sense of civic pride. This exchange was seen as reinforcing Bhaktapur's identity as a self-reliant city, free not only from international influence but also from intervention from within Nepal. Furthermore, the rejection of international funding meant that fewer actors and bureaucratic processes have been involved in the disbursement of funds, and the independent management of individual reconstruction projects has meant shorter implementation timelines.

The rejection of international funding in Bhaktapur follows from the ongoing impact of past interventions in the city. Two of these, in particular, have had lasting influence, namely the decade long Bhaktapur Development Project (BDP) and a series of restoration projects undertaken by UNESCO and UNDP. An urban regeneration intervention funded by the Federal Government of Germany, the BDP was planned and executed in three phases from 1974 to 1983 and remains a flagship of urban regeneration initiatives in South Asia. However, despite its lasting influence on heritage management and urban planning in Nepal, it has also been criticised for its top-down approaches to Bhaktapur's built fabric, livelihoods, and building systems which would often ignore pre-existing modes of construction, repair, and care (Silva, 2015). Nevertheless, both sets of interventions, the BDP and the UNESCO-UNDP projects, have set standards and systems for heritage management that continue to reflect in contemporary bureaucracy in the city. Simultaneously, their legacy is viewed with a degree of distrust amongst local communities, influencing local attitudes to 'foreign' influence whether it be in the form of funding or expertise. The BDP nevertheless remains one of the central reasons why heritage has become so inextricably linked with Bhaktapur's identity and economy and why heritage has featured so centrally in post-disaster recovery narratives of the city (Arora, 2022).

Since 2018, once reconstruction processes of heritage were finally systematised in Bhaktapur, local satisfaction with the progress and approach to reconstruction of heritage has been high. International heritage networks and organisations have, however, expressed dissatisfaction over the mechanics of the process, the lack of rigorous documentation, and adherence, or lack thereof, to notions of authenticity and historicity. Heritage experts, both within Nepal and internationally (UNESCO World Heritage Centre, 2021), have voiced their concerns about the speed, materials, and technologies being used in Bhaktapur's reconstruction and about a lack of adherence to UNESCO's *World Heritage Convention* and other international heritage charters (Box 15.1). Reconstruction of homes and non-listed community buildings have had to adhere to

---

Box 15.1 Local planning and funding mechanisms

The local approach to reconstruction of heritage in Bhaktapur has received extensive media attention in the wake of the 2015 Gorkha Earthquake, highlighting the involvement of communities in the active decision-making processes for individual projects and the maintenance of transparency in the way budgets are utilised. Bhaktapur's municipality has made it a rallying point in recent years that most of its individual reconstruction projects for heritage sites have been executed with less resources than estimated, in part because of a trend of locals volunteering labour and donating materials. Each project is managed by a locally constituted user committee (also known as local consumer committee) which consists between 10 and 12 member representatives from the vicinity who agree to oversee the disbursement of funds, the execution of reconstruction activities and the commissioning of labour and materials for individual projects (Figure 15.2).

Bhaktapur, and indeed across Nepal, particularly within Newar communities have had traditional systems for centuries that were in place for the upkeep and maintenance of

218 *Vanicka Arora*

Figure 15.2 Ongoing reconstruction of the Vatsala Durga Temple in Bhaktapur in 2019.

temples and community places. These systems were attributed to the *guthi,* a patrilineal, locally emplaced community, where members of the community would volunteer and be charged with the responsibility to perform various social and cultural functions, bolstered by varying degrees of participation by the wider community. Three primary types of *guthi* are those centred around a particular deity or festivity, *guthis* responsible for birth and funerary rites, and *guthi*s linked to neighbourhoods or locality-based activities, including music and crafts (Toffin, 2005). *Guthis* were endowed with land and monetary grants as a form of religious merit-making, but also as philanthropic endeavours, by the privileged castes and classes of Newar society and were therefore well-funded institutions that could undertake large-scale repairs, upgrades, and post-disaster reconstruction of places, that is, prior to their nationalisation in 1964. The Guthi Sansthan marked the nationalisation of the *guthi*s and, many scholars have argued, their eventual decline, as localised practices gave way to national institutions (Chapagain, 2008; Gutschow, 2017; Weise, 2018).

As a way to counter the rapidly declining efficacy and sustainability of *guthis* after their nationalisation during the Shah-Rana regime, and to remedy the competitive contract system that had been instituted in Bhaktapur leading to dissatisfaction in the way projects were managed the BDP in 1979 instituted a new form of locally-embedded organisational system. Under the Local Development Committees (LDC), as they were initially named, residents were recruited and facilitated to organise themselves into representative groups for individual projects and were responsible for delegating resources and reviewing project progress, effectively taking partial ownership of the ongoing interventions under the third phase of the BDP.

While the BDP and other international heritage-based interventions effectively introduced European principles of conservation to Bhaktapur, which included a focus on written documentation, drawing, and discussions of heritage value, they also brought with them several systems of project management and community organising. And though LDCs, now referred to as local consumer committees or local user committees, have gone through several iterations and changes since the BDP, they have resonated strongly with the political and social context of Bhaktapur, feeling almost contiguous with the traditional *guthi* system of heritage management.

Following the earthquake, most reconstruction projects involving listed monuments and public buildings have been undertaken by the Bhaktapur municipality. The public tendering system, used for most nationally funded projects, where independent contractors make financial and technical bids for individual heritage reconstruction projects, is uniformly shunned by the municipality. Instead, it relies on a system of local project management and construction oversight via the constitution of local consumer committees. Despite the initial critique of the local consumer committee model during the tenure of the BDP, which focused on how this system of heritage management superseded pre-existing management systems such as the *guthi*, this model has become increasingly popular in Bhaktapur. It has been lauded for encouraging local ownership and participation in the conservation and management of heritage, as well as, for creating a transparent system of managing funds and reducing corruption.

Following the earthquake, the Department of Archaeology (DoA), through the nationalised system of reconstruction of heritage, was responsible for 18 reconstruction projects within the city, executed using the standard contractor bidding system, while local consumer committees oversaw over 124 projects. Even so, both the national and local levels of governance monitor and evaluate each other.

Each reconstruction project that is located within the notified UNESCO World Heritage Site required approval from the DoA, but similarly, the municipality closely scrutinised the DoA's work. Both levels of governance are in constant negotiation with each other and also with international heritage institutions. Nevertheless, the municipality exerts the greatest amount of governmental control on heritage reconstruction as well as on urban development.

heritage-based codes and guidelines, though the pace of disbursal of funding and the overall implementation of these projects have been much slower than for listed heritage buildings (Arora, 2020; Suji et al., 2020).

Reconstruction of built heritage is only one avenue for the dissemination of international aid and relief in Nepal. Nevertheless, it is a significant arena for the enactment of international politics. Previously, aid and relief in Nepal were dominated by the West, taking on what many have argued is a neo-colonising influence and articulating new forms of international politics (*Disaster Studies Manifesto*, 2019). However, aid and relief politics have seen a shift as well, with non-Western aid actors playing increasingly influential roles in post-disaster response and recovery processes following the earthquake. Neighbouring countries like China and India, along with Japan, have invested heavily in the reconstruction of heritage to gain strategic influence politically (Daly & Feener, 2016). The politics of aid and assistance for Nepal's built heritage indicates not only pre-existing differentials in international geo-political structures but also a shift in priorities with respect to the position of heritage in the political economy of Nepal itself. The central role played by heritage in receiving international assistance highlights its growing significance as a sector in post-disaster recovery planning.

In terms of informing international debates on heritage reconstruction, the post-disaster landscape of Nepal has been extremely influential. Both the recently drafted *Culture in City Reconstruction and Recovery (CURE) Framework* (UNESCO & World Bank, 2018) and the *ICOMOS Guidance on Post Trauma Recovery and Reconstruction in World Heritage Sites* (ICOMOS, 2017) make explicit references to reconstruction of various heritage sites in Nepal, observing how notions of authenticity and historicity are challenged in practice, and acknowledging that the scope of reconstruction itself needs to be re-examined. The post-earthquake landscape of Nepal has come to represent the pivotal role played by built heritage in recovery processes and its contribution not only as economic asset and tourism infrastructure but also as a vital contributor to recovering national identity.

Conclusion

Whether it be locally, nationally, or globally, heritage has been a critical sector for post-disaster recovery in Nepal. However, recovery for heritage has been uneven, weighted significantly towards UNESCO World Heritage Sites and nationally listed monuments in the metropolitan region of Kathmandu Valley. Rural heritage, vernacular settlements, smaller temples, monasteries, and community heritage have received far less attention, both by international funding agencies as well as in the national plans for recovery. Unlisted heritage buildings and settlements with vast amounts of historic buildings have decayed following the earthquake, with large swathes replaced by standardised reinforced cement concrete houses with little or no sense of place. Within Bhaktapur, where the attention towards heritage has been consistent and there exist robust systems for reconstruction, traditional houses have nevertheless been replaced by Neo-Newar house types that only bear a superficial resemblance to the pre-existing housing stock in the city. The heritage by-laws of the city which have been used along with financial incentives following the 2015 Gorkha Earthquake fail to account for the rich diversity of built heritage that exists in the city in the form of historic houses and neighbourhoods.

It is important to understand the politics of heritage in the so-called Global South and how it can be leveraged to address larger geo-political agendas even following disasters. For instance, a great degree of emphasis is currently placed on heritage in the majority world being deliberately destroyed in active conflict zones, environmental risks such as climate change, urban floods, cyclones, and natural hazards such as earthquakes are framed as being increasingly dangerous to heritage sites as well. These trends illustrate not only the persistent view of heritage being a 'non-renewable' resource (Layton et al., 2001; Daly & Winter, 2012; Holtorf, 2014) but also that the heritage of the Global South is somehow at greater risk and in need of international intervention and assistance.

With respect to disaster risk discourses, Bankoff describes this view as the 'Western discourse on vulnerability' (Bankoff, 2001) and argues that it consistently reinforces existing patterns of neo-colonisation, where disasters have replaced 'tropicality' creating a 'moral imperative' for western intervention. For instance, in the wake of 2015 Gorkha Earthquake, there have been recurring discussions on placing the UNESCO World Heritage Sites identified in the Kathmandu Valley on the 'World Heritage in Danger' Lists, suggesting a continuing perception of heritage protection in Nepal as contingent upon international aid and expertise. Nonetheless, as Winter suggests, it is not a simple matter of describing the interventions in a West versus Non-west binary, but rather recognising that heritage practices and discourses within Asia are plural, nuanced, and often the outcome of complex negotiations at multiple scales (Winter, 2014). This recognition and acknowledgement of heritage as not just an economic resource or infrastructure or national symbol has much further to go to be assimilated in all scales of disaster recovery planning and discourse.

Key Learning Points

- Post-disaster reconstruction of built heritage is the venue for enacting political agendas and aspirations across local, national, and global scales of governance.
- Reconstruction of heritage is central not only to recovering livelihoods and physical infrastructures but also acts as a means of reclaiming collective memory and rebuilding identity.
- Heritage is not neutral or inactive, it is a process of valuation that involves power which is often substantially restructured during disasters.

Notes

1 Specific heritage sites received extensive media coverage following the Earthquake. For instance, Rajopadhyaya (2019) argues that Dharahara, a towering structure in Kathmandu received far more attention that the Kasthamandap which was an older structure intricately linked with the origins of Kathmandu as a capital, owing to the former's association with Nepal's national identity. Similarly, Apil et al. (2019) have observed that the UNESCO World Heritage Sites in the Kathmandu Valley received far greater attention than non-listed historic places in rural hinterland.
2 These numbers are regularly updated.

References

Allain, C., Copithorne, J., Eaton, J., Jigyasu, R., Selter, E., & Tandon, A. (2015). *Overview report of the Nepal cultural emergency crowdmap initiative*. Retrieved from https://www.icomos.org/images/DOCUMENTS/Secretariat/2015/Nepal/Nepal_Cultural_Emergency_Crowdmap_Initiative_Overview_ReportFinal.pdf

Apil, K., Sharma, K., & Pokharel, B. (2019). *Reconstruction of heritage structures in Nepal after 2015 Gorkha Nepal earthquake*. Paper presented at the Proceedings of the 12th Canadian Conference on Earthquake Engineering, Quebec, Canada.

Arora, V. (2020, 24/04/2020). Five years on from the earthquake in Bhaktapur, Nepal, heritage-led recovery is uniting community. *The Conversation*. Retrieved from https://theconversation.com/

five-years-on-from-the-earthquake-in-bhaktapur-nepal-heritage-led-recovery-is-uniting-community-136255

Arora, V. (2022). Reconstruction of heritage in Bhaktapur, Nepal: Examining tensions and negotiations between the 'local' and the 'global'. *Disaster Prevention and Management: An International Journal, 31*(1), pp. 41–50. doi:10.1108/DPM-03-2021-0093

Arora, V. (2022). Reconstructing Memory and Desire in Bhaktapur, Nepal. In: Linder, B. (eds) "Invisible Cities" and the Urban Imagination. Literary Urban Studies. Palgrave Macmillan, Cham. https://doi.org/10.1007/978-3-031-13048-9_16

Bajracharya, M., & Michaels, A. (2017). 'Religious' approaches to heritage restoration in post-earthquake Kathmandu. *Material Religion, 13*(3), pp. 379–381. doi:10.1080/17432200.2017.1335085

Bankoff, G. (2001). Rendering the world unsafe: 'Vulnerability' as Western discourse. *Disasters, 25*(1), pp. 19–35. doi:10.1111/1467-7717.00159

Bhattarai, S. (2018, 1 June 2018). Clash of cultures in Bhaktapur. *Nepali Times*. Retrieved from https://www.nepalitimes.com/banner/clash-of-cultures-in-bhaktapur/

Brosius, C., & Michaels, A. (2020). Vernacular heritage as urban place-making: Activities and positions in the reconstruction of monuments after the Gorkha earthquake in Nepal, 2015–2020 (The case of Patan). *Sustainability, 12*(20), p. 8720. https://doi.org/10.3390/su12208720

Chapagain, N.K. (2008). *Heritage conservation in Nepal: Policies, stakeholders and challenges.* Retrieved from University of New Mexico UNM Digital Repository: https://digitalrepository.unm.edu/cgi/viewcontent.cgi?article=1025&context=nsc_research

Chmutina, K., Jigyasu, R., & Okubo, T. (2019). Editorial for the special issue on "securing future of heritage by reducing risks and building resilience". *Disaster Prevention and Management: An International Journal, 29*(1), pp. 1–9. https://doi.org/10.1108/DPM-02-2020-397

Coningham, R., Acharya, K. P., Davis, C. E., Kunwar, R. B., Tremblay, J. C., Schmidt, A., LaFortune-Bernard, A. (2016). Post-disaster rescue archaeological investigations, evaluations and interpretations in the Kathmandu Valley World Heritage Property (Nepal): Observations and recommendations from a UNESCO mission in 2015. *Ancient Nepal, 191*, 72–92.

Daly, P. T., & Feener, M. R. (Eds.). (2016). *Rebuilding Asia following Natural Disasters: Approaches to Reconstruction in the Asia-Pacific Region.* Cambridge: Cambridge University Press.

Daly, P.T., & Winter, T. (2012). *Routledge Handbook of Heritage in Asia.* London and New York: Routledge.

Department of Archaeology, Government of Nepal (2016*). Guidelines on Conservation and Reconstruction of Earthquake Affected Heritage 2072 BS (2016).* Kathmandu: DoA Nepal.

Department of Archaeology, Government of Nepal. (2016). *Salvaging, screening and inventorying of carved wooden elements of Hanuman Dhoka Palace Complex.* Retrieved from Kathmandu: http://unesdoc.unesco.org/images/0024/002468/246816E.pdf

Gautam, D. (2017). Seismic performance of World Heritage Sites in Kathmandu Valley during Gorkha seismic sequence of April-May 2015. *Journal of Performance of Constructed Facilities, 31*(5), 0601–7003. https://doi.org/10.1061/(ASCE)CF.1943-5509.0001040

Gutschow, N. (2017). Architectural heritage conservation in South and East Asia and in Europe: Contemporary practices. In K. Weiler & N. Gutschow, eds. *Authenticity in Architectural Heritage Conservation: Discourses, Opinions, Experiences in Europe, South and East Asia,* Berlin: Springer, pp. 1–71.

Holtorf, C. (2014). Averting loss aversion in cultural heritage. *International Journal of Heritage Studies,* pp. 1–17. https://doi.org/10.1080/13527258.2014.938766

Hutt, M. (2015). Heritage, continuity and nostalgia. *Cultural Anthropology-Hot Spots.* Retrieved from: https://culanth.org/fieldsights/heritage-continuity-and-nostalgia.

ICOMOS. (2017). *ICOMOS guidance on post trauma recovery and reconstruction for World Heritage Cultural Properties (Working document).* Retrieved from http://openarchive.icomos.org/id/eprint/1763/

Joshi, R., Tamrakar, A., Magaiya, B. (2021). Community-based participatory approach in cultural heritage reconstruction: A case study of Kasthamandap. *Progress in Disaster Science, 10.* https://doi.org/10.1016/j.pdisas.2021.100153

Kathmandu Valley Preservation Trust. (2017). Retrieved from: https://kvptnepal.org/

Layton, R., Stone, P.G., Thomas, J., & Hall, M. (2001). *Destruction and Conservation of Cultural Property*. London: Routledge.

National Planning Commission Government of Nepal. (2015). *Post Disaster Needs Assessment*. Retrieved from Kathmandu: https://www.nepalhousingreconstruction.org/sites/nuh/files/2017-03/PDNA%20Volume%20A%20Final.pdf

National Reconstruction Authority, Government of Nepal. (2016). *Post Disaster Recovery Framework*. Kathmandu: Government of Nepal

National Reconstruction Authority, Government of Nepal. (2021). Evaluation of Socio-economic Impacts of Reconstruction in Nepal. Retrieved from Kathmandu://efaidnbmnnnibpcajpcglclefindmkaj/https://www.hrrpnepal.org/uploads/media/EISReportFinalRevisedbywps4Dec2021_20211210121321.pdf

Pant, S. (2020, January 15). Bhaktapur's famed Nyatapola receives post-earthquake facelift. *The Kathmandu Post*. Retrieved from https://kathmandupost.com/art-culture/2020/01/15/bhaktapur-s-famed-nyatapola-receives-post-earthquake-facelift

Prajapati, S. (2018, March 1). Bhaktapur shows the way by rebuilding itself. *The Nepali Times*. Retrieved from https://www.nepalitimes.com/here-now/bhaktapur-shows-the-way-by-rebuilding-itself/

RADIX: Radical Interpretations of Disasters. (2019). Power, prestige and forgotten values: A disaster studies manifesto, (2019) Retrieved from https://www.radixonline.org/manifesto-accord

Rajopadhyaya, A. D. (2019). Debating identity: Reflections on coverage of Dharaharā and Kāṣṭhmaṇḍap post Gorkha-Earthquake 2015. *Bodhi: An Interdisciplinary Journal*, 7, 67–104.

Rimal, P., & Rajbhandari, S.M (2021, 7 October 2021), In reconstruction, local ownership goes a long way, https://www.recordnepal.com/in-reconstruction-local-ownership-goes-a-long-way

Sharma, K., Kc, A., & Pokharel, B. (2022). Status and challenges of reconstruction of heritage structures in Nepal after 2015 Gorkha, Nepal earthquake. *Heritage & Society*, 15(1), pp. 89–112. doi:10.1080/2159032X.2022.2126229

Shneiderman, S., Baniya, J., & Billon, P. L. (2020, 24 April 2020). Learning from disasters: Nepal copes with Coronavirus pandemic 5 years after earthquake. *The Conversation*. Retrieved from https://theconversation.com/learning-from-disasters-nepal-copes-with-coronavirus-pandemic-5-years-after-earthquake-134009

Shrestha, R. (2021). Community led post-earthquake heritage reconstruction in Patan–issues and lessons learned. *Progress in Disaster Science*, 10, 100156. https://doi.org/10.1016/j.pdisas.2021.100156

Silva, K.D. (2015). The spirit of place of Bhaktapur, Nepal. *International Journal of Heritage Studies*, 21(8), 820–841. https://doi.org/10.1080/13527258.2015.1028962

Suji, M. (2020). *Discourse of post-earthquake heritage reconstruction: A case study of Bhaktapur Municipality*. Paper presented at the Annual Conference of Social Science Baha, Kathmandu.

Suji, M., Limbu, B., Rawal, N., Subedi, P.C., & Baniya, J. (2020). *Reconstructing Nepal: Bhaktapur – heritage and urban reconstruction*. Retrieved from Kathmandu, Nepal: https://soscbaha.org/wp-content/uploads/2020/04/reconstructing-nepal-bhaktapur.pdf

Toffin, G. (2005). From kin to caste: The role of the guthis in Newar society and culture. In S.S. Baha, ed. (Producer), *The Mahesh Chandra Regmi Lecture*, Lalitpur: Social Science Baha.

UNESCO World Heritage Centre. (1979). *Consideration of nominations to the World Heritage List*. Retrieved from Paris: UNESCO.

UNESCO World Heritage Centre. (2016). *State of conservation report*. Retrieved from Paris: https://whc.unesco.org/en/soc/?soc_start=2016&soc_end=2016&action=list&id_search_state=113&. Paris: UNESCO.

UNESCO World Heritage Centre. (2017). *Mission report: Kathmandu Valley (Nepal)* Retrieved from Paris: https://whc.unesco.org/en/list/121/documents/

UNESCO World Heritage Centre. (2021). *State of conservation of properties inscribed on the World Heritage List* (WHC/21/44.COM/7B.Add). Retrieved from Paris, France: https://whc.unesco.org/en/sessions/44com/documents

UNISDR (2015). *Sendai framework for disaster risk reduction 2015–2030*. United Nations World Conference on DRR, Sendai.

Weise, K. (2018). *Authenticity and the Safeguarding of Cultural Heritage in Nepal*. Paper presented at the Revisiting Authenticity in the Asian Context (Rome: ICCROM-CHA Conservation Forum Series 2).

Weise, K., Gautam, D., & Rodrigues, H.F.P. (2017). Response and rehabilitation of historic monuments after the Gorkha earthquake. In D. Gautam & H.F.P. Rodrigues, eds. *Impacts and Insights of the Gorkha Earthquake*, San Diego: Elsevier, pp. 65–94.

Whelpton, J. (2005). *A History of Nepal*. Cambridge: Cambridge University Press.

Winter, T. (2014). Beyond eurocentrism? Heritage conservation and the politics of difference. *International Journal of Heritage Studies*, *20*(2), pp. 123–137. https://doi.org/10.1080/13527258.2012.736403

UNESCO, & World Bank. (2018). *Culture in city reconstruction and recovery*. Retrieved from Paris: http://hdl.handle.net/10986/30733

16 'Dark Heritage'

Landscape, Hazard, and Heritage

Jazmin Scarlett, Miriam Rothenberg, Felix Riede, and Karen Holmberg

What Is Dark Heritage?

The presentation and representation of death, dying, and the dead are prominent facets of human cultures (Roberts & Stone, 2014). The study of tragedies, disasters, catastrophes, and calamities – and their tangible and intangible legacies – has been termed 'dark heritage'. Heritage refers to practices, things, and places that are passed down between generations and viewed as having value worth being perpetuated and/or preserved. Heritage can be either tangible (physical) or intangible (ideational) and can include traditions, objects, monuments, language, landscapes, music, and more. The label 'dark heritage' is applied to legacies of this kind whose origin is often unwanted, dissonant, uncomfortable, and contested. It often adheres to sites and landscapes as people go through the processes of grieving, questioning, understanding, and (un) acceptance. Iconic locales of dark heritage include prisons, concentration camps, battlefields, and crash sites.

Academic interest in dark heritage arose out of the field of dark tourism studies in the 1990s and 2000s (e.g., Lennon & Foley, 2000; Sharpley, 2009) and the two fields have remained closely linked. This is because, although acutely traumatic to those impacted by the precipitating tragedy, sites of dark heritage are also often alluring and many have become popular tourism destinations (Hooper & Lennon, 2017). Often referred to as 'dark sites', locations of dark heritage include those related to conflict, violence, genocide, tragedy, incarceration, and disaster. It is well recognised that they are continuously constructed and reconstructed into 'meaningful' places (Sather-Wagstaff, 2011). From the perspective of tourism studies, dark heritage has thus been defined as a means of acknowledging dark sites and their connection to atrocity as places of remembrance, education, or entertainment (Foley & Lennon, 1996).

Dark tourism research has focused on the experiences of visitors, researcher reflections upon visiting dark sites, and the perspectives of communities living in proximity to dark tourism destinations (e.g., Dunkley, 2017; Knudsen, 2017; Kulcsár & Simon, 2015; Strange & Kempa, 2003). The literature overwhelmingly describes sites of tragedies that are entirely human-induced – locations of genocide or battle, abandoned prisons and gulags, sites of murder and assassination, and sites associated with the Atlantic Slave Trade such as plantations and shipping ports (Hooper & Lennon, 2017). The continuing influence of dark tourism studies on dark heritage studies, while stimulating research in the latter field, has created a number of gaps and biases as to what comprises dark heritage and how it is understood. This chapter explores three overlapping facets of dark heritage that have received thus far only limited attention: dark heritage created by natural hazards, dark landscapes, and the value or importance of dark heritage beyond tourism.

Scholars of resilience and disaster have long recognised that there are no purely 'natural' disasters (O'Keefe et al., 1976) – although natural processes create hazards, these hazards only become 'disasters' when they impact a human community to a degree beyond that which the community is prepared to encounter (e.g., Hewitt, 1997; Hoffman & Oliver-Smith, 2002; Oliver-Smith, 1999; Riede & Sheets, 2020). In other words, a disaster occurs when a natural hazard 'surpasses the adaptation of a culture to its natural and technological environment' due to factors such as a lack of local knowledge of the hazard, weak social ties, and especially social inequality (Rothenberg, 2021: 8).

Academic research has only recently begun to approach the 'dark geocultural heritage' of disasters (notably Scarlett & Riede, 2019). Dimitrovski et al. (2017) argue that the difference between an 'ordinary' heritage site and a dark heritage site is determined by the 'shade of darkness' – a spectrum with one end categorised by sites of *deliberate* death (e.g., genocide) and the other end marked by experiences and events without any special interest in death itself, and perhaps even 'accidental' tragedies. By this metric, the dark heritage of natural hazards and disasters falls at the latter end of the spectrum, but this, in some ways, trivialises the force of their impacts, as we illustrate below, especially for those hazards which do indeed precipitate disasters.

In many cases, 'darkness' is only one aspect among many that are important to the heritage of a particular location or landscape, and other aspects of cultural heritage remain important (Koskinen-Koivisto & Thomas, 2016). A handful of studies engage with volcanoes and volcanic eruptions as focal points of dark tourism, as, for example, on Montserrat and at the sites destroyed by the 79 CE 'Pompeii eruption' of Italy's Mount Vesuvius (e.g., Skinner, 2018a, 2018b). However, for Pompeii, Herculaneum, and other associated sites, the cataclysm and destruction of the eruption is often only the hook that pulls tourists in; in both the academic literature and the curated tourist experience, it is the well-preserved traces of Roman life that are the primary emphasis. As we discuss below, the heritage of other natural hazards and disasters whose physical impacts are less dramatic or long-lasting have received even less attention, but these, too, warrant being approached as dark heritage.

When considering the dark heritage of hazard, it becomes necessary to expand the focus from individual sites to entire integrated landscapes. Unlike many socially-precipitated instances of dark heritage, which can be localised at a single site, natural hazards often take place at a landscape scale. Furthermore, as we describe below, many natural hazards originate from the landscape itself. The Pompeii eruption is a case in point for both arguments: the precipitating hazard originated from Mount Vesuvius, and its impacts were felt not only in the urban sites of Pompeii and Herculaneum but in dozens of other settlements, villas, and agricultural lands across the Campanian Plain. For this reason, the dark heritage of hazard must also be a dark heritage of landscape.

The shift to a landscape perspective also becomes a call for dark heritage studies to reach outward and away from dark tourism studies. Tourism is constrained by the need to identify discrete sites of visitation, but, as we argue, hazards affect both individual sites and the socio-ecological fabric of the landscapes surrounding them. Ideas from dark tourism remain useful for approaching the dark heritage of hazards, but they are not sufficient on their own. This, then, opens the door for conceiving of heritage beyond tourism and bringing in perspectives from other disciplines. For example, Dimitrovski et al. (2017: 696) acknowledge that dark sites 'could be perceived in a broader local, regional or national historical context', that is, that their importance to individuals and communities may extend well beyond touristic value. This statement echoes perspectives on the materiality of trauma from the fields of anthropology, archaeology, and memory studies, which often engage primarily with the non-touristic aspects of dark heritage (e.g., Crossland, 2002; Lydon & Ryan, 2018; McAtackney, 2014; Tumarkin, 2005).

By placing hazard at its centre, this chapter explores the heritage values associated with the land, covering the role of tourism, agriculture, livelihoods, and memory. We address themes such as cyclicity, (in)frequency, trauma, and how high-magnitude events impact a given society's relationship to its heritage, education, and emotional salience. The land and hazards are examined in the context of dark heritage, focusing on the role of destruction in both the built environment and the natural landscape. In the concluding section, we discuss traumatic heritage and approaches to it, trauma without death, and take-away lessons on dark heritage and disaster risk management. Based on the expertise of the authors, many of our examples revolve around volcanic eruptions, and our thinking touches on discussions occurring in the geosciences, archaeology, and anthropology, but we find that these examples and disciplinary perspectives have relevance for a broader range of natural hazards and heritage landscapes.

Landscape and Heritage

In anthropology and related disciplines, a distinction is made between land and landscape. Ingold explains that landscape 'is not "land", it is not "nature", and it is not "space"' (1993: 153). Instead, he argues, 'the landscape is the world as it is known to those who dwell therein, who inhabit its places and journey along the paths connecting them' (1993: 156). From this perspective, the landscape is a cultural construct, one that synthesises land, space, nature, and *culture* by relating them to human actions and perceptions. In this 'socio-symbolic' view, landscape is not only an intrinsic part of the human experience, but in fact, only exists as constituted through human experience, action, and perception (Knapp & Ashmore, 1999). The physical aspect of the landscape, furthermore, comprises both 'natural' forms and the 'built environment', that is, 'the products of human activity' including 'any physical alteration of the natural environment, from hearths to cities, through construction by humans' (Lawrence & Low, 1990: 454). Given the pervasiveness of human landscape modifications, a clear distinction between a natural and cultural or built environments is, in reality, rarely possible (cf. Boggs, 2016).

The anthropological understanding of landscape-as-cultural differs markedly from the way in which the term is used most frequently in the physical sciences. The latter definition of 'landscape' is one that distinguishes clearly between humans and the physical and environmental spaces they inhabit. From this disciplinary perspective, 'land' refers to the portions of the Earth's surface not covered by water, and 'landscape' refers to features of the land associated with the 'natural' environment and sometimes the built environment as well. Landscapes can be described as geographic, ecological, and/or topographic, and they develop around geological, biological, and climatic processes. This is not to say that landscape becomes merely a backdrop to human activity – humans rely on the landscape for resources and livelihoods, it commonly influences their behaviour and belief, and it is also frequently modified by human action, as Anthropocene studies have made abundantly clear (e.g., Crutzen, 2006; Vitousek et al., 1997). While the first part of this chapter used the term 'landscape' with deliberate ambiguity – we ensured that both meanings of the term were applicable in each instance – the rest of the chapter uses the term on its own to refer to landscape-as-a-synthesis-of-culture-and-nature, and we qualify the term with adjectives when we intend to refer to a scientific form of landscape (e.g., 'geological landscape', 'environmental landscape', and, most frequently, simply 'natural landscape').

The heritage values associated with natural and cultural landscapes commonly differ by culture and location, and can change significantly through time. With regard to agriculture, for example, beyond the production of food, agricultural practices and landscapes hold multiple purposes tied to heritage values including aesthetic and cultural values, traditional agricultural practices and landscapes (e.g., terraces), family farming, hydrogeological risk protection, connections to climate change, and agrobiodiversity and soil conservation (Barthel et al., 2013; Santoro et al., 2020). The physical heritage values associated with landscapes include resources that can be extracted or relied upon for subsistence (including agriculture), locations for settlement, refuge from violence and natural hazards, routes of connection between sites or communities, and control over territory (Knapp & Ashmore, 1999). Ideationally, landscapes may have aesthetic value, religious or spiritual significance, and political or territorial dimensions. A crucially important heritage value of landscape is its mnemonic capacity. Landscape features may be used by communities as touchstones of cultural memory, as places imbued with history and folklore, values and lessons, and traditional ecological knowledge (including of hazards and resources). Through repeated visitation and engagement with these places, these heritage values are perpetuated and passed down between generations (Basso, 1996).

Examples exist from around the world of entire landscapes that have been given statuses of protection and designations of significance. Officially designating a landscape as significant and protected heritage is a process that can occur at any level from the local to the international, resulting in the creation of state, regional, and national parks; wildlife reserves and sanctuaries; forest reserves; and specific landscape features (e.g., mountains, valleys, islands) that are marked as being important places of natural or geocultural heritage. Generally, the first step in protecting a landscape through policy, law, regulations, or heritage designations is identifying and assessing the importance of the landscape, but this can be challenging (Scazzosi, 2004). Nevertheless, there has been much success in large national and international bodies protecting natural heritage, including the US National Park Service – founded in 1916 and protecting 85 million acres across 423 distinct areas, serving nearly 300,000,000 visitors per year, and employing nearly 300,000 employees and volunteers (National Parks Service, undated) – and the UNESCO World Heritage Convention (founded 1972). Of the 1,154 properties on the UNESCO World Heritage List at the time of this writing, 218 were marked as being primarily 'natural' sites and 39 of mixed natural and cultural importance (UNESCO World Heritage List, undated).

Having now defined both 'dark heritage' and 'landscape', we can begin to put the two together. Heritage values, be they geological or cultural, dark or not, differ both by context (cultural, regional, and national) and by scale (from the individual to the societal). In an echo of the two different approaches to landscape, cultural heritage, and geoheritage are often approached separately – and there can be good reasons to do so – but in many cases, they overlap. Cultural heritage refers to the tangible (e.g., objects, buildings, and monuments) and intangible (e.g., traditions, knowledge, foodways, music) facets of a culture that have salience to those who are part of that culture or that they choose to present outwardly to represent themselves. Geoheritage refers to sites or areas that contain geological features of significant scientific, educational, cultural, or aesthetic value.

To demarcate instances of intersection between these two forms of heritage, Reynard and Giusti have proposed the term 'geocultural heritage', which they define as 'the links that exist between geoheritage and various forms of culture' (2018: 159). Examples of geocultural heritage abound, and the form that the linkages between natural and cultural heritage take can vary widely. A few representative examples include cultural sites of geological hazards (e.g., Roman Pompeii and Herculaneum, or Bronze Age Akrotiri on the island of Thera), geological places of religious or ceremonial importance (e.g., sacred Maya caves and cenotes in the Yucatán Peninsula (Palka 2014), or the anthropomorphised landscape features that constitute many Andean 'huacas' (Bray, 2015), Merapi and its many cultural and spiritual facets (Dove, 2008; Seeberg & Padmawati, 2015; Holmberg, 2022), and landscapes of resource extraction (e.g., ancient Roman gold mines in Spain (Reher et al., 2012)). Importantly, geoheritage, cultural heritage, and geocultural heritage

'Dark Heritage' 229

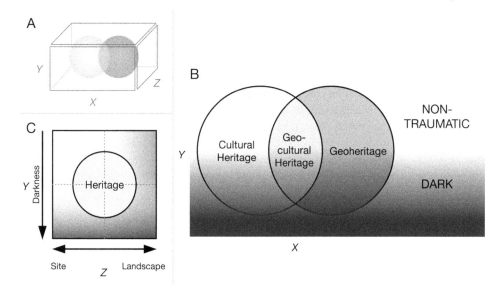

Figure 16.1 The two overlapping spheres of cultural heritage and geoheritage can be imagined to sit within a three-dimensional box (A). Geocultural heritage exists where the two spheres overlap on the x-axis; the y-axis grades from non-traumatic forms of heritage at the top to dark heritage at the bottom through various degrees of darkness; and the z-axis represents spatial scale, from the site level to the landscape level. Looking down the z-axis shows that all three forms of heritage represented here can also be dark heritage (B). Viewing the heritage box down the x-axis makes it clear that spatial scales and degree of darkness are both continuua and they allow for non-traumatic heritage sites, dark heritage landscapes, and everything in between (C). (Image by Miriam Rothenberg).

can all be linked with traumatic or catastrophic events, thus also becoming dark heritage; and they can also all occur at individual sites or across broad landscapes (Figure 16.1).

A key concept when approaching the intersection between landscapes and dark heritage is the idea of the 'traumascape'. The originator of the concept, Maria Tumarkin, describes traumascapes as 'not simply material locations of traumatic events, [but] physical places constituted by experiences of particular events and their aftermath' (2019: 5). Traumascapes are places marked by trauma – whether from a disaster, violence, or some other form of tragedy – that elicit emotions and memories among those who experienced or are otherwise connected to the precipitating event. They can be overwhelming and riddled with triggers that force the past into the present in painful and intrusive ways. Importantly, these triggers can be situated in the materiality of these sites, in the physical traces of the traumatising event (Tumarkin, 2005).

Many of the landscapes described in this chapter could be categorised as traumascapes, but the concept has rarely, if ever, been applied to geoheritage. None of the descriptions of the 257 UNESCO sites of natural or mixed importance emphasise dark or traumatic heritage, even when tragedies have indeed occurred at these sites, and even mentions of natural hazard are rare (UNESCO World Heritage List, undated). Directly addressing darkness and trauma in landscapes of natural and geocultural heritage creates avenues for education and scholarship. In developing public-facing materials for heritage sites and landscapes, emphasising the experiences of those who contended with a hazard is a way to bring together concepts of natural and cultural heritage, and thus to broaden perspectives on human-environment interactions. Focusing on disasters, as opposed to divorcing hazard processes (geoheritage) from cultural heritage, may also open the possibility for research collaborations, or at least minimise conflicts between natural and social scientists over these areas (see the recent debate over volcanological deposits at Pompeii: Devlin, 2019; Osborne, 2019; Scandone et al., 2019).

Dealing with landscapes of dark heritage, however, is also sensitive and complicated because of the weight of the trauma and the many often-conflicting narratives and perspectives that develop around it. This can create challenges for heritage professionals who are tasked with dealing with such a landscape, as there may be conflicting voices advocating for how (and whether) to conserve and share it. An example of this is the debate about what to do with the dark geocultural heritage of Montserrat's volcanic Exclusion Zone, as is described in Box 16.1.

Box 16.1   The dark heritage of Montserrat's Volcanic 'Exclusion Zone'

Montserrat is a small volcanic island located within the Lesser Antilles Volcanic Arc between Guadeloupe, Antigua, and Nevis. In July 1995, the Soufrière Hills Volcano, situated in the south-central portion of the island, suddenly became active, with explosive and dome-forming eruptions that have continued intermittently ever since. By April 1996, Montserrat's entire southern half had been permanently evacuated, including its capital town of Plymouth, and in June 1997, a dome collapse resulted in 19 fatalities, the only direct casualties from the volcano to date. In 2010, the volcano entered a period of relative quiescence, but much of the island remains in the so-called volcanic 'Exclusion Zone', an area demarcated by the government as being unsafe for public access, whose spatial footprint changes based on the volcanic hazard level. Today, Montserrat's volcanic ruins are engaged with by residents of the island, returning members of the large Montserratian diaspora community, volcanologists, and disaster tourists to Plymouth and its surroundings.

The volcanic crisis has subjected Montserrat's landscapes to a wide array of geological hazards. Ashfall events, pyroclastic flows, earthquakes, lahars (volcanic mudflows), and other high magnitude hazards have all been common in periods of high volcanic activity. Lahars, landslides, acid-based corrosion, and erosion have continued into less active periods. The island has also experienced meteorological hazards during this time, several hurricanes and strong storms with high winds. These processes along with human actions have transformed the built environment of Plymouth and the other settlements in central and southern Montserrat into extensive ruinscapes. Both acute moments of hazard-based destruction and slower process of dereliction have played a role in this transformation – with strict rules for who can enter the Exclusion Zone and under what circumstances, rebuilding and restoration are impossible and even undamaged buildings cannot be kept up and gradually fall into ruin.

The ruinscapes of the Exclusion Zone display many material traces of hazard processes, such as burial by volcanic ash, incursion of boulders and lahar deposits, burning, and damaged and missing roofs, to name just a few. They act as material reminders of the eruptions, including traumatic experiences of uncertainty, displacement, life in shelters, and loss of life and livelihoods. The ruined buildings are also visually prominent in the landscape, even from afar, highlighting the ongoing trauma of residents being unable to return to homes and home villages even decades after the crisis began and attracting tourists eager to explore them (Fig. 16.2).

Figure 16.2 Modern ruins and volcanic deposits on the outskirts of Plymouth, 24 July 2018.
Source: Photograph by Miriam Rothenberg.

Even in times of higher volcanic risk, Plymouth, sometimes advertised as the 'modern Pompeii of the Caribbean', has drawn adventure-seeking disaster tourists to the island. Illegal tourism to the Exclusion Zone has put both the tourists and potential rescue workers at risk, and has resulted in numerous instances of disrespectful and intrusive treatment of homes and properties. Since 2014, however, there has also existed a successful program for local taxi drivers to take tourists to Plymouth safely and legally. One question, however, is how sustainable disaster tourism will be in the long term. As these landscapes become more overgrown and less identifiable over time, retaining their allure for disaster tourists would likely require ongoing effort to maintain them at a particular degree of dereliction and visibility rather than either letting them decay into unrecognisable rubble or allowing for their erasure through rebuilding and reoccupation.

Discussions about the future of southern Montserrat highlight the contested nature of its dark heritage. Some residents want to be able to return to their homes and farmland despite the continuing risk and volcanic uncertainty, but others want to abandon the area completely and focus on life in the north. There have also been debates and informal proposals to designate the volcano and its surroundings a 'geopark' or to apply for UNESCO World Heritage status, in the hope that doing so would encourage tourism to the island and bring in funds (both in general and for conservation purposes). Some residents, however, have expressed the view that such a designation would require that these areas be preserved in their current state, thus ending the possibility of ever returning or rebuilding. These problems of future land use and tourism emphasise the complexity of the volcanic crisis, the multitemporal nature of landscapes and ruins, and the continuing uncertainty regarding volcanic risk (Rothenberg, 2021).

Heritage and Natural Hazards

Natural hazards are extreme events caused by environmental and geological processes which can become disasters when they disrupt human activities, cause injury and casualties, or damage property and the built environment. The relationship between hazards and heritage depends on the cyclicity, frequency, and the magnitude of the hazard. The terms 'risk', 'vulnerability', and 'resilience' have multiple definitions depending on the discipline and the hazard, but in the context of this chapter, their basic definitions suffice. 'Risk' is the likelihood and potential severity of a hazard to cause harm. 'Vulnerability' is defined as the factors that make something susceptible to a hazard, whilst 'resilience' is the capacity to cope and recover from the impact of hazard.

Natural hazard processes – whether geological, meteorological, or biological/ecological – are part of the landscape, and different landscapes will contain different hazards. Hurricanes, floods, and sinkholes are hazards that are not uncommon in sub-tropical, coastal, and karstic Florida; whereas it is earthquakes, landslides, and blizzards that mark tectonically-active, mountainous, and high-altitude Tibet. Cultures are more likely to be vulnerable to a hazard when it is either not one that is common in their region, or it occurs at a more severe magnitude than they have adapted to prepare for. Parallel examples of the former include the February 2021 North American cold wave and the summer 2022 European heat wave. In both cases, temperatures that were highly unusual for the regions created severe disruptions and generated casualties and deaths due to a lack of appropriate infrastructure (e.g., heating and air conditioning systems, and overloaded electricity grids), disproportionately affecting the most vulnerable portions of the populations (Doss-Gollin et al., 2021; Kaplan, 2022).

Unusually high-magnitude examples of otherwise common hazards can also surpass local adaptation, as was the case with Hurricane Maria in Puerto Rico and the northern Caribbean islands in September 2017. While tropical cyclones are common in this region and there exist robust systems for hurricane preparedness and management, Hurricane Maria was one of the strongest Atlantic hurricanes ever recorded, striking only a week after Hurricane Irma had damaged infrastructure and drained resources. Among its catastrophic effects on Puerto Rico were thousands of immediate and indirect casualties, severe damage to the built environment, the total destruction of the electric grid with some areas lacking electricity for months, flooding and landslides, the loss of agriculture, the destruction of transportation and healthcare infrastructures, and the displacement of thousands of individuals for months (Rodriguez & Mora, 2020).

Not all natural hazard processes leave clear traces in the landscape, and not all of the physical products of a hazard persist over long periods of time. In general, geological hazard processes (i.e., volcanoes, earthquakes, landslides) are more directly linked to the creation of prominent, long-lasting features in a landscape than meteorological hazards (e.g., tornadoes, hurricanes, droughts). The former has the power to change the physical topography by moving or even creating large swathes of new rock and Earth through the emplacement of solidified lava flows, the creation of scarps, and the rising, falling, cracking, and offsetting of the land's surface. These changes may be so dramatic as to be designated as sites of geoheritage worthy of conservation and tourism development such as Arxan-Chaihe Volcano Area, Inner Mongolia (Wang et al., 2014) or Morne Trois Pitons National Park in Dominica (Colmore, 2018).

Meteorological hazards are just as intrinsic to environments and landscapes as geological hazards, but their physical imprints on the natural landscape are often either temporary or less distinct than those of the former category. There are, of course, exceptions in the form of erosion, changed coastlines, or areas of formerly dry land that become permanently flooded, but, in general, distinctive landforms linked with meteorological hazards are rare. Nevertheless, these hazards can be damaging to lifeways and livelihoods and traumatic to those who live through them, and they can also be very destructive to the built environment, requiring extensive rebuilding or repairs. Traces of such events in the landscape, therefore, may be subtly present as ruins of destroyed or damaged buildings, collections of structures or repairs whose materials and architecture all date to a narrow period of time (months or years) after a meteorological disaster, or other changes to buildings associated with the hazard and its immediate aftermath. Finally, even when there is no severe damage to the built environment or when the damage has been repaired, meteorological disasters can be commemorated in the landscape with plaques or the construction of memorials and monuments (Figure 16.3a and b). These markers become dark heritage, acknowledging the trauma of the disaster and preserving its memory in the landscape, and they may be used to educate about past and future hazards.

Thus far we have focused primarily on the 'negative' aspects of hazards, but it is important to remember that they may just as easily have 'positive' consequences for a society. One positive example could be civilisations in Ancient Egypt utilising the annual cyclic floods of the Nile for agriculture – a practice, however, that left the population of the Nile Valley vulnerable from any disruption of that cycle as agricultural practices largely centred on the need for the flooding (Manning et al., 2017). Another example is that of agricultural practices around Mt. Bromo volcano in Indonesia, where farmers know to wait three years following an eruption to take advantage of the newly added enriching minerals (Bachri et al., 2015).

Figure 16.3a Two plaques installed on the historic market house in Providence, RI, marking high water lines associated with the hurricane of 1938 and the great gale of 1815. Although the physical damage from both storms is no longer present in the landscape, the events have been commemorated in the built environment through the instalment of markers like these throughout Providence and Rhode Island.

Source: Photograph by John F. Cherry, used with permission.

Figure 16.3b A so-called 'flood pillar' on the old harbour in Ribe, Denmark. Each band corresponds to a major flood that occurred in historical times. This is an example of a Northern European tradition commemorating traumatic coastal and River flood events in this way.

Source: Photograph by Hjart (CC BY-SA 4.0).

Natural hazards and disasters may also have religious significance or be used to encourage or enforce specific behaviours or morals within a society. Native Hawaiians view the duality of disasters – being both 'good' and 'bad' – through a spiritual lens. Kīlauea on Hawaii produces devastating eruptions – most recently in 2018 – but for Native Hawaiians and their belief in the Fire Goddess Pele, these eruptions are viewed as the Goddess 'reclaiming' her lands, which has been described as 'Paradise Tax' (Scarlett, 2014). Another example of religious associations with natural hazards is in the frequent attribution of earthquakes in Early Modern Europe and its colonies to God's punishment for sinful populations. The 1746 earthquake and tsunami in Lima (Walker, 2008), the 1755 'Great Lisbon earthquake' (Bassnett, 2006), and the 1692 earthquake that sank Port Royal, Jamaica, are all events that were viewed at the time as divine punishment for wickedness. Reverend E. Heath, a survivor of the latter, even wrote in its immediate aftermath, 'I hope by this terrible Judgment, God will make them reform their lives, for there was not a more ungodly People on the Face of the Earth' (Heath Rev, 1692). Thus, just as landscapes can hold lessons, so too can hazards, with their dark heritage being employed to enforce normative or religious behaviour and encourage social cohesion.

The response to these Early Modern earthquakes also demonstrates that, within particular worldviews, hazards may be blamed on human behaviour. While the above explanation would have had salience within its cultural context, it is not one that is supported by modern Western science. That is not to say, however, that human behaviour cannot promote or strengthen hazards. The natural hazards described in this chapter are all processes that have occurred for millions of years, but as humans continue to transform the landscape on ever-increasing scales, we have not always taken into consideration the impact our actions may have on natural hazard processes. Deforestation and intensive agriculture can both increase erosion rates to a devastating degree; dam building can prevent necessary floods and increase the risk of higher magnitude ones; destroying local grasslands can lead to drought and desertification (as in the Dust Bowl of the 1920s), and so on. A particular irony is that some attempts to control natural hazards and decrease local vulnerability to them may have the opposite result. The construction of artificial levees to prevent small floods can both make larger floods more damaging than they would otherwise have been and increase vulnerability to them (Burton & Cutter, 2008), and suppressing forest fires in fire-prone ecosystems can allow for the accumulation of flammable fuels, meaning that fires are much more difficult to control when they do break out (Calkin et al., 2015).

Finally, it is important to note both the impacts of colonialism and the effects of anthropogenic climate change on 'natural' hazards and resilience. Modernity, including the past 500 years of European colonialism and the subsequent development of capitalism, has altered Earth's landscapes and ecosystems on an unprecedented scale in human history (Pomerantz, 2009). Colonialism has disturbed native ecologies, erased Indigenous forms of traditional ecological knowledge, altered landscapes through intensive resource extraction, and institutionalised widespread inequalities. These factors and more have significantly increased the prevalence and strength of certain hazards (e.g., deforestation leading to slope destabilisation and landsliding) and have increased human vulnerability to them. Now, due to continued heavy resource extraction and populations experiencing 'unprecedented' impacts from natural hazards, heritage values which traditionally acknowledged and embraced hazards' negative and positive aspects together are diminishing.

Anthropogenic climate change is another consequence of European colonialism and the transition to a largely capitalist world system, and it, too, affects how humans engage with hazard. Many meteorological hazards are increasing in severity and/or frequency (e.g., hurricanes, droughts, floods, fires) or are becoming common in areas where they were previously unknown or infrequent. For example, wildfires occurring from natural means (such as lightning strikes) can be both dangerous and beneficial to some biomes' biodiversity (Burkle et al., 2015). However, with the increase in human-related wildfires and a changing climate, the wildfire regime is changing to the detriment of landscapes and environmental processes (Mariani et al., 2022).

The tangible impacts of anthropogenic climate change on heritage include physical threats to existing heritage sites (Sesana et al., 2021) but also the likelihood that an increase in the number and intensity of hazards will create more dark geocultural heritage in the coming decades. There are discussions about the preservation of cultural heritage in relation to climate change and other natural hazards (Harvey & Perry, 2015; Rico, 2016; Sorriso-Valvo, 2008), and parallel to this, the re-purposing of buildings that are dark sites (e.g., Pendlebury et al., 2018). Conceptually, however, there also seems to be an important shift taking place in terms of where to lay the blame for disasters. Whereas disasters are already known to occur from the combination of human and natural processes, many natural processes are now undeniably conditioned by human actions. Thus, the way recent disasters are described in popular discourse either tilts the balance of responsibility more towards the human or finally begins to acknowledge that the human-environment distinction is largely artificial.

Destruction as Process

Destruction plays a key but complicated role in the creation and conceptualisation of the dark heritage of hazards. We become attached to the places with which we interact, and the destruction of these places can be its own form of trauma distinct from (but related to) the trauma of surviving the destructive event or the lives lost to it. Natural hazard processes can damage and destroy buildings and infrastructure in a matter of seconds, but people will mourn the loss of homes and other places of personal or cultural significance for much longer. The physical presence or absence of destruction in the landscape may affect how the traumatic event is remembered and commemorated. Traces of a disaster may remain present as ruined structures and damaged infrastructure, or they may be erased and rewritten through restoration, preservation, reuse, or mitigation efforts. The question of 'what to do with the ruins of a disaster' is thus closely tied to the experiences of those with a connection to that disaster and how they wish to remember or forget it. As a result, discussions surrounding this question are deeply emotionally entangled and can quickly become fraught, as perspectives differ, narratives are contested, and conflicting desires are expressed.

Time intersects with destruction in several interesting ways. First, ruination can occur either suddenly or slowly due to the nature of the hazard. An earthquake or hurricane may create widespread destruction in a matter of minutes or hours, but the longer-term processes dereliction caused by abandonment is no less important in the creation of 'ruinscapes', that is, landscapes defined by the presence of ruins and ruination. Examples of such longer-term, landscape-wide abandonment include evacuated areas around active volcanoes or nuclear disaster sites, decisions made not to return to settlements damaged by floods or fires, and abandonment due to pre-disaster economic and subsistence strategies no longer being viable (e.g., a prolonged drought disrupting agriculture). While abandoned locations may not have experienced acute or violent destruction, they nevertheless express material traces of dereliction and ruination due to hazard and can thus equally become traumascapes.

A second intersection between time and destruction can be seen in the variable time spans between a hazard event and partial or full recovery of the affected area. Volcanic eruptions and hurricanes can destroy large areas of the landscape making them uninhabitable to humans, fauna, and flora, and recovery is not always as easy as simply 'rebuilding'. It can take time for the many interconnected systems at play in the landscape to rebound. Ecosystems impacted by natural hazard processes can recover in their own time and may not necessarily need the intervention of humans but rebuilding the built environment and re-establishing or reworking cultural connections to the landscape take a concerted effort. Furthermore, the alteration of many natural hazard regimes due to anthropogenic climate change can interfere with the ability of ecosystems to recover and adapt due to changes in the cyclicity, frequency, and magnitude of disruptions.

The third way in which destruction and time come into contact is through the distinctly nonlinear temporality of traumascapes which uncomfortably thrust the memories and trauma of a hazard event into the present. Scholars of ruination have described modern ruins as 'uncanny' (Edensor, 2005: 25) and 'untimely' (Yablon, 2009) places where pasts and presents collide (DeSilvey & Edensor, 2012; Pétursdóttir & Olsen, 2014). They are *'lieux de mémoire'*, which both store and trigger memories (Nora, 1989; Pálsson, 2012; Pétursdóttir, 2012), and they can serve as mnemonic objects for the transmission of the experience of trauma between generations. Destruction and ruination create material traces of traumatic hazard events, and thus their presence in the landscape provides a reminder of how and why things have changed from what they once were. A particularly poignant example of this form of dark heritage can be seen in the post-Katrina built environment of New Orleans. Here, the painted marks left on buildings by search-and-rescue teams and the flood residue that stuck to objects formed a 'patina' that linked the built environment to the disaster and residents' experiences with it long after the acute effects had subsided, repeatedly pulling these memories into the present as people went about their daily lives (Dawdy 2016).

Landscapes of destruction not only call the past into the present but force a reckoning with the future. Destruction may lead to acts of remembrance and memorialisation, as well as debates about what to do with these new dark sites – should they be abandoned, rebuilt, preserved, marketed to tourists, etc.? Deciding whether and how traumatic stories are told, and by whom, is complicated due to their inherently multivocal and contested nature. There may be conflicting opinions both within and among affected groups as to what is the 'best' or most desirable course of action – for example, some people may wish to rebuild and return to life as it was before a disaster, whereas others might want to forget the event and start over (see Box 17.1 for an example from Montserrat).

Destruction can spur people into action, for better or for worse. The destruction of one feature of the landscape and what is prioritised as 'valuable' may negate or neglect the attention of other features in the landscape. However, events that are destructive in the landscape can not only destroy but also create heritage. For example, in Banda Aceh, Indonesia, greatly devastated by the 2005 Indian Ocean tsunami, the destruction shone a light on heritage creation associated with (re)construction. There, a future-oriented perspective was taken that intentionally acknowledged and connected with the cultural heritage of buildings long lost, such as a building constructed in 1205 CE (Rico, 2016).

The destruction of a landscape or particular features within it can generate shock and spur discussions and policy changes with the aim of preventing future disaster and decreasing vulnerability to a particular hazard. For example, high-profile oil spills can generate calls to end our reliance on fossil fuels due to their environmental damage on wildlife and vegetation in the short-term, and on climate change in the long-term. This is a smaller scale version of discussions being had around global climate change, the vulnerabilities that it increases, and the calls to address it now that its effects are being felt. Destruction of the landscape – whether precipitated directly from human action or resulting from an environmental hazard – can both destroy existing heritage and create new landscape features that can become tied to dark heritage. For example, 'Big Hole' in South Africa is the remnants of a diamond mine which has destroyed and replaced a pre-existing landscape feature. Although the heritage associated with the previous feature may have been erased by mining, and Big Hole is comparatively young, it already has its own dark heritage: like many other mines scattered across the African continent landscape, it is a product of over-extraction by European colonial empires which subjected Africans to terrible work conditions, and the memory of this industry remains present in the gaping chasm (Calvão, 2011; Varanda & Cleveland, 2013).

Conclusion

Dark heritage is traumatic heritage, and disasters have been widely documented as causing symptoms of post-traumatic stress disorder and other mental health conditions in survivors (Neria et al., 2007), including 'eco-anxiety' and climate grief (e.g., Robbins & Moore, 2013). Because sites and landscapes of dark heritage can hold memories and triggers for those who experienced the trauma, disaster risk reduction strategies need to include considerations of dark heritage and the situatedness of trauma in the landscape. Because trauma can be long-lasting and situated in places and things, policies and recommendations that are supposed to protect those susceptible to disasters should include plans for contending with the built environment and physical landscape over the long term, and not exclusively on the immediate physical and mental health needs of the human population. Support structures need to exist both for rebuilding and protecting the tangible heritage of the built environment and natural landscape, and for recovering from trauma, which can be intangible, 'invisible', persistent, and in many places stigmatised. To diminish or ignore dark/traumatic heritage in disaster risk reduction strategies is to fundamentally misunderstand how trauma functions, and doing so can also create an environment of mistrust and disengagement with these strategies, which would also limit their efficacy by more traditional metrics.

As an example of change that is needed, 'no deaths' is seen as a success in disaster risk reduction terms, but of course, there can be long aftermaths and lingering trauma. Volcanic eruptions such as Chaitén, Chile, in 2008 and La Soufriére, St. Vincent, in 2021, caused no deaths, but their environmental impacts, disruption, and upheaval to daily life had a profound effect. However, in both examples, the people impacted are veering away from the 'dark' component and wanting to promote itself as a place of regeneration and renewed geological awareness. In the case of Chaitén, the locals created their own museum space for a place of remembrance and education. For La Soufriére, communities in the 'red zone' (highest area of risk that was evacuated in 2021) were engaged in a community project to document and photograph changes to the surrounding environment as a process of healing and advocacy that these communities are still in the 'recovery' phase. Thus, culturally sensitive approaches are required, which make managing and recovering from trauma an integral part of disaster risk reduction strategies. This will help ensure that no one is 'left behind' in the recovery phase, a goal that is important for its own sake and to help reduce future inequality and vulnerability to hazard.

What does dark heritage mean for disaster risk reduction and disaster risk management? Ultimately, it means adjusting perspectives on past, present, and future hazard events to account for a community's relationship with destruction, death, and trauma. Dark heritage spaces around the world already utilise destruction and death as conversation starters, calls for justice, avenues for addressing inequalities, and agents of change. However, the contested nature of traumatic heritage means that dark sites and landscapes necessarily have different meanings for different groups of people. Negotiating between the desires and experiences of different groups is a necessary, if difficult, task for heritage professionals, especially when these are wrapped up in local or national politics, colonial legacies, and other power dynamics, as they inevitably are.

Examining the links between dark heritage and disaster risk management also requires questioning what 'acceptable losses' are, whether in terms of monetary value, infrastructure, property or environmental damage, or human and animal lives. Acceptable losses are often defined in very concrete, measurable, and quantifiable terms, but dark heritage is always at least somewhat abstract. In order to properly account for the abstract and intangible impacts of a disaster, disaster management professionals must therefore become more abstract in their thinking as well. Most of the examples we have described in this chapter describe traumatic events that leave their marks upon the physical landscape. However, even if all physical traces of such an event have been erased, the landscape can still serve as a touchstone for lingering traumatic memories, as can individual structures and artefacts (be they physical objects or online on social media).

One way to add such abstract thinking to disaster risk reduction is to understand that dark heritage operates on multiple timescales, beginning from at least the moment that an event occurs (if not before) and continuing for generations, even potentially overlapping with the dark heritage of other disasters. This, too, describes landscapes, as they develop and change through time. Spaces of dark heritage can be used like forensic crime scenes to unveil these multitemporal layers and examine *how exactly* a hazard event turned into a disaster, not just physically, but emotionally and experientially. Their analysis can illuminate the natural reactions and responses of those impacted, which provides useful information for adjusting systems of disaster risk reduction and decreasing vulnerability to hazards more broadly. Scholars, disaster risk reduction organisations, and heritage professionals should all remain aware that death, destruction, and trauma sit within many landscapes and act over long and varying timescales; that hazard processes blur the line between the human and the environment; that traumatic events and processes can be highly contested; and that together these create complex and locally specific heritage milieux. While there can be no 'one-size-fits-all' approach to contending with heritage and landscapes of hazard, it is clear that the experiences of trauma and their long-term effects should be acknowledged and incorporated – not diminished, downplayed, or ignored – when formulating and implementing disaster risk reduction policies. Finally, we also stress that heritage, through its value and associated actors (e.g., knowledge-holders, museums), can play an important role in commemoration, reconciliation, and healing.

Key Points

- Dark heritage is uncomfortable, difficult, unwanted, dissonant, and often contested;
- Cultural heritage, geoheritage, and geocultural heritage can all also be dark heritage;
- A goal would be that the dark heritage of hazards and disasters needs to be considered at a landscape scale as much as at a site or object scale;
- Natural hazards come from the landscape and landscapes are culturally constituted, thus the dark heritage of disasters is contextualised within local landscapes and local cultures;
- When it comes to dark heritage, disaster management scholars and practitioners should not forget that they are also dealing with heritage, and heritage professionals must remember that they are also dealing with the effects of past, present, and future trauma and disaster;
- Any individual or organisation entering a community that has recently experienced trauma should be aware that the processing of trauma and a decision of what to do with its legacy requires both space and time.
- Finally, the experiences of death, destruction, and trauma differ at all scales from the personal to the national, creating multiple strongly-felt narratives around any disaster. For both heritage and disaster management professionals, navigating between different narratives and interest groups comes with many difficult ethical and emotional challenges. Communities that have experienced trauma should be allowed to decide for themselves how to construct their own post-disaster dark heritage – it is part of the process of grieving and (un)acceptance.

References

Bachri S., Stötter J., Monreal M., and Sartohadi J. (2015) The calamity of eruptions, or an eruption of benefits? Mt. Bromo human-volcano system: A case study of an open-risk perception. Natural Hazards and Earth System Sciences. Vol. 15(2). Pg. 277–290.

Barthel S., Crumley C.L., and Svedin U. (2013) Bio-cultural refugia – safeguarding diversity of practices for food security and biodiversity. Global Environmental Change. Vol. 23(5). Pg. 1142–1152. DOI: https://doi.org/10.1016/j.gloenvcha.2013.05.001

Bassnett S. (2006) Faith, doubt, aid and prayer: The Lisbon earthquake of 1755 revisited. European Review. Vol. 14(3). Pg. 321–328.

Basso (1996) Wisdom Sits in Places: Landscape and Language Among the Western Apache. Albuquerque, NM: University of New Mexico Press.

Boggs C. (2016) Human niche construction and the anthropocene. In Emmett R.S. and Lekan T. (eds.) In Whose Anthropocene? Revisiting Dipesh Chakrabarty's 'Four Theses'. RCC Perspectives 2: Transformations in Environment and Society. München: Rachel Carson Centre. Pg. 27–31.

Bray T.L. (ed.) (2015) The Archaeology of Wak'as: Explorations of the Sacred in the pre-Columbian Andes. Boulder, CO: University Press of Colorado.

Burkle L.A., Myers J.A., and Belote R.T. (2015) Wildfire disturbance and productivity as drivers of plant species diversity across spatial scales. Ecosphere. Vol. 6(10). Pg. 1–14.

Burton C.G. and Cutter S.L. (2008) Levee failures and social vulnerability in the Sacramento-San Joaquin Delta area, California. Natural Hazards Review. Vol. 9(3). Pg. 136–149.

Calkin D.E., Thompson M.P., and Finney M.A. (2015) Negative consequences of positive feedbacks in US wildlife management. Forest Ecosystems. Vol. 2(9). DOI: http://doi.org/10.1186/s40663-015-0033-8

Calvão F. (2011) When boom goes bust: Ruins, crisis and security in megaengineering diamond mining in Angola. In: Brunn S. (ed.) Engineering Earth. Dordrecht: Springer.

Colmore C.S. (2018) The Caribbean's geotourism potential and challenges: A focus on two islands in the region. Geosciences. Vol. 8(8). Pg. 273. DOI: https://doi.org/10.3390/geosciences8080273

Crossland Z. (2002) Violent spaces: Conflict over the reappearance of Argentina's disappeared. In Schofield J., Johnson W.G. and Beck C.M. (eds.) Matériel Culture: The Archaeology of Twentieth-Century Conflict. London: Routledge. Pg. 115–131.

Crutzen P.J. (2006) The 'Anthropocene'. In: Ehlers E. and Krafft T. (eds.) Earth System Science in the Anthropocene: Emerging Issues and Problems. Berlin: Springer. Pg. 13–18.

Dawdy (2016) Patina: a Profane Archaeology. Chicago: University of Chicago Press.

DeSilvey C. and Edensor T. (2012) Reckoning with ruins. Progress in Human Geography. Vol. 37(4). Pg. 465–485.

Devlin H. (2019) Pompeii row erupts between rival scientific factions [online]. The Guardian. https://www.theguardian.com/science/2019/jul/22/pompeii-row-erupts-between-rival-scientific-factions [accessed 20/10/2022]

Dimitrovski D., Senić V., Marić D., and Marinković V. (2017) Commemorative events at destination memorials-a dark (heritage) tourism context. International Journal of Heritage Studies. Vol. 23(8). Pg. 695–708.

Doss-Gollin J., Farnham D.J., Lall U., and Modi V. (2021) How unprecedented was the February 2021 Texas cold snap? Environmental Research Letters. Vol. 16. DOI: https://doi.org/10.1088/1748-9326/ac0278

Dove M.R. (2008) Perception of volcanic eruption as agent of change on Merapi volcano, Central Java. Journal of Volcanology and Geothermal Research. Vol. 172(3–4). Pg. 329–337.

Dunkley A. (2017) Analysing impact of dark tourism experiences on everyday life. In: Hooper G. and Lennon J.J. (eds.) Dark Tourism: Practice and Interpretation. London: Routledge.

Edensor T. (2005) Industrial Ruins: Spaces, Aesthetics and Materiality. Oxford: Berg.

Foley M. and Lennon J.J. (1996) Editorial: Heart of darkness. International Journal of Heritage Studies. Vol. 2(4). Pg. 195–197.

Harvey, D. and Perry J. (2015) Heritage and climate change: the future is not the past. In: Havery D. and Perry J. (eds) The future of heritage as climates change: loss, adaptation and creativity. London: Routledge. Pg. 3–22.

Heath Rev L. (1692) A full account of the late dreadful earthquake at Port Royal in Jamaica; Written in two letters from the minister of that place. From Aboard the Granada in Port Royal Harbour, June 22, 1692. London: Printed for Jacob Tonson, and sold by R. Baldwin. John Carter Brown Library, Providence, RI.

Hewitt K. (1997) Regions of Risk: A Geographical Introduction to Disasters. London: Addison Wesley Longman.

Hoffman S.M. and Oliver-Smith A. (eds.) (2002) Catastrophe and Culture: the Anthropology of Disaster. Santa Fe, NM: School of American Research Press.

Holmberg K. (2022) Merapi and its dynamic 'disaster culture'. In: Gertisser R., Troll V.R., Agung Nandaka I.G.M., and Ratdomopurbo A. (eds.) Merapi Volcano: Geology, Eruptive Activity and Monitoring of a High-Risk Volcano. Berlin & Heidelberg: Springer-Verlag.

Hooper G., and Lennon J.J. (2017) Dark Tourism: Practice and Interpretation. London: Routledge.

Ingold T. (1993) The temporality of the landscape. World Archaeology. Vol. 25(2). Pg. 152–174.

Kaplan S. (2022) Europe just had its hottest summer on record [online]. The Washington Post: https://www.washingtonpost.com/climate-environment/2022/09/08/europe-record-hot-summer-extreme-heat/ [accessed 20/10/2022]

Knapp A.B. and Ashmore W. (1999) Archaeological landscapes: Constructed, conceptualised, ideational. In Ashmore W. and Knapp A. (eds.) Archaeologies of Landscape: Contemporary Perspectives. Oxford: Blackwell Publishers. Pg. 1–30.

Knudsen B.T. (2017) Experiencing dark heritage live. In: Hooper G. and Lennon J.J. (2017) Dark Tourism: Practice and Interpretation. London: Routledge. Pg. 186–198.

Koskinen-Koivisto E. and Thomas S. (2016) Lapland's dark heritage: Responses to the legacy of world war II. In Silverman H., Waterton E. and Watson S. (eds.) Heritage in Action: Making the Past in the Present. New York: Springer. Pg. 121–133.

Kulcsár E. and Simon R.Z. (2015) The magic of dark tourism. Management and Marketing Journal. Vol. 13(1). Pg. 124–136.

Lawrence D.L. and Low S.M. (1990) The built environment and spatial form. Annual Review of Anthropology. Vol. 19. Pg. 453–505.

Lennon J.J. and Foley M. (2000) Dark Tourism: The Attraction of Death and Disaster. London and New York: Continuum.

Lydon J. and Ryan L. (2018) Remembering the Myall Creek Massacre. Sydney: NewSouth Publishing.

Manning J.G., Ludlow F., Stine A.R., Boos W.R., Sigl M., and Marlon J.R. (2017) Volcanic suppression of Nile summer flooding triggers revolt and constrains interstate conflict in ancient Egypt. Nature Communications. Vol. 8. DOI: https://doi.org/10.1038/s41467-017-00957-y

Mariani M., Connor S.E., Theuerkauf M., Herbert A., Kuneš P., Bowman D., Fletcher M.-S., Head L., Kershaw A.P., Haberle S.G., Stevenson J., Adeleye M., Cadd H., Hopf F., and Briles C. (2022) Disruption of cultural burning promotes shrub encroachment and unprecedented wildfire. Frontiers in Ecology and the Environment. Vol. 20(5). Pg. 292–300.

McAtackney L. (2014) An Archaeology of the Troubles: The Dark Heritage of Long Kesh/Maze Prison. Oxford: Oxford University Press.

National Parks Service. Frequently asked questions [online]: https://www.nps.gov/aboutus/faqs.htm [accessed 20/10/2022]

Neria Y., Nandi A., and Galea S. (2007) Post-traumatic stress disorder following disasters: A systematic review. Psychological Medicine. Vol. 38(4). Pg. 467–480.

Nora P. (1989) Between memory and history: les lieux de mémoire. Representations. Vol. 26. Pg. 7–24.

O'Keefe P., Westgate K., and Wisner B. (1976) Taking the naturalness out of natural disasters. Nature. Vol. 260. Pg. 566–567. DOI: https://doi.org/10.1038/260566a0

Oliver-Smith A. (1999) What is a disaster?': Anthropological perspectives on a persistent question. In Oliver-Smith A. and Hoffman S.M. (eds.) The Angry Earth: Disaster in Anthropological Perspective. New York: Routledge. Pg. 18–34.

Osborne H. (2019) Pompeii archaeologists 'committing vandalism to volcanology' by destroying history of Vesuvius eruption, scientists claim [online]. Newsweek. https://www.newsweek.com/popmpeii-archaeologists-vandalism-vesuvius-eruption-1449676 [accessed 20/10/2022]

Palka (2014) Maya Pilgrimage to Ritual Landscapes: Insights from Archaeology, History, and Ethnography. Albuquerque, NM: University of New Mexico Press.

Pálsson, G. (2012) These are not old ruins: A heritage of the Hrun. International Journal of Historical Archaeology, 16, pp. 559–576.

Pendlebury J., Wang Y.-W., and Law A. (2018) Re-using 'uncomfortable heritage': The case of the 1933 building, Shanghai. International Journal of Heritage Studies. Vol. 24(3). Pg. 211–229.

Pétursdóttir Þ. (2012) Concrete matters: Ruins of modernity and the things called heritage. Journal of Social Archaeology. Vol. 13(1). Pg. 31–53.

Pétursdóttir Þ. and Olsen B. (2014) An archaeology of ruins. In Olsen B. and Pétursdóttir (eds.) Ruin Memories: Materialities, Aesthetics and the Archaeology of the Recent Past. London: Routledge. Pg. 3.

Pomerantz K. (2009) Introduction: World history and environmental history. In Burke E. III and Pomerantz K. (eds.) The Environment and World History. Berkeley: University of California Press.

Reher G.S., López-Merino L., Sánchez-Palencia F.J., and López-Sáez J.A. (2012) Configuring the landscape: Roman mining in the *conventus asturum* (NW *Hispania*). In Kluiving S.J. and Guttman-Bond E.B. (eds.) Landscape Archaeology between Art and Science: From a Multi- to an Interdisciplinary Approach. Landscape & Heritage Studies Proceedings. Amsterdam: Amsterdam University Press. Pg. 127–136.

Reynard E. and Giusti C. (2018) The landscape and the cultural value of geoheritage. In Reynard E. and Brilha J. (eds.) Geoheritage: Assessment, Protection, and Management. Amsterdam: Elsevier. Pg. 147–166.

Rico T. (2016) Constructing Destruction: Heritage Narratives in the Tsunami City. New York: Routledge.

Riede F. and Sheets P. (2020) Framing catastrophes archaeologically. In Riede F. and Sheets P. (eds.) Going Forward by Looking Back: Archaeological Perspectives on Socio-Ecological Crisis, Response, and Collapse. New York: Berghahn Books. Pg. 1–14.

Robbins P. and Moore S.A. (2013) Ecological anxiety disorder: Diagnosing the politics of the anthropocene. Cultural Geographies. Vol. 20(1). Pg. 3–19.

Roberts C. and Stone P.R. (2014) Dark tourism and dark heritage: Emergent themes, issues and consequences. In Convery I., Corsane G., and Davis P. (eds.) Displaced Heritage: Responses to Disaster, Trauma and Loss. Woodbridge: Boydell Press.

Rodriguez H. and Mora M.T. (2020) Hurricane Maria: Disaster response in Puerto Rico. In Oxford research encyclopedia of politics [online]. DOI: https://doi.org/10.1093/acrefore/9780190228637.013.1609

Rothenberg M.A.W. (2021) Community and corrosion: A contemporary archaeology of Montserrat's volcanic crisis in long-term comparative perspective. PhD dissertation in the Joukowsky Institute for Archaeology and the Ancient World, Brown University, Providence, RI.

Santoro A., Venturi M., and Agnoletti M. (2020) Agricultural heritage systems and landscape perception among tourists. The case of Lamole, Chianti (Italy). Sustainability. Vol. 12. DOI: https://doi.org/10.3390/su12093509

Sather-Wagstaff J. (2011) Heritage that Hurts: Tourists in the Memoryscapes of September 11. Walnut Creek, CA: Left Coast Press Inc.

Scandone R., Giacomelli L., Rosi M., and Kilburn C. (2019) Preserve Mount Vesuvius history in digging out Pompeii's. Nature. Vol. 571. DOI: https://doi.org/10.1038/d41586-019-02097-3

Scarlett J.P. (2014) 'Paradise tax': The price Hawaiians are prepared to pay for living near volcanoes [online] The Conversation: https://theconversation.com/paradise-tax-the-price-hawaiians-are-prepared-to-pay-for-living-near-volcanoes-34375 [accessed 25/04/2022]

Scarlett J.P. and Riede F. (2019) The dark geocultural heritage of volcanoes: Combining cultural and geoheritage perspectives for mutual benefit. Journal of Geoheritage. DOI: https://doi.org/10.1007/s12371-019-00381-2

Scazzosi L. (2004) Reading and assessing the landscape as cultural and historical heritage. Landscape Research. Vol. 29(4). Pg. 335–355.

Seeberg J. and Padmawati R.S. (2015) Between the queen of the South Sea and the spirit of Mount Merapi – Political and cosmological dimensions of the Central Java earthquake in 2006. In: Riede F. (ed.) Past Vulnerability. Volcanic Eruptions and Human Vulnerability in Traditional Societies Past and Present. Aarhus: Aarhus University. Pg. 23–37.

Sesana E., Gagnon A.S., Ciantelli C., Cassar J., and Hughes J.J. (2021) Climate change impacts on cultural heritage: A literature review. Wiley Interdisciplinary Reviews: Climate Change. Vol. 12(4). Pg. e710. DOI: https://doi.org/10.1002/wcc.710

Sharpley R. (2009) Shedding light on dark tourism: An introduction. In Sharpley R. and Stone P.R. (eds.) The Darker Side of Travel: The Theory and Practice of Dark Tourism. Bristol: Channel View Publications. Pg. 3–22.

Skinner J. (2018a) Plymouth, Montserrat: Apocalyptic dark tourism at the Pompeii of the Caribbean. International Journal of Tourism Cities. Vol. 4(1). Pg. 123–139.

Skinner J. (2018b) The smoke of an eruption and the dust of an earthquake: Dark tourism, the sublime, and the re-animation of the disaster location. In Stone P.R., Hartmann R., Seaton A.V., Sharpley R., and White L. (eds.) The Palgrave Handbook of Dark Tourism Studies. London: Palgrave Macmillan. Pg. 125–150.

Sorriso-Valvo M. (2008) Natural hazards and natural heritage: Origins and interference with cultural heritage. Geografia Fisica e Dinamica Quaternaria. Vol. 31. Pg. 231–237.

Strange C. and Kempa M. (2003) Shades of dark tourism: Alcatraz and Robben Island. Annals of Tourism Research. Vol. 30(2). Pg. 386–405.

Tumarkin M. (2005) Traumascapes: The Power and Fate of Places Transformed by Tragedy. Carlton, Australia: Melbourne University Press.

Tumarkin M. (2019) Twenty years of thinking about traumascapes. Fabrications. Vol. 29(1). Pg. 4–20.

UNESCO World Heritage Convention (undated). World Heritage List [online]: https://whc.unesco.org/en/list/ [accessed 20/10/2022]

Varanda J. and Cleveland T. (2013) Unhealthy relationships: African labourers, profits and health services in Angola's colonial-era diamond mines, 1917–75. Medical History. Vol. 58(1). Pg. 87–105.

Vitousek P.M., Mooney H.A., Lubchenco J., and Melillo J.M. (1997) Human domination of Earth's ecosystems. Science. Vol. 277. Pg. 494–499.

Walker C.F. (2008) Shaky Colonialism: The 1746 Earthquake-Tsunami in Lima, Peru, and Its Long Aftermath. Durham, NC: Duke University Press.

Wang L., Tian M., Wen X., Zhao L., Song J., Sun M., Wang H., Lan Y., and Sun M. (2014) Geoconservation and geotourism in Arxan-Chaihe volcano area, inner Mongolia, China. Quaternary International. Vol. 349. Pg. 384–391.

Yablon N. (2009) Untimely Ruins: An Archaeology of American Urban Modernity, 1819–1919. Chicago: University of Chicago Press.

Section IV
Moving Forward

17 Arts and Other Cultural Expressions as Tools for Disaster Risk Management

Claudia González-Muzzio, Claudia Beatriz Cárdenas Becerra, and Bernadette Esquive

Could art and culture can be considered as a mechanism for disaster risk management and disaster risk reduction? Is it something that is done spontaneously or is it deliberate?

The definition of what is art is controversial although it is usually associated with a creative activity: 'traditionally, artworks are intentionally endowed by their makers with properties, often sensory, having a significant degree of aesthetic interest, usually surpassing that of most everyday objects' (Adajian, 2022: n.p.). In Spanish, the dictionary of the Real Academia Española (2022) states that art is a 'Manifestación de la actividad humana mediante la cual se interpreta lo real o se plasma lo imaginado con recursos plásticos, lingüísticos o sonoros'.[1] In English, 'something that is created with imagination and skill and that is beautiful or that expresses important ideas or feelings'. Regardless of the aesthetic sense (or not) that art has, it expresses ideas and feelings through a creative process, a manifestation of human beings.

Culture is also a contested concept, and its notion is 'dynamically changing over time and space – the product of ongoing human interaction' (Skelton & Allen, 1999: 4). However, we can take UNESCO's definition of culture as 'the set of distinctive spiritual, material, intellectual and emotional features of society or a social group, that encompasses, not only art and literature, but lifestyles, ways of living together, value systems, traditions and beliefs' (UNESCO, 2001). According to Daskon and Binns (2009, in Kulatunga, 2010), culture is related with livelihood and opportunities for development.

The 'culture of prevention' appeals to the attitude and actions of human beings that allow us to avoid and reduce those situations in life that could cause harm or damage. In this context, Anthony Oliver-Smith stated that cultural factors influence behaviour of people when facing a hazard (Oliver-Smith, 1996).

When confronting a threat, people not only consider the danger but also give priority to factors like social values, religious beliefs, traditions, and attachment to place (Kulatunga, 2010: 4). However, not always these beliefs prevent people from suffering or being hurt when a disaster occurs.

But preventive behaviour is prevalent, it is then that we could speak of a culture that allows people to avoid situations and actions that cause damage and losses to the social collective. The culture of prevention, according to Luis Bruzón Delgado (CECC/SICA Culture Coordinator), 'requires a series of values such as precaution, responsibility, altruism, social awareness, and cooperation' (CECC/SICA & USAID, 2018).

The Hyogo Framework for Action (2005–2015) recognised in its Priority for Action 3: 'use knowledge, innovation and education to build a culture of safety and resilience at all levels', and that traditional and indigenous knowledge and cultural heritage should be incorporated among the information provided, taking into account cultural and social factors (UNISDR, 2005). Likewise, the Sendai Framework for DRR in its Priority 1 states regarding its implementation: 'to ensure the use of traditional, indigenous and local knowledge and practices, as appropriate, to complement scientific knowledge in disaster risk assessment and in the development and implementation of policies, strategies, plans and programmes of specific sectors, with a cross-sectoral approach, which should be tailored to localities and to the context' (United Nations, 2015: 15) (for more details, see Chapter 3).

In this chapter, we will focus on how artistic and other cultural expressions, including those coming from traditional knowledge, despite may not be sufficient to address the risks faced by today's society, indubitably are a mechanism to support disaster risk management, focusing on prevention, risk reduction, response, and/or recovery; in a way that contribute to increase resilience of the population.

We Will Not Forget

The examples (see also Box 17.1) below show how through art and culture it is possible to activate memory and remembrance of the consequences of past events in order to prevent future disasters.

The Fundación Proyecta Memoria

Proyecta Memoria Foundation arose in Concepción after the earthquake and tsunami on 27 February 2010: a group of architects and students decided to re-signify debris from the catastrophe and particularly those from destroyed heritage buildings, returning them to the community in public spaces as 'elements of memory'. Subsequently, the foundation focused on the development of artistic and educational projects to remember past disasters, to account for their consequences and their imprint on history, with 'alerts of memory' and 'hypocentres of memory'

Box 17.1   Research and education on disasters through art. DESARTES.[3] CIGIDEN's Arts and Disaster Unit

CIGIDEN[4] is a research centre promoted by four Chilean universities[5] led by Pontifical University Catholic of Chile, which was founded as a result of the earthquake and tsunami occurred in the country in 2010.

DESARTES, its unit of Arts and Disasters, seeks to generate a bridge between scientific research and artworks that address disasters in Chile. Through artistic exhibitions, such as ALUVIÓN, artists reflect and produce works based on research. For example, Gutiérrez and Briceño's work, *Viene bajando la Quebrada* (2022), is a spatial and sound installation that explores the 'sensory ethnography' of a CIGIDEN study, recreating the sound of alluvium and activating itself according to the years of alluvium recorded in the Quebrada de Macul from 1900 to the present. It focused on both recurrence of a natural hazard and auditive memory and emotions of it through sound (CIGIDEN, 2022).

DESARTES aims to coordinate, produce, and disseminate projects based on art and culture for the study and intervention in disasters. The unit's programme comprises three lines of activities: interdisciplinary meetings, seminars on art projects, and circulation of art works (www.desartes.cl).

Figure 17.1 'Hypocentres of memory'.
Source: Proyecta Memoria (2016).

(Figure 17.1). They argue that reconstruction is not only material and memory is a tool for recovery, healing, and prevention (Proyecta Memoria, n/d).

The initiatives of Proyecta Memoria, throughout ludic and artistic interventions, have the intention of bringing to memory past events and raising awareness, in a country with recurrent disastrous events. It seeks that through the knowledge of these causes and effects, the population will avoid repeating mistakes and reproducing vulnerabilities, contributing to the prevention of future disasters.

Songs and Murals after the 1939 Earthquake

On 24 January 1939 a 7.8 (Mw) earthquake severely affected the city of Chillán and others in the centre and south of Chile, with around 25,000 people were killed. It is the earthquake with the highest number of human losses recorded in this country (Figure 17.2).

Among other expressions, songs were composed to remember the event, the loss of loved ones and the hope for reconstruction, as well as recognising the importance of support and aid received from neighbouring countries (Figure 17.3).

> Van a escuchar los sentidos versos y arreglos de este vals, como un responso sobre el sueño grave de los muertos, en la tragedia que nunca se ha de olvidar.[2]
> (opening recitation of the waltz 'Bajo Los Escombros de Chillán', by Victor Acosta)

Although this theme is not well known today, perhaps because waltz is no longer a predominant musical expression, there are others that remain to this day as a symbol of a monumental process of reconstruction.

Figure 17.2 Chillan after the earthquake, 1939.
Source: Zig-Zag (1939: 36).

The Mexican government, in an act of solidarity promoted by the poet Pablo Neruda, sent muralists David Alfaro Siqueiros and Xavier Guerrero to the city of Chillán with the mission of create and paint murals for the Escuela México, an educational building funded by the Mexican government (Figure 17.4). The Escuela México, in the national context, acquires significance for being one of the first works built by the *Sociedad Constructora de Establecimientos Educacionales*, an institution created during the process of reconstruction the earthquake by the government of Pedro Aguirre Cerda, whose motto was 'to govern is to educate'.

Its relevance is also widely acknowledged by the community as an identity symbol of the city, which recognises its historical and touristic importance. The building is protected by the municipality in its land use planning instrument. Likewise, the murals 'Mexico to Chile' by Xavier Guerrero and 'Death to the Invader' by David Alfaro Siqueiros were declared national monuments in 2004. They were restored before and after the 2010 earthquake by Mexican restorers and conservators, thanks to an agreement between Mexico and Chile.

In Mourning and Recovery

The following experiences exemplify how the arts can contribute to processes of emotional recovery after an emergency, to building a sense of community, to share the burden of distress as well as for political advocacy.

The Spring Festival is a celebration initiated by students from the secondary school of Dolores in the Department of Soriano, Uruguay, in 1960, to memorialise the beginning of spring. It has

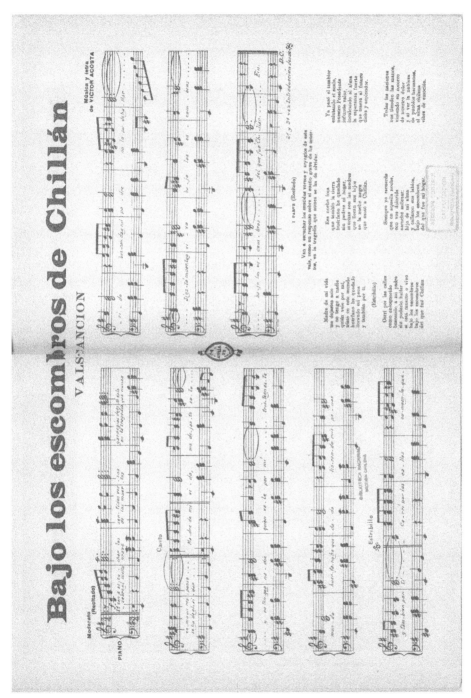

Figure 17.3 'Bajo los escombros de Chillán (vals-canción)' (Under the debris of Chillán: waltz-song).

Source: Biblioteca Nacional de Chile (n/d).

Figure 17.4 Mural 'Mexico to Chile' by Xavier Guerrero.

become the main event in the locality, with the collaboration of students, their families, and the teaching staff gaining national significance.

On 15 April 2016, a tornado with winds of approximately 250 to 330 km/h hit the town of Dolores; 5 people were killed, nearly 200 injured, and over 9,000 people affected. This was considered the second most devastating tornado in the country; it literally crossed the city as it formed on the outskirts of the urban area (La Diaria, 2016).

Despite the destruction, the emotional distress, and widespread economic impacts of the event, a few months after the disaster, the locality and its inhabitants organised and celebrated the Spring Festival (Figure 17.5), spending less resources on its production than in previous

Figure 17.5 Spring festival celebrated in 2016.
Source: fiestanacionaldelaprimavera.com (2016).

years and obtaining higher profits, which directly contributed to the emotional recovery of the community for rebuilding their lives and the city.

> 'The city has revived and is pleasantly surprised. People decorated the front of their houses with balloons, flowers and butterflies, there are loudspeakers with music playing in the neighbourhoods and the "gurisada" took to the streets to rehearse choreographies. This did the community a lot of good', explained one of the organisers.
>
> (Rojas, 2016)

La Marca del Agua and Volver a Habitar: flooding in the city of La Plata, 2013

In Argentina, the city of La Plata, capital of Buenos Aires province, was severely affected by a flood on 2 April 2013, considered the worst in its history, after it rained nearly 400 mm in a few hours. Approximately 35% of the city houses were affected, with severe damage to critical infrastructure, such as the refinery and some hospitals, among others; approximately 100 people died due to causes related to the disaster. In La Plata, a flood was not unexpected as the city has been affected by several serious flooding throughout its history.

After the event, a series of initiatives were developed with the aim of utilising art 'as a way of generating meanings around the flood in conjunction with the city's neighbours and in different urban spaces' (Capasso & Muñoz, 2016: 85). *La marca del agua*[6] and *Volver a habitar*[7] are initiatives analysed by Capasso and Muñoz (2016) as social practices based on cultural expressions that seek to reconstruct the 'community fabric', both arising from collectives made up mainly of artists: editors, photographers, designers, videographers, and interviewers from the city of La Plata. They also sought to make social and institutional structures uncomfortable as a political action.

The collective 'Volver a Habitar' made murals in seven neighbourhoods affected by the flood in collaboration with the neighbours (Figure 17.6). The description on their Facebook page reads: 'This is a collective work project to foster new spaces of encounter and creation. It is an open-air community museum, conceived as a meeting space for muralists, painters, graffiti artists and neighbours. It is a museum for those from below'. This collective was organised out of the need to give meaning to the catastrophe by creating artistic works that produce a collective memory, contributing to increase solidarity and social ties among neighbours.

'La marca del agua', another artistic collective also organised after the flood to deal with the catastrophe yet characterised by a different kind of artistic practice. It originated from a 'poetic action' that broke into the municipal Book Fair at the Pasaje Dardo Rocha Cultural Centre. The group had the twofold explicit objective of making themselves visible as local publishers excluded from the fair, and the purpose of making visible and denouncing the official silence about the catastrophe two months after the event. Public lectures of poems both by artists and 'common people' have been their means of expression in different public events so far (Capasso & Muñoz, 2016).

The most important artistic event took place on 2 April 2014, one year after the flood. On that date, a meeting of self-organised cultural collectives from civil society took place. This action, called 'Desbordes. Colectivo de Colectivos', organised activities such as sculpture and photography exhibitions, an open radio, with a square as a meeting place. With the aim of denouncing the concealment regarding some facts related to the flood (i.e., the real number of deaths and the responsibility of the government in the lack of necessary hydraulic works to reduce the risk of flooding), artistic and communication strategies were the tools

Figure 17.6 'Volver a habitar' by DAN, licensed under CC BY-NC-SA 2.0.

chosen to make visible, denounce what happened, and generate 'noise' in the urban space. (Capasso & Muñoz, 2016). The experience brought together more than 30 local collectives, repeating the event each year since then, along with other events such as photograph exhibitions (Figure 17.7) with both a political purpose of denounce and to promote collective art expressions.

Figure 17.7 Exhibition 'Seguimos inundadxs: archivos e imaginaciones del agua común'. (We are still flooded: Archives and imaginations on common water).
Source: SADO Colectivo Fotográfico (2019).

Art Therapy, Some Experiences and Applications

Art therapy has been employed in different places as a means for the treatment of post-traumatic stress in people affected by disasters with a wide range of geographic applications. Chilcote (2007) employed art therapy with children that survived the 2004 tsunami in Sri Lanka. Many of these children were not provided with the means to express their grief – and it is through art that they were given such an opportunity, through a process comprising a series of sessions which allowed therapists gain children confidence, open their feelings, and helping them mourning and process losses. This author remarks on the importance of interventions culturally sensitive which means to understand local culture, education systems, and rituals of expressing grief.

The following cases show the potential of art therapy to contribute not only to treat post-traumatic stress but also to promote collective recovery after disasters and contribute to reconstruction processes.

1 The Rogue Foundation: 'I am Haiti' initiative

On 12 January 2010, an earthquake in Haiti left more than 200,000 dead and caused massive destruction in the capital Port-au-Prince and other population centres in a country already immersed in a chronic crisis. Children, who were deeply affected, were not always given the required attention and needs such as education was postponed in favour of other needs such as shelter, food, or medical care in disaster contexts. The case of the Marantha School and Orphanage is interesting in this sense.

The Rogue Foundation is an organisation that aims to empower children in conflict zones and economically challenged environments around the world through art and creativity.[8] All projects of the Rogue Foundation are self-funded by events and art exhibitions. More than 200 images inspired the 'I am Haiti' exhibition, with art pieces made by Marantha's students. In addition, the children's artworks were later commissioned to be made into luxury scarves (Alegría Tejeda, 2018). All the sales funded the rebuilding of the school, destroyed by the earthquake. Children felt responsible and empowered for successfully rebuilding their school because they were the authors of artwork sold to fund it. Today, Marantha is recognised as one of the best schools in Port-au-Prince with students scoring higher than the national average.

The most recurring topics were houses children dreamed about for their families, along with flowers and trees for the new city and houses.

2   Art therapy in Manta after the earthquake in Ecuador, 2016

On 16 April 2010, an earthquake 7.8 (Mw) hit Ecuador, and especially the province of Manabí, where Manta, city is located. This is the biggest and most populated urban area of the province, and one of the most severely affected. After the event, there were several interventions by artists who brought their art to hundreds of people in shelters, some of which were collected and analysed by Henríquez Coronel and Navas Guzmán (2017). The authors highlight the fact although initial response of people was of certain rejection because they were expecting material donations from 'visitors'. But the artistic interventions achieved changes in the state of mind, especially through the integration of children through drawing and singing as well as the engagement of the older people. Artistic initiatives have some basic conventions: not to intervene, not to victimise, focus on listening, not to make the artist visible but to seek the involvement of the people (Henríquez Coronel & Navas Guzmán, 2017).

The Department of Culture of the Universidad Laica Eloy Alfaro de Manabí carried out six initiatives involving dance, theatre, music, puppets, and other activities carried out by a variety of artists, from painters, singers, musicians, dancers, actors and actresses, street artists and even a chess player, who contributed with their art to the collective proposals organised by the university congregating groups and individuals from different parts of the world.

Another initiative was proposed by the Centro de Artes La Trinchera, a cultural space created by the Fundación Cultural La Trinchera, for the city of Manta. They self-organised after the earthquake acting as a unifying force to create what was called the artistic caravan 'Arte por la Vida' (Art for Life). They decided to propose dramatic plays. The script was based on two characters who were elderly and had already lived through several earthquakes, by sharing their experiences on previous earthquakes they remembered that people could rebuild the city and move forward. The work also integrated a diversity of artists such as musicians, jugglers, dancers, actors, and photographers.

3   Ay, María! And other artistic initiatives in Puerto Rico after the hurricane Maria

This play was prepared in response to the effects of Hurricane Maria in Puerto Rico in 2017. Artists could not work, just like many of the population in the country, due to lack of electricity, water, and other services that caused the interruption of most of activities at national level (Alegría Tejeda, 2018). The production house Ipso Facto mobilised a group of actors and actresses that performed in the 78 municipalities of the island in a 'mobile theatre', building a space for reflection and healing through a collective creation based on people's experiences of the hurricane.

¡Ay, María! was one of many artistic proposals that emerged on the island after the hurricane. For example, the Museo de Las Américas, located in Old San Juan, three months after the disaster held a call for artists in all disciplines and the community in general called Catarsis: re/constructing after María (Museo de las Américas, 2018), to express their feelings about the effect of Hurricane María on their lives. 'We will have the opportunity to share in one place our works, stories and feelings to somehow purge, purify and release our feelings caused by this event', said López Vilella. (Sin Comillas.com, 2017).

Grandparents Said: Legends and Songs that Save Lives

Traditional knowledge considers memory as a vehicle for making historical practices visible and giving them sustainability through transfer to other generations (Tironi & Molina, 2019). Although the recognition and incorporation of local knowledge in mitigation, management, and recovery plans remains a challenge, the following cases present legends and songs laid on traditional knowledge that contribute to build a culture of disaster prevention and have saved lives during disasters. 'They are informal knowledge that, nevertheless, are based on the accumulation and systematisation of concrete experiences with changing natures. They are collections that are often passed orally from generation to generation, or that are simply an immanent part of the worlds that peoples, collectives, and communities construct and inhabit' (Tironi & Molina, 2019: 5).

Kai-Kai and Treng-Treng,[9] and Other Legends and Songs

The story of Kai-Kai and Treng-Treng is considered by some authors the myth of origin of Mapuche people (Bengoa, 2000; ; Díaz, 2007). It tells about the constant conflict among two snakes, representing water and land. Kai-Kai lives in the sea and wants to get rid of humanity, and Treng-Treng, is another giant snake that lives on land, camouflaged in the mountains, who loves and protects people (Figure 17.8a and 17.8b). The struggle remains in the form of tsunami, earthquakes and volcanic eruptions. One day, Kai-Kai began to flood the territory, but Treng-Treng allowed people to seek refuge on the hills and began to grow.

When the great tsunami after the earthquake of Valdivia in 1960 took place, many people saved their lives by evacuating to the mountains they called ten-ten or treng-treng. 50 years later, the voluntary evacuation of all residents of the coastal settlement of Tirúa (Bio-Bío Region) and the residents of Puerto Saavedra (Araucanía Region) in Chile to the surrounding highlands after the earthquake occurred on 27 February 2010 was remarkable and took place prior to any official evacuation notice, resulting in no fatalities due to the tsunami in those settlements. On the contrary, the evacuation order was given too late, contributing to the death toll in other

Figure 17.8 'Simple' and 'complex' figures representing the two snakes Kai-Kai and Treng-Treng.
Source: Mege R. (1991).

areas of the country. Both towns have a high percentage of Lavkenche population (Mapuche people from the coast and lakes).

The transmission of ancestral knowledge can be considered as a form of risk communication, and disaster risk prevention, expresses through a process based on an exchange of information between different actors in society with the purpose of sharing meaningful messages (Cadierno Gutiérrez, 2018). In this case, a collection of stories called '*piam*' (also present in some artistic works) it refers both to a preventive behaviour and the recognition of a territory where some places have a role for protection in emergency situations (Mege R, 1991; Cadierno Gutiérrez, 2018; Molina Camacho et al., 2018; CIGIDEN, 2021).

It is very interesting to note that in other places in the world this kind of tales passed from generation to generation have also allowed people to protect themselves after a tsunami. For example, after the Indian Ocean tsunami in 2004, the Simeulue communities successfully survived, whereas migrants and tourists who did not have any knowledge about past disasters occurred in the area were severely affected (Arunotai, 2008). Knowledge on tsunamis was gained through a set of traditional stories which are commonly about events in the past and carry an historical as well as a moral message. Some of them referred to an earthquake and tsunami that affected Simeulue island in 1907 and are acquired by the population in the form of lullabies mothers and grandmothers sing to infants and young children (Arunotai, 2008; Sutton et al., 2021). Likewise, the Moken tribe, in the Surin islands, Thailand, were aware of tsunami also through stories regarding laboon or 'seventh wave' according to what if the laboon comes, they have to go to the mountains or head to deep water. This knowledge helped them to respond accordingly when the tsunami occurred without warning in 2004 (Smillie, 2014).

Box 17.2   Hazards and disasters in music, a tool for disaster risk reduction

An interesting discussion on how songs and music can contribute (or not) to disaster risk reduction and how their use has evolved through history by means of a literature review and the analysis of lyrics and descriptions is the one developed by Bob Alexander (2012).

These artistic expressions have been employed in many ways: 'education tools, as history, as coping and relief mechanisms, and as expressions of disaster risk capacities' (Alexander, 2012: 131).

Music and songs are an effective tool for DRR as it is easier to remember and other kind of teaching materials. For that reason, it is being used in different parts of the world such as Australia, USA, Indonesia, Bangladesh) to learn about hazards, know what to do in case of disasters, climate change, and remembering past events.

Alexander also highlights the work of the non-profit Music for Relief established by artists and music industry professionals for disaster relief and mitigation. It was founded by the band Linkin Park in 2005, and it 'aims to support immediate and long-term disaster relief with a primary goal of making a powerful and sustainable impact in highly affected areas' (www.musicforrelief.org).

He also discusses on how songs in popular music 'representing hazards and disasters seemed ill-suited to be the communication medium' for DRR because, in many cases, they 'cherishes the tragic and the heroic, not long-term good practices' for reducing vulnerability and other root causes of disasters (Alexander, 2012: 139). Sutton et al. (2021) also consider music and songs as one of many means of risk communication.

Some Reflections

Traditional stories and songs, strongly linked to the territory and people's experience at it, allow a better understanding of the nature of the hazards that may affect a specific place, especially with reference to earthquakes and tsunamis but also to other hazards such as floods and volcanic eruptions (Thrush & Ludwin, 2007; Srivastava, 2012). They have proved to be an effective way to communicate risks and prepare people to evacuation in case of emergency, although both types of knowledge are needed for effective DRR (Mercer et al., 2012).

Given the intertwining of communities with their territory and surrounding nature, considering culture as part of the dialectic process between nature and society is key to understand how they relate to each other, and for promoting community development strategies for a sustainable future.

Arts and other cultural expressions have the potential to access diverse audiences when installing issues of risk, disaster, and environmental damage (Alexander, 2012) for preventing damage and losses when an event occurs, reducing risks, and contributing to recovery and reconstruction processes. In the same manner, artistic practices as cultural manifestations, contribute to support values such as solidarity, empathy, attachment to place, and sense of community, which allow the population and places to recover in a sustainable way both physically and emotionally, configuring at the same time elements of memory that will allow reducing the risk to future generations. Artistic production and practices not only allow the reconstitution of community meanings but also of their affective ties contributing to the processes of signification of disasters.

Most of the examples provided have a collective origin and involve a wide range of exponents including painters, actors and actresses, poets, writers, sculptors, musicians, and others. Involvement of the 'audience' is considered crucial and desirable, they not only 'receive' art as a gift but also contribute to creations with their experiences, feelings, and hopes, thus avoiding to consider them as 'victims'.

Emergent groups and behaviours are common after disasters and contribute to face their effects (Quarantelli, 1994; Provitolo et al., 2011). As in other cases, regarding artistic collectives, some of them emerge immediately after the emergency to help people to cope with post-traumatic effects of the event, while others appear later in response to an identified need or with a political agency to raise awareness on what is perceived as insufficient action from the government to reduce vulnerabilities or promote sustainable reconstruction through artistic activism (Mouffe, 2007). While most of those groups dissolve soon after the disasters, others shift their focus to foster disaster prevention and risk reduction in the long term.

Key Points

- There are different ways in which people process disasters and the traumas they produce. Art is one of them. Providing ways for people to express their feelings can help them overcome difficulties caused by disastrous events.
- Reconstructing after a disaster is not just a matter of rebuilding constructions; memory is a tool for recovery and social cohesion after adverse events. It helps to effective learning and to prevent repeating dangerous behaviours in the future contributing to be better prepared.
- Culture, cultural and artistic activities are often postponed behind other needs, however, they are crucial for people, individually and as a collective, to overcome difficulties and obstacles that communities face.
- We should pay attention to the grandparents' experiences. They can help save our lives. Past generations have learned through experience, it is something we should consider for the future.

Notes

1 'A manifestation of human activity through which the real is interpreted or the imagined is shaped with plastic, linguistic or sonorous resources' (authors' translation).
2 *You will listen to the heartfelt verses and arpeggios of this waltz, like a dirge over the grave sleep of the dead, in the tragedy that must never be forgotten.*
3 A contracted term from Desastres and Artes (Disasters and Arts), would be 'Disart'.
4 Centro de Investigación Integrada para la Gestión del Riesgo de Desastres.
5 Pontificia Universidad Católica de Chile, Universidad Técnica Federico Santa María, Universidad Andrés Bello, and Universidad Católica del Norte.
6 'The mark of water'.
7 'Return to inhabit'.
8 'I am Haiti experience' by Rogue Films can be watched on https://vimeo.com/319630312
9 Kai-Kai and Treng-Treng or Cai cai vilu and ten ten are some of the given names to a story some consider the myth of origin of Mapuche people and/or a legend on the relation among land and water.

References and Suggested Lectures

Adajian, T. (2022). The definition of art, in Edward N. Zalta (ed.), The Stanford Encyclopedia of Philosophy (Spring 2022 Edition). Available online at https://plato.stanford.edu/archives/spr2022/entries/art-definition/.

Alegría Tejeda, P. (2018). El arte como herramienta de intervención comunitaria después de un desastre natural: El caso del Huracán María en Puerto Rico. Trabajo Final de Máster en Mediación Artística: Arte para la transformación social, la inclusión social y el desarrollo comunitario, Barcelona, Spain: Universitat de Barcelona.

Alexander, B. (2012). Hazards and disasters represented in music, in J.C. Gaillard & I. Kelman (Eds.), Handbook of Hazards and Disaster Risk Reduction, Taylor and Francis, New York, pp. 131–141.

Arunotai, N. (2008). Saved by an old legend and a keen observation: The case of Moken sea nomads in Thailand, in R. Shaw, N. Uy and J. Baumwoll, eds., Indigenous Knowledge for Disaster Risk Reduction: Good Practices and Lessons Learnt from the Asia-Pacific Region, UNISDR Asia and Pacific, Bangkok, pp. 73–78.

Bengoa, J. (2000). Historia del pueblo mapuche. Siglos XIX y XX, LOM Ediciones, Santiago de Chile.

Boen, T. & Jigyasu, R. (2005). Cultural considerations for post-disaster reconstruction: Post Tsunami challenges. Available online at http://www.adpc.net/irc06/2005/4-6/TBindo1.pdf

Cadierno Gutiérrez, J. (2018). Transmisión de conocimiento ancestral lavkenche sobre terremotos y tsunamis y su implicancia en la reducción del riesgo de desastres. Revista de Estudios Latinoamericanos sobre Reducción del Riesgo de desastres REDER, 2 (2), 16–27.

Capasso, V.C. & Muñoz, M.A. (2016). Arte después de la inundación. Dos casos de procesamiento de la dislocación después de la catástrofe. Política y Cultura, primavera, 45, 79–98.

CECC/SICA & USAID. (2018). Foro 'Reducción del riesgo de desastres, cultura y resiliencia comunal'. Ciudad de Panamá. Available online at https://ceccsica.info/sites/default/files/docs/INFORME%20FINAL%20FORO%20DE%20CULTURA%20Y%20RRD.pdf

Chilcote, R. (2007). Art therapy with child tsunami survivors in Sri Lanka. Estados Unidos: Art Therapy: Journal of the American Art Therapy Association, 24 (4), pp. 156–162. Available online at https://files.eric.ed.gov/fulltext/EJ791439.pdf

CIGIDEN (2021), Instalación artística-científica reproduce las experiencias y memorias de las comunidades Lafquenches durante el terremoto de 1960. 27 October 2021. Available online at https://www.cigiden.cl/instalacion-artistica-cientifica-reproduce-las-experiencias-y-memorias-de-las-comunidades-lafkenches-durante-el-terremoto-de-1960/ [accessed 28 May 2022].

CIGIDEN (2022) Instalación artística conmemora aluviones ocurridos en Quebrada de Macul desde 1900 hasta hoy. Available online at https://www.cigiden.cl/instalacion-artistica-conmemora-aluviones-ocurridos-en-quebrada-de-macul-desde-el-1900-hasta-hoy/

Consejo de Monumentos Nacionales. (n/d). Murales de la Escuela México. Available online at https://www.monumentos.gob.cl/monumentos/monumentos-historicos/murales-escuela-mexico

Daskon, C. & Binns, T. (2009). Culture, tradition and sustainable rural livelihoods: Exploring the culture-development interface in Kandy, Sri Lanka. Community Development Journal, 45 (4), 494–517.

Díaz, J.F. (2007). El mito de 'Treng-Treng Kai-Kai' del pueblo Mapuche. Universidad Católica de Temuco. Revista CUHSO, 14 (1), 43–53.

Fiesta Nacional de la Primavea (2016). Me contaron de una ciudad pequeña. Available online at https://www.fiestanacionaldelaprimavera.com/2016/10/me-contaron-de-una-ciudad-pequena/

Henríquez Coronel, P. & Navas Guzmán, L. (2017). La expresión artística como vehículo para la recuperación emocional en caso de desastres naturales. Terremoto Ecuador 16 Abril 2016. II CONGRESO INTERNACIONAL DE INVESTIGACIÓN EN ARTES VISUALES, ANIAV 2017. Available online at https://www.researchgate.net/publication/320248032_La_expresion_artistica_como_vehiculo_para_la_recuperacion_emocional_en_caso_de_desastres_naturales_Terremoto_Ecuador_16_abril_2016 [accessed 28 May 2022].

Kulatunga, U. (2010). Impact of culture towards disaster risk reduction. International Journal of Strategic Property Management, 4, 304–313.

La Diaria. (2016). Fiesta de la Primavera en Dolores, 21 October 2016. Available online at https://ladiaria.com.uy/articulo/2016/10/fiesta-de-la-primavera-en-dolores/

Mege R, P. (1991). La imagen de las fuerzas: ensayo sobre un mito mapuche. Boletín del Museo Chileno de Arte Precolombino, 5, 9–22.

Mercer, J., Gaillard, J.C., Crowley, K., Shannon, R., Alexander, B., Day, S. & Becker, J. (2012). Culture and disaster risk reduction: Lessons and opportunities. Environmental Hazards, 11 (2): 74–95, DOI: 10.1080/17477891.2011.609876

Molina Camacho, F., Constanzo Belmar, J. & Inostroza Matus, C. (2018). Desastres naturales y territorialidad: El caso de los lafquenche de saavedra. Revista de Geografía Norte Grande, 71, 189–209.

Mouffe, C. (2007). Artistic activism and agonistic spaces. Art & Research: a Journal of Ideas, Contexts and Methods, 1 (2), 1–5. Available online at https://chisineu.files.wordpress.com/2012/07/biblioteca_mouffe_artistic-activism.pdf

Museo de las Américas. (2018). CATARSIS: RE/CONSTRUYENDO después de María. Available online at https://www.museolasamericas.org/catarsis.html

Nacional de Chile, B. (n/d). Bajo los escombros de Chillán [música]: vals-canción [para canto y piano]. Available online at https://www.memoriachilena.gob.cl/602/w3-article-99489.html

Oliver-Smith, A.S. (1996). Anthropological research on hazards and disasters. Annual Review of Anthropology, 25, 303–328.

Provitolo, D., Dubos-Paillard, E. & Müller, J. (2011). Emergent human behaviour during disaster: Thematic versus complex systems approaches. Proceedings of EPNACS 2011 within ECCS'11Emergent Properties in Natural and Artificial Complex Systems Vienna, Austria – 15 September 2011, 47–57.

Proyecta Memoria. (n/d). Fundación Proyecta Memoria. www.proyectamemoria.cl

Proyecta Memoria. (2016). Con masivo ejercicio de memoria Chillán conmemoró el terremoto de 1939. Available online at http://proyectamemoria.cl/?p=2324

Quarantelli, E.L. (1994). Emergent Behaviors and Groups in the Crisis Time Periods of Disasters. University of Delaware, Disaster Research Center. Boulder, USA

Real Academia Española. (2022). Diccionario de la Lengua Española. RAE. Actualización 2022. Available online at https://dle.rae.es/arte

Rogue Films. (2019). Rogue Foundation's 'I Am Haiti' Art Empowerment Project. Available online at https://vimeo.com/319630312

Rojas, D. (2016). Dolores florece tras la tragedia. Tradicional Fiesta de la Primavera se hizo en la ciudad devastada, con 15.000 participantes. El País, Uruguay. Available online at https://www.elpais.com.uy/informacion/dolores-florece-tragedia.html

Sin Comillas.com. (2017). El Museo de las Américas presenta: Catarsis: re/construyendo después de María. Available online at https://sincomillas.com/el-museo-de-las-americas-presenta-catarsis-re-construyendo-despues-de-maria/

Skelton, T. & Allen, T. (Eds.). (1999). Culture and Global Change, Routledge, London.

Smillie, S. (2014). The Last Sea Nomads: Inside the disappearing world of the Moken. *The Guardian*, UK., p. 56. Available online at https://www.theguardian.com/news/2014/dec/15/the-last-sea-nomads-new-guardian-shorts-book-extract

Srivastava, S. (2012). Managing indigenous and scientific knowledge for resilience building. Journal of Advances in Management Research, 9 (1), 45–63.

Sutton, S.A., Paton, D., Buergelt, P., Sagala, S. & Meilianda, E. (2021). Nandong smong and tsunami lullabies: Song and music as an effective communication tool in disaster risk reduction. International Journal of Disaster Risk Reduction, 65, 102527, ISSN 2212-4209. Available online at https://doi.org/10.1016/j.ijdrr.2021.102527.

Thrush, C. & Ludwin, R. (2007). Finding fault: Indigenous seismology, colonial science, and the rediscovery of earthquakes and tsunamis in Cascadia. American Indian Culture and Research Journal, 31 (4), 1–24.

Tironi, M. & Molina, F. (2019). Desastres desde abajo. El rol de los saberes locales en la gestión del riesgo de desastres. Available online at https://www.cigiden.cl/wp-content/uploads/2021/07/2019.-Tironi-M-Molina-F-Desastres-desde-abajo-.pdf

UNESCO. (2001). UNESCO Universal Declaration on Cultural Diversity. UNESCO, Paris.

United Nations Office in Disaster Risk Reduction (UNISDR) (2005). Hyogo Framework for Action 2005–2015: Building the Resilience of Nations and Communities to Disasters, Geneva.

United Nations Office in Disaster Risk Reduction (UNISDR). (2015). Sendai Framework for Disaster Risk Reduction 2015–2030. The United Nations, Geneva. Available online at https://www.preventionweb.net/files/43291_sendaiframeworkfordrren.pdf

Volver a Habitar. (2013). Barrio barrio que tienes el alma inquieta de un gorrión sentimental. Available online at https://www.facebook.com/volverahabitar/photos/a.321136158016416/361454740651224/

Zig-Zag. (1939). Calle de Chillán tras el terremoto de 1939. Zig-Zag, 2 February 1939, p. 36. Photograph available online at https://www.memoriachilena.gob.cl/602/w3-article-71049.html

18 'Planning for Disasters Facing Heritage at Risk: Ethics and Epistemes'

Fallon Samuels Aidoo

Introduction

Chronic risks and acute threats to cultural heritage – both tangible and intangible, buildings and landscapes, properties, and places – are addressed by planners of disaster reduction, response, and recovery around the world. Plans that protect cultural heritage from disasters propose to strengthen and/or lengthen relationships forged between people, place and the resources that course through them. Still, much variability can be found in the objectives, imperatives, incentives, and foci of these plans. Moreover, who is responsible for their conception, execution, revision, and communication remains a matter of debate at every level of risk management. Planning discourse, scholarship, and pedagogy reflects the vastness and vitality of mitigation, adaptation, and sustainability of heritage – thus, making any summary thereof merely a sectional cut through climate action and a call for more substantive reflection on the feasibility, sustainability, effectiveness, and appropriateness of these actions and agents.

This introduction to planning for disasters focuses on the interplay between the inquiries, interventions, and investments of preservationists operating at different scales and for distinct institutions. To do so, it highlights the connectivity and reciprocity – and lack thereof – between global/international bodies planning disaster risk reduction; national organizations that generate, amplify, and/or evaluate plans for climate adaptation, hazards mitigation, disaster recovery, and emergency management; and the diverse local and regional bodies that apply governmental and non-governmental capital and capacity (private, philanthropic, grassroots, etc.) to cultural heritage risk assessment and management. The national context highlighted here – the United States – serves as a touchstone for comparison of countervailing approaches and critical reflection on these actors and actions.

An Anglo Lens: Limitations and Opportunities

English-language inquiry into vulnerability, adaptability, and recoverability of diverse heritage pales in comparison to the scale, scope, and maturity of investigation into heritage at risk in the many languages spoken and written globally. Yet, the language and its framing of 'heritage at risk' underscores examination and execution of plans for disaster reduction, response, and recovery by the United Nations International Strategy for Disaster Reduction (UNISDR, now UNDRR) and the World Bank Global Facility for Disaster Reduction and Recovery (GFDRR). National heritage trusts and agencies of the United States, United Kingdom, and British commonwealths (e.g., Australia) and colonies (former and current) thus play an outsized role in setting the terms and pace of investigation, intervention, and investment. The National Trust for Historic Preservation, the Canadian Trust, and the Heritage Trust (UK), for example, contribute

DOI: 10.4324/9781003293019-23

not only to the themed studies of federal land/marks managers like the US National Park Service but also to the global Climate Heritage Network, regional heritage conservancies, state historic preservation offices, local preservation commissions, and even neighbourhood and commercial heritage districts. Yet, their relationships to researchers and responders to disaster risk often waver, and in regards to environmental concerns, has waned as social issues took the stage (e.g., The National Trust for Historic Preservation disbanded its Green Lab researchers). Waves of comprehensive and comparative investigation of environmental action or policy amongst preservation planners precede or follow long lapses in inquiry.

The instability and infancy of English-language inquiry create silences within global discourse on cultural heritage vulnerability and sustain existing absences of cultural heritage stewards and scientists in global discourse on disasters. Social vulnerability inquiries, both qualitative and quantitative, connect race, gender, and class to measures of risk (Fothergill & Peek 2004; Zhang & Peacock 2009; Oliver-Smith 2010; Peacock et al. 2014; Sun & Faas 2018). These studies not only recognize 'there's no such thing as a natural disaster', to quote Neil Smith's (2006) essay, but also identify recovery and reconstruction practices, policies, and plans that generate inequities across social categories. In the process of generalising from a singular disaster, however, these assessments can minimize or overlook inequities that precede and endure the event, or power dynamics and differences that planning of disaster response and recovery may generate or affirm. Social vulnerability indices, Jacobs (2019, p. 34) adds, 'place the problem of disasters and inequity at the feet of being Black, being poor and being a woman as opposed to recognising racist, sexist and classist structures.' Both analyses and critiques of social vulnerability shed light on preservation policies or protection practices that reinforce or reproduce what Rivera (2022) and Hsu et al. (2015) call 'procedural vulnerability.'

The case study of this chapter returns to the potentialities of equity planning for disaster vulnerability. First, a few frameworks favoured by planners of risk reduction and disaster recovery undergo examination. These metrics and mindsets, which make up the status quo of preservation planning, range from sustainability indices to 'heritage at risk' theories but revolve around tangible resources and measurable risk not the intangibility and immeasurability of some cultural resources and risks to them. The vulnerability, adaptability, and sustainability of organizations planning preservation and protection of cultural heritage are equally important, yet far fewer frameworks and formulas focus on their capacities and capital. This chapter closes by highlighting these more marginalized formulations of cultural heritage at risk of disaster that account for racialized risks and organizations, specifically scholarship and strategies focused on blackness and indigeneity of heritage sites and their stewards. What cultural heritage managers of the Global South and North know about disaster risks and how they generate, retain, apply, and revise their knowledge warrant more attention than provided here, but this essay prompts critical engagement with how and why procedural and social vulnerability persist as wicked problems despite growing resistance to disaster colonialism amongst grassroots and governmental planners.

Heritage at Risk: A Functional but Faulty Framework for Planning Disaster Mitigation and Reduction

At a glance, planning scholars strive to reduce disaster risks to cultural heritage through their own networks of knowledge production and dissemination. Their scholarship specifically (a) theorizes, critiques, and enhances comprehensive, contextual, and, to a lesser extent, community-devised adaptation and mitigation measures; (b) models the impact of these measures on the vulnerability and adaptability of cultural heritage; (c) measures the effectiveness and costs of these diverse approaches to disaster risk management; (d) assesses adoption of and adherence to plans for disaster risk reduction in distinct contexts and cases; and (e) evaluates access to and affordability of planning for risk reduction amongst different institutions and individuals. The robustness of this research on heritage at risk can be inaccessible to the very organizations responsible for rendering models and metrics applicable and actionable. However, the concept of 'heritage at risk' and calculations thereof have cut through language barriers and crossed epistemic boundaries that otherwise limit risk knowledge-sharing.

A global concern with disasters' impact on cultural heritage took off in 1990 with the launch of the UN International Decade for Natural Disaster Reduction (IDNDR). The initiative synthesized language concerning cultural heritage affected by extreme weather events and provided a dedicated framework for raising questions about the fate of cultural heritage at risk of loss. Before IDNDR, few major publications had considered loss as the result of disasters (Council of Europe 1985; Jones 1986; Feilden 1987), but acknowledgement of and attention to different types of threats and vulnerabilities developed in the decade that followed (Thiel 1992; Spennemann & Look 1998). The initiative ultimately yielded an ICCROM-sponsored manual *Risk Preparedness: A Management Manual for World Cultural Heritage* (Stovel 1998), which recognizes 'disasters are a rarity and thus a phenomenon of limited interest, [sic] and the scale of their destruction far outweighs the cumulative impacts of daily wear and tear with which conservation management is more often concerned' (Rico 2014, p. 162). More frequent and ferocious extreme events in the 21st century – some weather related, others warfare related – compelled planners of disaster preparedness to develop understanding of 'heritage at risk' by climate change as much as economic, social, and political change.

In the 21st century, the concept of 'heritage at risk' has come to stand for vulnerability – a more complex characterization of environmental, economic, socio-cultural, technological, and material conditions of intangible and tangible heritage that encompasses assumptions and assessments of sensitivity and exposure to harm, and the capacity of the heritage to withstand, cope, and adapt to those hazards. Heritage experts at the helm of archaeological and architectural networks developed the meaning and metrics through international meetings and national forums throughout the 1990s, the UN International Decade of Natural Disaster Reduction. The meetings and publications that followed transmitted the concept to national trusts and local governments with the power to regulate practices and implement policies at the community or regional scale. Likewise, the global conferences that followed broached the topic of risk management faced by local bodies grappling with the destructive and damaging effects of climate change (Ross 2021; Rossler 2006). Expert judgment, social consensus, and state sanction thus characterized the 2005 Kyoto Declaration on the Protection of Cultural Properties, Historic Areas and Their Settings, and other mobilising statements of international bodies.

On occasion, these convenings and compendiums captured the knowledge of community and cultural organizations planning for disasters and their reduction. For instance, Trinidad Rico (2014) discusses coastal communities that planned and put into effect cultural regulations – some that restrict specific agricultural, industrial, and fishery practices while others incentivize more constructive or corrective alternatives. Likewise, tourism-dependent cities within deltaic and nautical regions received recognition for their surgical and curatorial approaches to risk reduction such as a policing portfolio that allows some profitable yet harmful practices identifiable as cultural heritage or associated with culture bearers to continue while restricting others (Rindrasih & Witte 2021; Balan 2022). On the whole, however, 'Authorized Heritage Discourse' (AHD) on disaster response and recovery managed by national and international organizations (Rico 2014, p. 158) has recognized little local stewardship of cultural heritage at risk and recognized even fewer cultural understandings of and responses to destruction, decay, and impermanence as pertinent to the production of knowledge and preparation for disasters.

Disasters in Torino (Italy), Kyoto (Japan), Haiti, New Orleans, LA and Mobile, AL (US) during 2004 and 2005 exposed the risks of assessing local disaster risks to cultural heritage from the proverbial 'ivory tower' of global organizations. UN and UNESCO communities that provided thought leadership on disaster risk assessment and management in the wake of these events produced and propagated a praxis of preparedness that revolved around the measurement and mutability of hazards. The 'threat' matrix devised by these experts in ecology, economics, conservation, and culture for World Heritage Sites and National Trusts provided an accurate accounting of environmental and ecological risks to the economic viability and material integrity of these places, but not community, parochial and heirs property called home in disaster-affected and disaster-prone regions (Cambrice 2014; Boger, Perdikaris & Rivero-Collazo 2019; Sou & Webber 2019). Although each matrix connected the integrity, significance, and value of tangible and intangible heritage at risk, their claims stopped short of capturing endangered and damaged systems of knowledge and action (e.g., evacuation routes, property databases) let alone calculating threats to cultures and communities that self-manage risk (Aka et al. 2017; Jon & Purcell 2018). Moreover, matrices – and the vulnerability and sustainability indices generated from them – dismiss or diminish the importance of systemic inequities in preparedness resources and regulatory protections, as well as the development of response systems and recovery plans, the deployment of risk reduction strategies, and the support of diverse community stewards and culture bearers. The absence of qualitative, historical, and systemic knowledge of heritage at risk makes narratives of disaster response, recovery, and resilience ever more valuable to the management of disaster risks but also more difficult to incorporate into planning for disasters due to take place.

The surveying of tangible and intangible heritage – from reconnaissance to research to remembrance – once constituted the common denominator between governmental and grassroots, structural and non-structural approaches to disaster risk management. However, models and matrices of heritage at risk paradoxically grow more extensive and less inclusive via geographical information systems (GIS), a data processing and mapping technology enabled by geolocatable devices (e.g., a smartphone) and geospatial tools (e.g., ground penetrating radar, LIDAR). Proprietary, governmental, and open-source versions of the technology such as ArcGIS, qGIS, and ATLAS, enable planners to amass and manage diverse knowledge of actualized and projected risks, hazards, damage, and harm; to model response or recovery from them; and to map out future actions and actualization. International nongovernmental organizations such as the World Monuments Fund, Kaplan Fund, and the Getty Conservation Institute as well as global corporations like ESRI and Google largely advance these mapping enterprises, prompting in some cases countermapping efforts (Roberts 2020b). GIS-driven cultural resource management at the scale and scope of disasters relies on digitization of existing documentation, collection of new heritage data, and the integration and combination of disparate documentation and data into a comprehensive knowledge base. But, neither the infrastructure nor the expertise to support sprawling digital technologies and analogue techniques of heritage knowledge consolidation are freely available worldwide despite recent investments in open-source GIS and open-access geoscience. Even when they are, environmental degradation and systematic disinvestment in historic buildings and sacred landscapes can preclude their presence in databases designed around exclusionary ideologies of material integrity (Blanks et al. 2021) and/or established techniques and emergent technologies of heritage documentation (Fortenberry 2019).

The resulting repositories of national, state, and local heritage for risk are neither geographically nor categorically inclusive of all heritage at risk, or even, cultural heritage on the frontlines of the world's most ferocious and frequent hazards. From tribal settlements developed by master planners of Taiwan to the Census Designated Places of Black settlement within the deltaic United States, considerable mapped heritage remains absent from the governmental registries of risk on which disaster preparedness planners depend (Huang 2018; Roberts 2020b).

Additionally, resistance to use and proliferation of digital heritage documentation tools and methods – a path taken by indigenous and marginalized stewards of endangered heritage – contributes to some 'null' values in heritage databases and GIS processes pertinent to disaster planning (Brown & Nicholas 2012; Roberts 2020a). Scholars that have constructed digital records of endangered heritage with marginalized, resource-deprived, and systematically disinvested stewards attribute absences in risk knowledge to epistemological issues: the universality of database technology and terminology undercuts community authorship and erases ethnic stewardship of cultural heritage; the expense of LIDAR scanners and GIS subscriptions that generate data precludes participation in related knowledge production and sharing practices by low-income heritage stewards; and databases coded to mine and process analogue documentation restrict the latter to forms of knowledge already sanctioned or utilized by state authorities such as measured drawings (González-Tennant & González-Tennant 2016; Sullivan, Nieves and Synder 2017; Ammon 2018; Minner 2018; ; Roberts 2019).

However misaligned with contemporary conceptions of diversity, equity, and inclusion, sprawling yet selective registers of disaster risks and impacts have pathed the way for planners to frame how, with whom, and for whom diverse individuals and institutions plan to manage cultural heritage. In many such cases, discourse amongst different governmental offices, consultants, and their expert or elite constituencies leads 'heritage at risk' evaluations to include only architecture that entrenched political economies and ecologies can retrofit, reconstruct, restore, and relocate. 'Authorized Heritage Discourse', as Laurajane Smith and other analysts of exclusionary cultural resource management accordingly find the prioritization of select sites, structures, landscapes, and ecologies for protection and preservation follows findable heritage data and fundable heritage management decisions (Smith 2007; Zhang and Lenzer 2020; Coutts 2021; Wendel 2009). By their account, 'heritage at risk' will continue to flourish as a framework for planning the reduction, avoidance, and mitigation of identifiable and measurable risks as long as digital but also political and economic infrastructure renders it actionable.

Translation – literal and conceptual – of indigenous planning theories and practices into international indices and insights increasingly functions as scaffolding for climate action even when social to procedural vulnerabilities limit indigenous agency. 'Disaster colonialism' corresponds to various acts of exploitation and erasure, including epistemic boundary work that delegitimizes indigenous knowledge of climate change and marginalizes cultural recovery and community repair (Rivera 2022a). Tribal, grassroots, charitable, and corporate organizations participate in these public discourses to varying degrees – measurable qualitatively by 'Arnstein's Ladder' of participation (Roberts 2019), legally by indigenous status (Huang 2018), or haphazardly by 'fuzzy' metrics of social inclusion (Chandrasekhar, Zhang & Xiao 2014; Dewi et al. 2019). Despite their epistemic and institutional engagements, subaltern and anti-colonialist theorists of disaster risk reduction, mitigation, and or adaptation remain unrecognized as planners and their work rarely appears in compendiums of reproducible risk reduction plans.

Reduction of a disaster's impact on human and non-human populations consistently but cursorily account for stewardship plans even while discounting said stewards that shape the survival of properties, places, cultures and communities. Plans that meet that threshold emphasize extant and emergent threats to the monumentality, integrity, and/or viability of national and local heritage – even when they incorporate indigenous culture in their matrices of knowledge and models of local action (Smith 2007; Huang 2018). In other words, the 'heritage at risk' framework activates equity planning of disaster preparedness and mitigation but also affirms the power and practices of planning organizations that perpetuate the risks they are entrusted to manage and distrust others to manage.

Planning Protection of Heritage at Risk – an Organizational Matter

Authorizers of heritage at risk – governmental and nongovernmental organizations with the capacity and capital to review, prepare, finance, and implement plans – can plan to reduce damage, redistribute harm, avoid losses, and/or adapt to hazards. Epistemological differences exist between these four approaches: damage reduction requires measurement of disaster impacts (and typically accounting of recovery costs); harm redistribution revolves around impact assessments; risk adaptation rests on observation of responsiveness to risks identified, projected, and experienced plus evaluations of adaptive capacity; and hazards avoidance requires evaluation of disastrous outcomes and acting on their relevance elsewhere and to others. In other words, entities empowered to preserve and protect cultural heritage can engage in these distinct modes, scales, and temporalities of planning mitigation of and adaptation to the risks of typhoons, wildfires, hurricanes, earthquakes, and tornados. But, planning scholarship shows that not many do – even in the United States, where federal and state declarations of a disaster trigger a flood of financial and human resources not only to identify and rectify damage to cultural resources but also to mitigate any additional adverse effects that intervention and investment in disaster recovery and resilience may bring to heritage already at risk (Box 18.1).

Box 18.1 Reducing the risk of washed-out Black heritage

Hurricane Katrina (2005) submerged thousands of single-storey, slab-on-grade homes in New Orleans, displacing the African Americans they disproportionately sheltered. More than 700 houses of nearly 1,100 such deluged houses in New Orleans's Pontchartrain Park Historic District (PPHD) underwent rehabilitation with financial aid and technical assistance from the Federal Emergency Management Agency and the Preservation Resource Center of New Orleans, respectively. At the same time, the historic district lost to demolition nearly 275 damaged but standing ranch houses built for 'coloured people' following passage of the US Fair Housing Act of 1949. These homeowners – nearly all the original homebuyers or their descendants – voluntarily sold their properties to Louisiana Land Trust in exchange for capital to invest in residency out of harm's way. Instead, retreat took the form of structures constructed above projected base flood elevations on cleared lots within PPHD – not by choice or by chance, but in compliance with grant and loan programs requiring Black participation in land buyout programs (*Road Home*) and Black facilitation of green home building and buying (*Build Back Better*).

In this case, neither the storm nor the state took away historic structures signifying African Americans' belonging in the city plus suburbia generally. Housing loss in PPHD points instead to both the storm and the state making space in the political economy of preservation in New Orleans for green building sponsors and stakeholders (i.e., housing manufacturers, insurers, builders, engineers, and designers but also state land trusts and city bond markets). Pontchartrain Park preservationists received $4.1M in funding for housing rehabilitation and community rebuilding; the contracts and covenants associated with these grants and loans from the Salvation Army, Kresge Foundation, Surdna Foundation, Greater New Orleans Foundation, and the Foundation for Louisiana, as well as the US Department of Housing and Urban Development,

Figure 18.1 Elevated EnergyStar® houses under construction in Pontchartrain Park five years after Hurricane Katrina as planned by Community Development Corporation (CDC) founders, descendants of the historic district's elderly homeowners, and prescribed by the CDC's funders, a collection of private, public, and philanthropic organizations.

Source: ATTRIBUTION: Infrogmation of New Orleans, CC BY-SA 2.0 <https://creativecommons.org/licenses/by-sa/2.0/.

restricted design choices and development options to housing typologies inaccessible, physically and financially, to many of Pontchartrain Park's elderly culture bearers: single-family, solar-powered dwellings with front and back steps up to entries above base flood elevation.

Determined not to leave community elders high and dry, descendants of Pontchartrain Park elders planned high-rise residential development and recruited supportive commercial enterprises for high-ground near the historic district (Fig. 18.1). The procedural vulnerability of these Black development and preservation planners – paternalistic planning and philanthropy – kept such refuge from materialising. Instead, charitable individuals and institutions propagated disaster colonialism by permitting and financing sustainable building stock in subsiding landscapes and subprime marketplaces. In New Orleans and elsewhere in the United States (e.g., Princeville, NC), in short, global patrons of climate mitigation and local planners of cultural preservation may co-produce heritage at risk as they relocate BIPOC community stewards to architecture designed to perch above measured, insurable risks.

US Planning for Resilient Colonial Heritage

Many planning organizations preparing for hazards to heritage in their jurisdictions of the United States issue plans that socialize and spatialize only the risks that these organizations are prepared to redress and reduce. Each of the 50 states within the US plus US territories has a unique capacity to access and apply knowledge of heritage at risk, and their own political economy for doing so (i.e., some eschewing actions requiring a 'big government' response or partnership with a private operator). Rather than flatten these differences, federal and state planning invites local managers of cultural heritage and climate hazards to turn their distinctiveness into a competitive advantage over peers in the fight for scarce resources (Berke et al. 2014). Scholars examining these programs find they reinforce resource allocations already made by states, such as state subsidies for local governments that join them in quantitatively and qualitatively accounting for the human, economic, and financial costs of mitigating hazards to historic assets and cultural resources (Rumbach, Bierbrauer & Follingstad 2020). Existing local and state governments may focus their resources on specific geographies such as 'Main Streets' (Appler & Rumbach 2016), 'colonias' (Rivera 2022), colonies (Cepeda 2002; Rumbach & Foley 2014), or unincorporated areas (Roberts 2020a, 2020b). Persistent mismatches between need and resources suggest systemic indifference towards heritage risk management experiences and expertise of these communities (Aidoo 2021; Nelson et al. 2022). The portfolio of relocation, resettlement, redevelopment, retrofit, and/or rehabilitation plans underway throughout the United States, its territories, and colonies in the Caribbean and South Pacific instead represent application of and adherence to hazards mitigation procedures and historic preservation policies confining the finances and flexibility of state and local planning to measurable hazards and manageable risks.

Considered fiscally prudent by federal authorities and state auditors, redundant resource allocation for disaster risk reduction favours *entrenched,* even exclusionary, thought and action over *emergent,* including equity planning frameworks. Exemplified by US programming of post-disaster recovery resources throughout the Gulf South following hurricanes of the 21st century (Katrina, Rita, Ike, Harvey, Maria), many organizations planning to redress social vulnerability have deprioritized and delegitimized reparative justice strategies proposed as alternatives to typical structural or non-structural "'solutions'" (Wendel 2009; Johnson 2011;). The capital, capacity, and authority to pursue these solutions lie out of reach of frontline communities of the Global North and South (Laska 2020; Sheller 2018) in many cases. Planning for biophilic and landscape urbanism, for instance, costs more than engineering coastal heritage protections such as seawalls and levees when public authorities already plan on administration and maintenance of the latter (Fleming 2019). Large-scale plans for non-structural management of heritage at risk that governmental authorities solicit and support, namely the Resilient by Design competition winners, also sit on the proverbial shelf (Holmes 2020) – never forming part of the National Flood Insurance Program or Community Ratings System by which the Federal Emergency Management Agency (FEMA) provides resources and guidance for owners, managers, and stewards of tangible cultural heritage. Alternatively, the feasibility of an emergent or marginal approach depends on its insurability by private sector metrics and markets – assurances to residential, commercial, and even industrial insurers and lenders that reducing disaster risks will pay off (i.e. prevent most costly damage).

Disaster risk reduction guidance – a relatively recent role for FEMA, long a disaster response agency – frames planning as private actions that increase the public's preparedness. FEMA policies and programs present 'preparedness' as multi-dimensional and variable but measured sporadically (e.g., every 5–10 years) either quantitatively or qualitatively (e.g., number of buildings fortified). For instance, FEMA neither compels nor counts periodic photographic documentation of heritage in recovery as valuable data on disaster risk reduction (Hendricks & Meyer 2021a).

Instead, FEMA's preparedness guidance solicits documentation not typically produced by heritage managers and supports expertise not typically responsible for heritage management (e.g., engineers). Not surprisingly, resulting plans prioritize structural solutions: elevating structures, regrading land, swapping materials of construction, and/or designing assets to sustain wind, building structures that dam floodwaters, and prescribing land use and development to contain wildfire and flood. Additionally, federal planners of flood and fire prevention support uprooting heritage exposed to disastrous hazards, specifically changing the location – X-Y coordinates (Phillips, Stukes, & Jenkins. 2012; de Vries and Fraser 2017) and the elevation (Aidoo 2021). Reduction of disaster risk in both cases depends on mobility of otherwise 'immovable heritage' relative to the geography of immutable hazards.

Ironically, capital critical to actualising recommended plans for disaster risk reduction – including hiring the professionals that FEMA suggests – largely follows disastrous weather events and flows from competitive programs administered at unpredictable intervals. Even then, recipients require matching funds – funds often sourced from foundations and charities with patron preferences and/or planning prescriptions such as specific locations for relocation or select designers of heritage retrofits (Johnson 2011; Aidoo 2021). The colour-blind priorities and land swap programs of international charities and global foundations such as Global Green and Surdna Foundation suggest their plans to apply knowledge of risks brewing in oceans, seas, skies, tectonic plates, and soil develops long before and dissipates long after disasters make landfall (Berglund & Loukaitou-Sideris 2016). Predictions, politics, and plans of these non-governmental organizations remain underrepresented in research on disaster risk management broadly (Adams 2013), even though FEMA, the US Department of Housing and Urban Development, the National Park Service, and other federal agencies advise and reward alignment with charitable formulations of and funding for reducing repetitive loss of heritage at risk.

Whether FEMA's hazards mitigation programs and participants that aim to reduce heritage at risk end up endangering cultures and communities remains an open question – a question driving litigation and investigation surrounding disaster risks affecting indigenous peoples and heritage. Plans perpetuating disaster colonialism can also take place within planning organizations reckoning with their role in the racialization of risks, the relational power of respondents to them, and even racial planning to reduce them or their impacts. How they define cultural heritage worth saving and risk management worth taking, i.e., perform risk assessment and risk reduction, is rarely reflected on let alone revised (Rumbach & Nemeth 2014; Rivera 2022). Thus, proliferation of planning reforms in the Black Atlantic in recent years warrants a closer look and case study.

Planning Reduction of Disaster Colonialism throughout the Black Atlantic

Planning practitioners across the globe – not just in the United States – learn from disasters via reflection, recalibration, and revision of obsolete, faulty, and forgotten plans without unlearning risk epistemologies rendering injustice. International organizations entrusted with 'world monuments' or less notable cultural heritage rely on spatial modelling – and the institutional and financial support for modelling the temporality, ferocity, and frequency of anthropic and environmental crises. The complexity of these calculations privilege standard-setting disaster risk reduction strategies substantiated by global scientific knowledge of hazards and vulnerability over context-specific local knowledge drawn from community-based actions (Wisner 2015; Rosa, Santangelo, & Tondelli 2021). From the UN's international strategy for Disaster Risk Reduction (UNDRR) and the World Bank's Global Facility for Disaster Reduction and Recovery (see Chapter 4) to the glocalization frameworks of NGOs such as the Surdna Foundation, conventional metrics of scientific understanding or technocratic expertise determine which vulnerability to redress, when, how, for how long, and with whom. In far more cases than scholars acknowledge, implementation of these plans for disaster risk reduction involves and/or relies on the capital raising infrastructure and the command-and-control structures of national trusts, international bodies, and global NGOs that can regulate and redress risks to public safety, private property, natural resources, and historic assets all at once (Alexander 2002b; Delfin and Gaillard 2008).

UNESCO, ICOMOS, and ICCROM interventions and investments in plan making for World Heritage Sites, national monuments, regional heritage, and local landmarks in partnership with national governments and trusts reinforce regulations and responsibilities (see Chapters 3 and 4). For instance, ICCROM, in partnership with the Smithsonian Institution, descended on Haiti in the wake of the devastating earthquake to shore up cultural institutions lacking staff and infrastructure to safeguard their collections. In this case of 'First Aid to Cultural Heritage' and others, first responders reinforce existing institutions and infrastructure of heritage recognition – even if the disasters reveal the exclusivity or insufficiency of these frameworks to account for probabilities of occurrence, timing, intensity, and interdependency on other hazards near, mid-, and/or long-term (e.g., whether high winds accompany heavy rain or precede a tornado).

The presence of international bodies may eclipse but not erase local narratives of heritage at risk or community-driven planning of risk reduction. Trinidad Rico's critical ethnography of indigenous heritage practices in the aftermath of a tsunami hitting Banda Aceh, Indonesia, in 2004 is but one of many accounts of post-colonial community organising and discourse that counters – but considers – the application of Western ideas concerning heritage at risk and globalized ideologies of risk reduction through expert engagement with heritage destruction and deconstruction. Such critical storytelling is accessible to both globe-trotting heritage professionals nor grounded governmental authorities via books on disasters that have greatly impacted disenfranchised populations and cultures. In them can be found diverse stewards engaged but rarely entrusted with management of risks to heritage. Thus, comparing their planning praxis to that of municipal planning commissions, regional authorities devoted to flood or fire mitigation, state departments of emergency management, national trusts and conservancies, and international protectorates for parks, trails, sites, and landscapes requires planners, and their critics, adopt a new framework for comparative and comprehensive studies of disaster risk reduction. Until then, planning practitioners and analysts must seek and synthesize insights from risk and heritage managers at opposite ends of the capacity, capital, and credibility scale, as well as the earth.

Conclusions

Since the early 1990s, cultural heritage stewards and stakeholders – both grassroots and governmental, capitalist and charitable – have sought for disaster risk reduction to take their experience and expertise into account. Yet, their reports, research, and reporting on resilient and vulnerable heritage remained, well into the 21st century, marginal and, in many cases, absent from disaster policy, programs, projections, and pedagogy (in the English language). The wide-ranging impacts of missed synergies between heritage preservation and disaster prevention have been noted elsewhere (Minner 2016), especially by analysts of geodatabases indexing vulnerability and resilience (e.g., Roberts 2020b) and other knowledge bases upon which recognition and redress of disaster risk depend (e.g., Hendricks & Meyer 2021; Hendrinks & Van Zante 202b). Discontinuities between those bearing and deploying knowledge of tangible and intangible heritage significantly – and arguably disproportionately (Rivera 2022; Jacobs 2021) – affect the hazards and risks faced by frontline communities across the colonized world, not just coasts of the United States (Peacock et al. 2019). These distinct and disparate epistemologies converge and conflict when disaster colonialism comes to a head, offering the chance to discontinue heritage and disaster discourses uncritical of actors and actions jeopardising vulnerable cultures and corresponding heritage. Doing so requires plans to reduce disaster risks to Black and brown bodies, both human and intellectual, that advance anti-colonial, anti-racist worldmaking – an enterprise well outside the scope of most preservation societies, trusts, and schools of thought. Rapidly increasing rates for repetitive loss of cultural heritage in the Global South and Global North ironically offer seemingly endless second chances, however. Repetitive loss, typically partial or piecemeal, offers recurring opportunities to demonstrate culturally competent and collectively constructed plans for disaster risk reduction can take root and be impactful.

When, where, and how established heritage trusts and emergent heritage funds make and meet risk reduction goals are examined at length here but warrant substantial investigation. The interdependency of sustainability and heritage praxis will depend, going forward, not only on cultural heritage managers learning from disaster management planned and practiced by their peers and participants in their projects – a benchmark outlined by Marie-Theres Albert et al. (2017). Planners will also need to take stock of the range of structural and non-structural strategies for disaster risk reduction, the rate at which they develop and replicate, and the places where conception, implementation, evaluation, and maintenance take place. 'Going Beyond' these frames for action calls for heritage managers planning changes to the historical frameworks that shape contemporary action: public narratives, patronage, political economies, and public administration. In short, preservationists planning to curtail risks to cultural heritage, to conserve cultures of stewardship, and to cultivate supportive political economies are effective at reducing disaster risks, and disastrous results only to the extent that they navigate established and emergent economies and epistemes of care.

Key Points

1 Transformation of US programs for hazards mitigation and disaster recovery, designed to systematize and support state planning, has proven instructive for those repeating past missteps and those repairing the physical damage and social harm such mistakes created. The US system of affirming, funding, implementing, and replicating the risk reduction strategies of heritage managers depends on these stewards and stakeholders adopting metrics of measurable action and techniques of universal applicability at odds with the spatial and civil epistemologies of risk that shape their knowledge of hazards, climate, mitigation, and adaptation. Climate adaptability amongst some cultures constitutes actions at odds with preservation principles and policies, while the preservation practices of some communities count as non-compliance with mitigation measures designed to reduce disastrous outcomes of an extreme event.

2 Producing discourses and planning actions that build partnerships between peer practitioners of disaster mitigation has not been the prerogative of local, regional, national, and international organizations seeking a role in disaster risk reduction. Scholars rightfully question whether organizations founded on the preservation of colonial heritage typologies, values, and stewards are capable of producing and propagating an anti-colonial praxis. Research on transformative organizations undergoing their own transformation (e.g., Brand 2020), many informed by Victor Ray's theory of racialized organizations (Ray 2008), suggest the risks of disaster resulting from repetitive use of expedient, universalising, white-washed formulations of heritage at risk will – in time – fuel reflection and redress of their reach, limits, resources, and impacts. Advocacy planning – practiced by preservationists but rarely recognized by scholars or stakeholders (Roberts & Kelly 2019) – speeds up the process. FEMA relinquishing its Katina lexicon of colour-blind urban greening (Campanella 2015; Rivera & Hendricks 2022) in favour of trauma-informed, community-driven managed retreat plans for resettlement of Black, Indigenous, and other people of colour (Nelson et al. 2022) represents one such step towards agnostic assessment and affirmation of competency and capacity to manage risks to heritage. Whether more have followed, and where they're headed, remain open questions.

3 Bellweathers of epistemological and organizational change in preservation planning can buttress and control systems, methods, and ethics of knowledge production that limit the scope and scale of change. The entities empowered to take up novel risk management approaches or engage long marginalized stewards tend to qualify as authorities to insurers, underwriters, and donors. The public, private, and philanthropic sectors not only support, through hazards mitigation grants and climate action plans, specific ways of knowing heritage at risk and the actors and actions protecting them from various hazards. They cultivate political economies of knowledge about hazards and heritage that positions select building typologies, documentation methodologies, and management technologies in disaster risk reduction – selections that sustain some institutions and initiatives but not others. Case in point: The National Trust for Historic Preservation once aimed to collect and compare data on heritage vulnerability and resilience nationwide with that of historic trusts at the local and international levels – and directed fundraising to its Green Lab accordingly. Now it funds climate action on endangered African American cultural heritage such as Pontchartrain Park through a donor-advised fund invested in saving places of significance to Black history. Charitable organizations, disaggregated but dedicated, fill the gap – suggesting the common denominator across national borders is the importance of NGOs and their patrons to planning the future of cultural heritage.

References and Suggested Reading

Adams, Vincanne. *Markets of Sorrow, Labors of Faith: New Orleans in the Wake of Katrina*. Duke University Press, 2013.

Aidoo, F.S. 'Architectures of mis/managed retreat: Black land loss to green housing gains.' *Journal of Environmental Studies and Sciences* 11, no. 3 (2021): 451–464.

Aka, Festus Tongwa, Gaston Wung Buh, Wilson Yatoh Fantong, Isabella Tem Zouh, Serges Laurent Bopda Djomou, Richard Tanwi Ghogomu, Terry Gibson et al. 'Disaster prevention, disaster preparedness and local community resilience within the context of disaster risk management in Cameroon.' *Natural Hazards* 86, no. 1 (2017): 57–88.

Albert, Marie-Theres, Francesco Bandarin and Ana Pereira Roders, Editors *Going Beyond: Perceptions of Sustainability in Heritage Studies No. 2*. Springer, Berlin, 2017.

Alcántara-Ayala, Irasema. 'Time in a bottle: Challenges to disaster studies in Latin America and the Caribbean.' *Disasters* 43 (2019): S18–S27.

Alexander, David. 'From civil defence to civil protection–and back again.' *Disaster Prevention and Management: An International Journal* 11, no. 3 (2002): 209–213.

Ammon, Francesca. 'Digital humanities and the urban built environment: Preserving the histories of urban renewal and historic preservation.' *Preservation Education & Research* 10 (2018): 11–31.

Appler, D. and A. Rumbach. 'Building community resilience through historic preservation.' *Journal of the American Planning Association* 82, no. 2 (2016): 92–103.

Balan, Nina. 'Climate Change and Urban Tourism: Preservation Strategies and Destination Management Policies in Coastal Cities. The Case of Venice.' PhD thesis. University of New Orleans, 2022.

Berglund, Lisa and Anastasia Loukaitou-Sideris. 'The road home: An examination of the successes And challenges of housing non-profits in New Orleans since Katrina.' *Journal of Civil Society* 12, no. 2 (2016): 121–140.

Berke, Philip, John Cooper, Meghan Aminto, Shannon Grabich and Jennifer Horney. 'Adaptive planning for disaster recovery and resiliency: An evaluation of 87 local recovery plans in eight states.' *Journal of the American Planning Association* 80, no. 4 (2014): 310–323.

Blanks, J., A. Abuabara, A. Roberts and J. Semien. 'Preservation at the intersections: Patterns of disproportionate multihazard risk and vulnerability in Louisiana's historic African American cemeteries.' *Environmental Justice* 14, no. 1 (2021): 1–13.

Boger, Rebecca; Perdikaris, Sophia; and Rivero-Collazo, Isabel, "Cultural Heritage and Local Ecological Knowledge under Threat: Two Caribbean Examples from Barbuda and Puerto Rico" (2019). Global Studies Papers & Publications. 6. https://digitalcommons.unl.edu/global/6

Britz, Johannes and Peter Lor. 'A moral reflection on the digitization of Africa's documentary heritage.' *IFLA Journal* 30, no. 3 (2004): 216–223.

Bronin, S.C. 'Law's disaster: Heritage at risk.' *Columbia Journal of Environmental Law* 46 (2020): 489.

Brown, Deidre and George Nicholas. 'Protecting indigenous cultural property in the age of digital democracy: Institutional and communal responses to Canadian First Nations and Māori heritage concerns.' *Journal of Material Culture* 17, no. 3 (2012): 307–324.

Richard Campanella, "A Katrina Lexicon," Places Journal, July 2015. Accessed 09 Oct 2023. https://doi.org/10.22269/150727.

Cepeda, Ricarda P. 'Disaster Mitigation and Recovery Planning for Historic Buildings: Guam as a Case Study.' PhD dissertation. Texas A&M University, 2002.

Chandrasekhar, Divya, Yang Zhang and Yu Xiao. 'Nontraditional participation in disaster recovery planning: Cases from China, India, and the United States.' *Journal of the American Planning Association* 80, no. 4 (2014): 373–384.

Colten, Craig E. 'Environmental management in coastal Louisiana: A historical review.' *Journal of Coastal Research* 33, no. 3 (2017): 699–711.

Coutts, Robert. *Authorized Heritage: Place, Memory, and Historic Sites in Prairie Canada*. University of Manitoba Press, 2021.

Council of Europe (1985) Convention for the Protection of the Architectural Heritage of Europe (Granada, 3.X.1985). Retrieved on January 20, 2006 from: http://conventions.coe.int/Treaty/en/Treaties/Html/121.htm

Crue, C. and J. D Robin Clark. 'Missing links: Connecting emergency management and the cultural heritage industry.' *Journal of Emergency Management* 8, no. 4 (2010): 9–16.

de Vries, D.H. and J.C Fraser. 'Historical waterscape trajectories that need care: The unwanted refurbished flood homes of Kinston's devolved disaster mitigation program.' *Journal of Political Ecology* 24, no. 1 (2017): 931–950.

Delfin Jr, Francisco G. and Jean-Christophe Gaillard. 'Extreme versus quotidian: addressing temporal dichotomies in Philippine disaster management.' *Public Administration and Development: The International Journal of Management Research and Practice* 28, no. 3 (2008): 190–199.

Dewi, Cut, Erna Izziah, Julie Meutia and Nichols. 'Negotiating authorized heritage discourse (AHD) in banda aceh after reconstruction.' *Journal of Architectural Conservation* 25, no. 3 (2019): 211–227.

Eggleston, Jenifer and Jen Wellock. 'The national flood insurance program and historic resources.' *National Trust for Historic Preservation's Preservation Leadership Forum Journal, High Water and High Stakes: Cultural Resources and Climate Change* 29, no. 4 (2015): 34–46.

Feilden, Bernard M. 1987. Between Two Earthquakes: Cultural Property in Seismic Zones. Rome; Marina del Rey, CA: ICCROM; Getty Conservation Institute. http://hdl.handle.net/10020/gci_pubs/between_two_english

FEMA P-1037: Reducing Flood Risk to Residential Buildings That Cannot Be Elevated, September 2015. https://www.fema.gov/sites/default/files/2020-07/fema_P1037_reducing_flood_risk_residential_buildings_cannot_be_elevated_2015.pdf

Fleming, Billy. 'Design and the green new deal.' *Places Journal* (2019).

Fortenberry, Brent. 'Digital documentation in historic preservation education and research: Prospects and perils.' *Preservation Education & Research* 11, no. 1 (2019): 81–116.

Gafford, Farrah D. '"It was a real village" community identity formation among black middle-class residents in Pontchartrain Park.' *Journal of Urban History* 39, no. 1 (2013): 36–58.

González-Tennant, Edward and Diana González-Tennant. 'The practice and theory of new heritage for historical archaeology.' *Historical Archaeology* 50, no. 1 (2016): 187–204.

Hendricks, Marccus D. and Michelle Annette Meyer. 'Modeling long-term housing recovery after technological disaster using a virtual audit with repeated photography.' *Journal of Planning Education and Research* (2021a). DOI: 10.1177/0739456X211002910.

Hendricks, Marccus D. and Shannon Van Zandt. 'Unequal protection revisited: Planning for environmental justice, hazard vulnerability, and critical infrastructure in communities of color.' *Environmental Justice* 14, no. 2 (2021b): 87–97.

Holmes, Rob. 'The problem with solutions'. *Places Journal*, July 2020. Accessed 28 Apr 2022. https://doi.org/10.22269/200714.

Horowitz, A.D. 'Planning before disaster strikes: An introduction to adaptation strategies.' *APT Bulletin: The Journal of Preservation Technology* 47, no. 1 (2016): 40–48.

Horowitz, Andy. *Katrina: A History, 1915–2015*. Cambridge, MA: Harvard University Press, 2020.

Hovanic, C. 'Stronger Than the Storm? Promoting the Post-Sandy Resilience of Historic Resources in New Jersey's Coastal Communities'. Doctoral dissertation. Columbia University, 2016.

Hunter, J. Robert. 'Insuring against natural disasters.' *Journal of Insurance Regulation* 12, no. 4 (1994).

Hsu, M., R. Howitt and F. Miller. 'Procedural vulnerability and institutional capacity deficits in post-disaster recovery and reconstruction: Insights from Wutai Rukai experiences of Typhoon Morakot.' *Human Organization* 74, no. 4 (2015): 308–318.

Huang, Shu-Mei. (2018) Heritage and postdisaster recovery: Indigenous community resilience. *Natural Hazards Review* 19(4). DOI: 10.1061/(ASCE)NH.1527-6996.0000308.

Jacobs, Fayola. 'Black feminism and radical planning: New directions for disaster planning research.' *Planning Theory* 18, no. 1 (2019): 24–39.

Jacobs, Fayola. 'Beyond social vulnerability: COVID-19 as a disaster of racial capitalism.' *Sociologica* 15, no. 1 (2021): 55–65.

Jigyasu, R. 'Reducing disaster risks to urban cultural heritage: Global challenges and opportunities.' *Journal of Heritage Management* 1, no. 1 (2016): 59–67.

Johnson, Cedric. 'Charming accommodations: Progressive Urbanism meets Privatizations in Brad Pitt's Make It Right Foundation.' In Johnson, Cedric, ed. *The Neoliberal Deluge: Hurricane Katrina, Late Capitalism, and the Remaking of New Orleans*. U of Minnesota Press, 2011.

Jon, Ihnji and Mark Purcell. 'Radical resilience: Autonomous self-management in post-disaster recovery planning and practice.' *Planning Theory & Practice* 19, no. 2 (2018): 235–251.

Jones, Barclay G. 'Experiencing loss.' In *Protecting Historic Architecture and Museum Collections from Natural Disasters*, pp. 3–14. Butterworth-Heinemann, 1986.

Laidlaw, P., D. H. Spennemann and C Allan. 'Protecting cultural assets from bushfires: A question of comprehensive planning.' *Disasters* 32, no. 1 (2008): 66–81.

Laska, Shirley. *Louisiana's Response to Extreme Weather: A Coastal State's Adaptation Challenges and Successes*. Springer Nature, 2020. https://link.springer.com/book/10.1007/978-3-030-27205-0#bibliographic-information

López-Marrero, Tania and Ben Wisner. 'Not in the same boat: Disasters and differential vulnerability in the insular Caribbean.' *Caribbean Studies* 40 (2012): 129–168.

Minner, Jennifer. 'Revealing synergies, tensions, and silences between preservation and planning.' *Journal of the American Planning Association* 82, no. 2 (2016): 72–87.

Minner, Jennifer. 'Open data flows, spatial histories, and visualizing the future of preservation.' *Preservation Education and Research* 10 (2018): 33–50.

Minner, Jennifer, Andrea Roberts, Michael Holleran and Joshua Conrad. 'A smart City remembers its past: Citizens as sensors in survey and mapping of historic places.' In *Crowdsourcing: Concepts, Methodologies, Tools, and Applications*, pp. 489–516. IGI Global, 2019. DOI: 10.4018/978-1-5225-8362-2.ch025.

Nedvědová, K. and R. Pergl. 'Cultural heritage and flood risk preparedness.' The International Archives of the Photogrammetry, Remote Sensing and Spatial Information Sciences XL-5-W2 (July 1, 2013): 449–51. https://doi.org/10.5194/isprsarchives-XL-5-W2-449-2013

Nelson, Marla, Renia Ehrenfeucht, Traci Birch and Anna Brand. 'Getting by and getting out: How residents of Louisiana's frontline communities are adapting to environmental change.' *Housing Policy Debate* 32, no. 1 (2022): 84–101.

Oliver-Smith, Anthony. *Defying Displacement: Grassroots Resistance and the Critique of Development*. University of Texas Press, 2010.

Olshansky, R.B. 'Planning after Hurricane Katrina.' *Journal of the American Planning Association* 72, no. 2 (2006): 147–153.

Peacock, Walter Gillis, Shannon Van Zandt, Yang Zhang, and Wesley E. Highfield. 'Inequities in long-term housing recovery after disasters.' *Journal of the American Planning Association* 80, no. 4 (2014): 356–371.

Peacock, Walter, Michelle Meyer, Annette, Shannon Van Zandt, Himanshu Grover and Fayola Jacobs. 'The adoption of hazard mitigation and climate change adaptation policies, programs, and actions by local jurisdictions along the Gulf and Atlantic coasts.' In: *The Routledge Handbook of Urban Disaster Resilience*, ed. Michael K. Lindell (London: Routledge, 2019), 109–143.

Phillips, B., P.A. Stukes and P. Jenkins. 'Freedom Hill is not for sale – and neither is the lower ninth ward.' *Journal of Black Studies* 43, no. 4 (2012): 405–426.

Quinlan, S.K. 'Climate Change and Cultural Heritage: Disaster Management under the Trump Administration'. Doctoral dissertation. Rutgers: The State University of New Jersey, School of Graduate Studies, 2019.

Ravankhah, M., Chmutina, K., Schmidt, M., Bosher, L. (2017). Integration of Cultural Heritage into Disaster Risk Management: Challenges and Opportunities for Increased Disaster Resilience. In: Albert, MT., Bandarin, F., Pereira Roders, A. (eds) *Going Beyond. Heritage Studies*. Springer, Cham. https://doi.org/10.1007/978-3-319-57165-2_22

Ravankhah, M., M. Schmidt and T. Will. Multi-hazard disaster risk identification for world cultural heritage sites in seismic zones. *Journal of Cultural Heritage Management and Sustainable Development* (2017). Vol. 7 No. 3, pp. 272–289. https://doi.org/10.1108/JCHMSD-09-2015-0032

Ray, V. A theory of racialized organizations. *American Sociological Review* 84, no. 1 (2019): 26–53. https://doi.org/10.1177/0003122418822335

Rico, Trinidad. 'The limits of a "heritage at risk" framework: the construction of post-disaster cultural heritage in Banda Aceh, Indonesia.' *Journal of Social Archaeology* 14, no. 2 (2014): 157–176.

Riddell, G.A., H. van Delden, H.R. Maier and A.C. Zecchin. 'Exploratory scenario analysis for disaster risk reduction: Considering alternative pathways in disaster risk assessment.' *International Journal of Disaster Risk Reduction* 39 (2019): 101230.

Rindrasih, Erda and Patrick Witte. 'Reinventing the post-disaster cultural landscape of heritage tourism in Kotagede, Yogyakarta, Indonesia.' *Journal of Heritage Tourism* 16, no. 2 (2021): 136–150.

Rivera, Danielle Zoe. 'Disaster colonialism: A commentary on disasters beyond singular events to structural violence.' *International Journal of Urban and Regional Research* 46, no. 1 (2022): 126–135.

Rivera, Danielle Zoe and Marccus D. Hendricks. Municipal undergreening: Framing the planning challenges of implementing green infrastructure in marginalized communities, *Planning Theory & Practice* (2022). DOI: 10.1080/14649357.2022.2147340

Rivera, Danielle Zoe, Bradleigh Jenkins & Rebecca Randolph (2022) Procedural Vulnerability and Its Effects on Equitable Post-Disaster Recovery in Low-Income Communities, *Journal of the American Planning Association*, 88:2, 220–231, DOI: 10.1080/01944363.2021.1929417

Rivera, Danielle Zoe and Marccus D. Hendricks. 'Municipal undergreening: framing the planning challenges of implementing green infrastructure in marginalized communities.' *Planning Theory & Practice* 23, no. 5 (2022): 807–811.

Roberts, Andrea R. '"Until the lord come get me, it burn down, or the next storm blow it away": The aesthetics of freedom in African American vernacular homestead preservation.' In. *Buildings & Landscapes: Journal of the Vernacular Architecture Forum* 26, no. 2 (2019): 73–97.

Roberts, Andrea R. 'Preservation without representation: Making CLG programs vehicles for inclusive leadership, historic preservation, and engagement.' *Societies* 10, no. 3 (2020a): 60.

Roberts, Andrea. 'The end of bootstraps and good masters: Fostering social inclusion by creating counternarratives.' *In Preservation & Inclusion*. ed. Erica Avrami. New York: Columbia University Press: Columbia Books on Architecture, 2020b.

Roberts, Andrea and Grace Kelly. 'Remixing as praxis: Arnstein's ladder through the grassroots preservationist's lens.' *Journal of the American Planning Association* 85, no. 3 (2019): 301–320.

Rosa, Angela, Angela Santangelo and Simona Tondelli. 'Investigating the integration of cultural heritage disaster risk management into urban planning tools. The ravenna case study.' *Sustainability* 13, no. 2 (2021): 872.

Ross, Susan M. 'Review: Perceptions of sustainability in heritage studies ed. by Marie-Theres Albert, and; Going beyond: Perceptions of sustainability in heritage studies No. by Marie-Theres Albert et al.' *Future Anterior* 18, no. 1 (2021): 147–153.

Rössler, Mechtild. 'World Heritage cultural landscapes: A UNESCO flagship programme 1992–2006.' *Landscape Research* 31, no. 4 (2006): 333–353.

Rumbach, Andrew, Anna Bierbrauer and Gretel Follingstad. 'Are we protecting our history? A municipal-scale analysis of historic preservation, flood hazards, and planning.' *Journal of Planning Education and Research* (2020): 0739456X20948592. https://journals.sagepub.com/doi/abs/10.1177/0739456X20948592.

Rumbach, Andrew and Dolores Foley. 'Indigenous institutions and their role in disaster risk reduction and resilience: Evidence from the 2009 tsunami in American Samoa.' *Ecology and Society* 19, no. 1 (2014): 19–28.

Rumbach, Andrew and Jeremy Németh. 'Disaster risk creation in the Darjeeling Himalayas: Moving toward justice.' *Environment and Planning E: Nature and Space* 1, no. 3 (2018): 340–362.

Saito, H. (Ed.). *Kobe/Tokyo International Symposium Risk Preparedness for Cultural Properties: Development of Guidelines for Emergency Response*. Chuo-Koron Bijutsu Shuppan, Tokyo, 1997.

Sewordor, Emefa, Ann-Margaret Esnard, Alka Sapat and Lorena Schwartz. 'Challenges to mobilising resources for disaster recovery and reconstruction: Perspectives of the Haitian diaspora.' *Disasters* 43, no. 2 (2019): 336–354.

Siegel, Peter E., Corinne L. Hofman, Benoît Bérard, Reg Murphy, Jorge Ulloa Hung, Roberto Valcárcel Rojas and Cheryl White. 'Confronting Caribbean heritage in an archipelago of diversity: Politics, stakeholders, climate change, natural disasters, tourism, and development.' *Journal of Field Archaeology* 38, no. 4 (2013): 376–390.

Sheller, Mimi. 'Caribbean reconstruction and climate justice: Transnational insurgent intellectual networks and post-hurricane transformation.' *Journal of Extreme Events* 5, no. 04 (2018): 1840001.

Smith, Laurajane. 'Empty gestures? Heritage and the politics of recognition.' In *Cultural Heritage and Human Rights*, pp. 159–171. Springer, New York, NY, 2007.

Smith, Neil. "There's no such thing as a natural disaster." *Understanding Katrina: perspectives from the social sciences* 11 (2006).

Sou, Gemma and Ruth Webber. 'Disruption and recovery of intangible resources during environmental crises: Longitudinal research on "home" in post-disaster Puerto Rico.' *Geoforum* 106, (2019): 182–192.

Spennemann, Dirk R. and David W. Look, eds. *Disaster management programs for historic sites*. US National Park Service, 1998.

Spennemann, D.H.R. 'Cultural heritage conservation during emergency management; luxury or necessity?.' *International Journal of Public Administration* 22, no. 5 (1999): 745–804.

Stovel, Herb. *Risk Preparedness: A Management Manual for World Cultural Heritage*. Roma: ICCROM. (1998).

Sullivan, Elaine, Angel David Nieves and Lisa M Snyder. "Making the Model: Scholarship and Rhetoric in 3-D Historical Reconstructions." In *Making Things and Drawing Boundaries: Experiments in the Digital Humanities*, edited by Jentery Sayers, 301–16. University of Minnesota Press, 2017. https://doi.org/10.5749/j.ctt1pwt6wq.38.

Sun, Lei and A. J. Faas. 'Social production of disasters and disaster social constructs: An exercise in disambiguation and reframing.' *Disaster Prevention and Management: An International Journal* 27, no. 5 (2018): 623–635.

Thiel, Charles C. 'A strategy for planning and protecting immovable heritage: Lessons from natural disaster planning.' *Materials Research Society Symposium J – Materials Issues in Art and Archaeology III, MRS Online Proceedings Library*. Volume 267 (1992): 149.

Wendel, Delia Duong Ba. 'Imageability and justice in contemporary New Orleans.' *Journal of Urban Design* 14, no. 3 (2009): 345–375.

Wisner, Ben. 'The disaster experts: Mastering risk in modern America.' *Disaster Prevention and Management* 24, no. 3 (2015): 417–419.

Woods, Clyde. *Development Drowned and Reborn: The Blues and Bourbon Restorations in Post-Katrina New Orleans*. Vol. 35. Athens: University of Georgia Press, 2017.

Zhang, Mengke and James H. Lenzer Jr. 'Mismatched canal conservation and the authorized heritage discourse in urban China: A case of the Hangzhou section of the Grand Canal.' *International Journal of Heritage Studies* 26, no. 2 (2020): 105–119.

Zhang, Yang and Walter Gillis Peacock. 'Planning for housing recovery? Lessons learned from Hurricane Andrew.' *Journal of the American Planning Association* 76, no. 1 (2009): 5–24.

19 New Technologies and Disaster Risk Management for Cultural Properties

Hirofumi Ikawa

Increasingly frequent and devastating disasters caused by natural hazards are being observed on a global scale. Cultural heritage sites of various types and locations must be safeguarded and made resilient to these hazards. The utilisation of technology in disaster preparedness is an indispensable aspect that must be considered. By implementing knowledge through applied engineering, technology can effectively mitigate disaster risks.

This study aims to furnish an overview, through case studies, of the utilisation of advanced technologies such as 3D documentation and machine learning in the management of cultural heritage during disasters.

3D Documentation and Disaster Risk Management

The international community recognizes the significance of 3D documentation in managing the risks faced by cultural heritage sites exposed to natural hazards, climate change, and conflicts. For instance, UNESCO, in partnership with ICONEM[1] and the General Directorate of Antiquities of Lebanon (DGA), has established a database of cultural heritage sites destroyed during the Syrian civil war, which has been digitized in 3D and includes images and videos. The growing demand for digital documentation of cultural heritage has led to an increasing number of projects by various research groups. This chapter will showcase examples of the use of 3D documentation for disaster prevention.

Overview of 3D Documentation Technology

There are two primary methods of recording cultural heritage in 3D: laser scanning and photogrammetry.

Laser scanning is a measurement technique that utilizes a laser beam to obtain the 3D spatial position of an object. The laser is rotated 360 degrees horizontally and vertically at a constant speed, and 3D coordinates are calculated from the distance to the object and the irradiation angle. By continuously irradiating a laser, the 3D coordinates of a large number of points on an object can be obtained quickly with an accuracy of 10 mm or less. The aggregate of these points is known as point cloud data. A laser scanner can acquire point cloud data for the object in front of it but not for the object behind it or in the back. Therefore, the laser scanner must be finely relocated to acquire point cloud data clear of obstacles. In recent years, a great variety of laser scanners have become available, not only those that are mounted on a tripod and operated, but also handheld types and even small ones mounted on a smartphone, thus greatly reducing the amount of labour involved in recording.

DOI: 10.4324/9781003293019-24

Figure 19.1 Various methods for 3D measurement, from left: Laser scanning, photogrammetry using drone photos, Lidar app.

Photogrammetry is a method of analysing digital images with software to create 3D models. Using the stereo camera principle, when an object is photographed from two different locations, the relative position and direction of the photographed points to the object are calculated from the positions of multiple feature points captured in both images. The feature points can then be obtained as point cloud data. When applied to a large number of images (in the hundreds to thousands), it can be applied to a wide and large object. Therefore, by flying the UAV at a certain height and taking a large number of photographs, extensive point cloud data can be generated. The accuracy of the resulting point cloud data is about 10 to 50 mm, depending on the flight and shooting conditions.

When precise point cloud data is required, laser scanning is the preferred option. Even complex shapes can be accurately documented, and point clouds acquired at multiple locations can be combined in software. On the other hand, photogrammetry is suitable for large, widespread, two-dimensional objects such as ground surfaces and walls. Therefore, when documenting cultural heritage sites in 3D, a workflow that combines laser scanning and photogrammetry is chosen to take advantage of each other's merits and compensate for their shortcomings. In other words, a laser scanner is installed on the ground for scanning, while a drone (UAV) captures images and creates a 3D model using photogrammetry, and the two are integrated in software (Figure 19.1).

Case of 3D Documentation for Disaster

3D documentation possesses several advantages for the management of disaster risks, as demonstrated by the following case examples: (1) the capacity for rapid and safe assessment of damage, (2) the utility of simulation for forecasting disasters, and (3) the potential for contributing to the reconstruction of lost structures.

The first advantage is its ability to allow for rapid and efficient damage assessment. Cultural heritage sites located in areas affected by earthquakes and conflicts are often at risk of collapse, and thus, require prompt and safe documentation without the need for proximity to the buildings. 3D documentation facilitates accurate documentation of buildings from a distance, for example, through the use of drones. For instance, in the case of a church that collapsed in stages following an earthquake in central Italy in 2016, the process of the collapsing church was documented in 3D, providing a valuable analysis of how the earthquake affected the structure of the building.[2] Additionally, during the 2017 seismic event in Lesbos, Greece, photogrammetric data acquired from aerial photographs (UAV) was amalgamated with point cloud data procured via terrestrial laser scanning (TLS) to furnish a comprehensive 3D documentation of a religious edifice affected by the earthquake (Chatzistamatis et al., 2018).

A further benefit is that the utilisation of 3D modelling in the preservation of cultural heritage sites allows for the simulation of geological characteristics and behaviour of the impacted region. These simulations can be utilized to evaluate the potential risk of landslides and seismic activity in the future. Examples include the simulation of damage resulting from the rockfall that occurred near the hamlet of Cortes de Paras, located 80 km southwest of Valencia on 6 April 2015,[3] and the geohazard forecasting in Choirokoitia, Cyprus, executed by PROTHEGO,[4] which employed field survey techniques to measure and document the extent of damage caused by natural hazards in cultural heritage sites.[5]

Lastly, the utilisation of photogrammetric techniques enables the creation of 3D models of lost cultural heritage sites. The 'Everyone's Shurijo Digital Restoration Project'[6] is an exemplar of this method. On 31 October 2019, a conflagration consumed nine buildings of Shuri Castle, including the main hall. Immediately following the fire, researchers established a website and solicited individuals worldwide to submit photographs of Shuri Castle. Over 2,000 images taken by various people at different times were compiled and employed to construct the model. Such endeavours are exemplified in the 3D reconstruction of the ruins of Palmyra, which were lost during the Syrian Civil War,[7] and the reconstruction of the ancient city of Mosul, a casualty of war, through 3D recording and the development of a digital archive and database of cultural heritage in the old city.[8]

Expected Future Developments

Many cultural heritage sites on UNESCO's World Heritage List encompass vast areas. To effectively monitor these sites, UNESCO collaborated with UNITAR-UNOSAT in 2015 to conduct surveillance of disasters and other occurrences utilising satellite imagery and produced a report on its efficacy (UNESCOUNITAR-UNOSAT, 2016). Presently, high-resolution satellite imagery is approaching the resolution range of aerial photography, and technologies for super-resolution, which converts images from low resolution to high resolution, are being realized (Wang et al., 2021). In other words, the necessary tools for more efficiently collecting images of a broader range of cultural heritage are now available. For instance, it is feasible to document cultural heritage in 3D during times of peace and compare it with its 3D appearance following a disaster in order to gauge the extent of damage or utilize it for reconstruction at another location.

Additionally, a variety of 3D laser scanning equipment has been developed and low-cost methods have been created to realize 3D laser scanning. In recent years, Lidar on the iPhone/iPad[9] and the development of various applications have set a new trend in the field of 3D documentation of cultural heritage (Figure 19.2). Although initially limited to small-scale objects such as movable cultural heritage, Lidar is gradually being applied to capture the cityscapes and topographic data. To obtain high-quality results, it is necessary to understand the characteristics

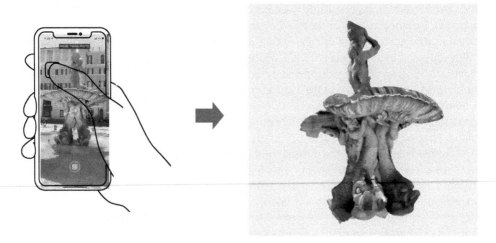

Figure 19.2 3D documentation using smartphone app.

of the application used, but there is an online community where members who use Lidar on smartphones to document cultural heritage share their results. Some of these results are quite remarkable, and by sharing their findings with one another, they have become a highly reliable means of documenting cultural heritage.

Machine Learning and Disaster Risk Management

Overview of Machine Learning and Cases of Using Machine Learning for Disaster Risk Management

Machine learning is a technology that enables computers to learn from vast amounts of data in order to extract rules and patterns from the data. This technology permits the creation of algorithms and models to perform tasks such as classification and prediction.

A study reviewing the case studies of the use of machine learning techniques in natural disaster management determined that the most appropriate areas for machine learning are monitoring, detection, and forecasting. Data sources utilized in this process include satellite imagery, social media, and so forth (Arinta & Andi, 2019).

Monitoring Cultural Heritage Sites Using a Combination of Machine Learning and Remote Sensing

Machine learning is increasingly being employed in disaster risk management in conjunction with remote sensing techniques. Remote sensing is a technology that enables the acquisition of images of the earth's surface from satellites, aircraft, and drones, and the extraction of relevant information. The challenge with remote sensing has been the analysis of vast amounts of data. For instance, in the event of a large-scale disaster, analysing high-resolution aerial images and identifying damaged buildings with the human eye is a highly labour-intensive and time-consuming task. As a result, researchers are attempting to use machine learning techniques to extract the shapes of buildings and features, such as indications of damage, that require attention. For example, there is a study that focused on seismic data and evaluated the performance of Convolutional Neural Networks (CNNs) for detecting visible structural damage from satellite images and aerial photographs. CNNs are a type of deep learning model that are particularly effective at analysing images and are widely used in computer vision tasks (Nex et al., 2019). The rapid development of machine learning has also led to its widespread use in disaster monitoring and damage assessment of cultural heritage sites over a wide area of the earth's surface.

Monitoring Cultural Heritage by Combining Machine Learning and Social Media

With the proliferation of smartphones, there have been experiments to gather text and images related to damaged buildings through social networking services and utilize them for disaster monitoring and damage assessment. Social media provides access to ground-level photographs of cultural heritage sites affected by disasters that are not obtainable from the aerial perspective provided by UAVs or satellite imagery. However, as cultural heritage photographs constitute a small percentage of the total images posted, it is necessary to extract relevant images from the vast quantity of images on social media. It is not feasible, of course, to perform this task manually. Therefore, machine learning techniques are employed to detect whether cultural heritage sites are affected by disasters from images posted on social media in the event of a disaster, excluding images that are unrelated to the affected cultural heritage sites (Kumar et al., 2020).

Risk Prediction of Cultural Heritage through Machine Learning

If the disaster risk associated with a cultural heritage site can be assessed, more appropriate management can be implemented, such as enhancing the level of monitoring and allocating financial resources in a more targeted manner. For example, the likelihood of natural hazards such as earthquakes and landslides can be statistically calculated from geological data, enabling the assessment of future disaster risks. Machine learning techniques can be used to identify high-risk areas for human factors such as theft, illegal construction, and vandalism by comparing parameters related to management in a region. For instance, a study assessed the risk of cultural assets in various provinces of China (Li et al., 2021).

Deterioration Monitoring of Cultural Heritage through Machine Learning

While monitoring methods employ relatively accessible big data such as satellite images and social media, there are efforts to monitor signs of damage and deterioration by capturing images of the condition of cultural heritage sites, such as buildings, and storing image data related to building damage in the cloud. By monitoring the structure of the heritage site, it is possible to prolong its remaining lifespan by predicting possible areas of damage. Examples include the extraction of damage to roof tiles or cracks in stone walls. Furthermore, data collected from material experiments and from the field can be combined with machine learning and effectively used to obtain more accurate prediction models. These models can be used to predict the compressive strength of masonry and repair mortars, possible damage scenarios for heritage buildings, assess seismic vulnerability, determine mechanical properties of materials, and several predictive applications for archaeological site surfaces due to weathering, material loss, bleaching, infiltration, algae growth, and moss deposition.

Issues in Implementing Machine Learning

Machine learning is a highly promising technology for monitoring cultural heritage, but there are several challenges that need to be addressed. Three of these challenges are outlined below.

The first challenge is how to collect data suitable for learning. One method of cultural heritage damage extraction is using supervised learning, which requires training data containing correct answers. The performance of an algorithm generated using training data in extracting features is defined by that training data. In other words, some types of training data may be effective at extracting tile breaks used in a particular time period or region but not for tile breaks from a different region or time period. This issue may also arise for photographs obtained from satellite, aerial, and UAV sources, and without the proper collection of training data, damage cannot be properly extracted (Gupta et al., 2019). Possible solutions to this challenge include the creation of a more global database or improved algorithms, but none have fully solved the problem.

One issue when using machine learning technology to detect building damage is determining the severity of the damage. To address this problem, a possible solution is to establish a framework for quantitatively evaluating the image. This framework can be used to determine the severity of the damage, such as cracks in a wall or missing stone, found by the machine learning technology. For instance, in the case of extracting tile damage from a roof image of the National Palace Museum in China, two different glazed tile photographs were automatically detected and cropped to assess the extent of damage (Wang et al., 2020). Similarly, the main façade of the Chapel Royal at Stirling Castle (Scotland) was used to validate automated damage detection and classification of ashlar masonry walls through machine learning, with missing stones extracted from a sparse point cloud of the wall surface (Valero et al., 2019). However, this approach is limited to planar objects such as roof surfaces and walls, and is not applicable to damage to axial structures such as columns and beams.

The last issue pertains to the development of relationships with the community. This issue is primarily concerned with the social implementation of advanced technologies, such as machine learning, within the community rather than solely with technological development. When using machine learning to monitor cultural heritage sites managed by the community, it is essential to involve the community in the process. This is because the community managers of these heritage sites may have concerns about the limitations of current monitoring methods and the costs of adopting new technologies. Even if machine learning is demonstrated to make the monitoring of cultural heritage more efficient, field managers and community representatives may question whether there are tangible benefits. When introducing a new technology to the community, it is necessary to demonstrate specific advantages, such as labour savings in comparison to traditional methods, in order to gain the trust of the community. Additionally, the implementation process should aim to share the objectives of the monitoring methods that machine learning aims to achieve and provide opportunities for community feedback to be incorporated.

Conclusion

This study introduces a case examination leveraging 3D documentation technology in the field of disaster risk management during the initial segment, followed by a case examination of the utilisation of machine learning technology in the latter half. It is evident that these technologies are progressively transitioning towards practical applications, transitioning from the laboratory and being employed to safeguard cultural heritage, buoyed by the influx of more potent and cost-effective devices entering the market in quick succession.

It is worth highlighting the ongoing discourse surrounding the handling of technology. Presently, we are involved in a project in Japan aimed at devising a system for the surveillance and

Figure 19.3 Distribution of cultural heritage in Japan and type of damage.

preservation of cultural assets through the utilisation of machine learning. There are approximately 5,000 nationally designated cultural assets in Japan, primarily situated in the Kinki region, specifically in Kyoto and Nara, however, a significant number are also dispersed throughout rural areas such as islands and mountainous regions. These cultural assets are annually susceptible to damage from natural disasters (Figure 19.3). The proprietors of each cultural asset submit damage reports to the Agency for Cultural Affairs through the municipalities in order for appropriate measures to be taken. However, properly inspecting cultural asset buildings scattered across the country poses certain challenges in the inspection process. One challenge is the lack of clear guidelines for the inspection cycles and poor data management of inspection results.

Currently, when a damage report is submitted, experts visit the cultural properties individually to diagnose the condition of each building, but most of the time, the results are not recorded in the data. Another challenge is the shortage of trained experts to properly diagnose the damage. It takes a lot of time to train them, and experienced technicians are aging. In Japan, there are 130 engineers who specialize in the repair of cultural property buildings, but due to the large number of cultural properties and remote areas, there is not enough manpower to cover all the regions.

To overcome these challenges, the Agency for Cultural Affairs and local prefectures have implemented a project called 'cultural property patrols' where residents of the community check the cultural properties of their area. While this is an important support for local cultural properties, the cultural property patrols are conducted by local volunteers and not by experts, making it difficult to accurately grasp the damage situation. To bridge the gap between the skills of experts and volunteers, using Machine Learning has been identified as a solution to monitor and maintain cultural properties effectively (Figure 19.4).

The findings of our endeavour have been showcased at international symposia,[10] and the concerns are succinctly outlined in this manuscript: the methodology for procuring the requisite data for instruction, the method of quantifying the deterioration, and the strategy for establishing a liaison with the constituency in which the monitoring will be conducted. Recent technological advancements have been remarkable, and these state-of-the-art technologies are being implemented on a regular basis in the realm of cultural heritage conservation, and there are steady transformations occurring in Japan as well. Nevertheless, in order to bridge the divide between those who innovate technologies and those who implement these technologies in the community, it is imperative to have a platform for reciprocal exchange of ideas and perspectives. Those who can bridge the divide between advanced technology and cultural heritage practice are in the highest demand today (Box 19.1).

Distribution and damage of cultural heritages

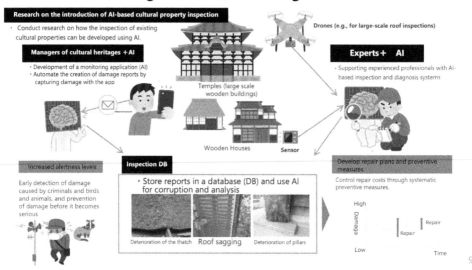

Figure 19.4 Development of a monitoring and conservation system for cultural properties using machine learning.

Box 19.1 Reconstruction of Mosul's ancient city and machine learning for monitoring

Cultural heritage sites face a growing risk of natural disasters, climate change, and conflicts. To manage these risks, 3D documentation technology and machine learning are increasingly being utilized for rapid and accurate documentation of cultural heritage sites. This case study examines the use of these technologies in disaster prevention and explores the challenges and potential solutions associated with their implementation.

The ancient city of Mosul, a casualty of war, has been undergoing reconstruction efforts through the use of 3D recording technology. A digital archive and database of cultural heritage in the old city have been developed to assist in the restoration and preservation of the site, particularly after the destruction of its main temple. This initiative highlights the potential of 3D documentation in post-conflict recovery and cultural heritage preservation.

Machine learning, particularly CNNs, can be employed to extract building shapes and features, such as damage indicators, from satellite images and aerial photographs. The rapid development of machine learning has led to its widespread use in disaster monitoring and damage assessment of cultural heritage sites. This case study explores the use of machine learning technology for monitoring cultural heritage and identifies three critical issues that need addressing: data collection, damage severity, and community relations. Addressing these issues is essential for enhancing the effectiveness of machine learning in cultural heritage preservation.

Key Points

- Cultural heritage sites face increased risks due to natural disasters, climate change, and conflicts, prompting the use of 3D documentation technology for effective management and preservation.
- The reconstruction of Mosul showcases the potential of 3D recording, digital archiving, and the development of databases for restoring war-damaged cultural heritage sites.
- Machine learning, particularly Convolutional Neural Networks (CNNs), shows promise in monitoring and assessing damage to cultural heritage sites using satellite imagery and aerial photographs.
- Challenges in implementing machine learning for cultural heritage preservation include data collection, damage severity assessment, and community relations, requiring clear guidelines and a well-trained workforce.
- Bridging the divide between technology innovators and community practitioners is crucial for successful adoption of advanced solutions in cultural heritage preservation, demanding a platform for reciprocal exchange of ideas and perspectives.

Notes

1. Iconem is a French start-up company founded in 2012 that develops digitization and 3D modelling technologies for cultural heritage.
2. Iconem is a French start-up company founded in 2012 that develops digitization and 3D modelling technologies for cultural heritage.
    ch of S. Agostino, located southeast of Amatrice, was damaged in stages by the August 24 earthquake, the October 30 earthquake, and the January 18 earthquake, with large parts of the church collapsing. After each earthquake, a 3D model of the church was created using a drone (UAV) for 3D documentation. The process of the church's collapse and the effects of the earthquakes on the building's structure are interestingly analysed (Chiabrando et al., 2017).
3. In this project, UAV photogrammetry and laser scanning were used to obtain a point cloud of the collapse area, and a three-dimensional (3D) simulation of the fall was performed to assess the risk to the village where the cultural heritage site is located (Sarro Trigueros et al., 2018).
4. PROTHEGO (PROTection of European Cultural HEritage from GeO-hazards) is a project that aims to make an innovative contribution to the analysis of geohazards in European cultural heritage areas. New space technology based on radar interferometry was applied to monitor European monuments and sites inscribed on the World Heritage List that are potentially unstable due to geohazards and to analyse the geological processes affecting their heritage characteristics.
5. This project combined InSAR ground-motion data on changes in the earth's surface with field data collected by UAV and photogrammetry to detect and analyse deformation phenomena in order to monitor and predict geologic hazards affecting cultural heritage sites (Themistocleous & Danezis 2020).
6. Shuri Castle Digital Reconstruction, https://www.our-shurijo.org/en/
7. The International Committee of Architectural Photogrammetry (CIPA) is making a technical contribution to the preservation of Syria's heritage by creating an open access database based on data collected by CIPA members through various projects in Syria in the years prior to the outbreak of the civil war in 2011. The CIPA project has created a useful database that can support protection, conservation, restoration, and reconstruction strategies. For example, the NewPalmyra project (www.newpalmyra.org), an open source project to reconstruct Palmyra, has created a 3D reconstruction of the destroyed Palmyra site.
8. The heritage of the Old City of Mosul was the victim of armed conflict with ISIS between 2013 and 2017, which resulted in extreme violence in the city center and undue damage to the historical and cultural heritage and dignity of the community. The Mousl project aims to understand the human and cultural heritage of Iraq's postwar damaged historical city of Mosul and documenting and preserving the heritage that is being lost. As part of this project, we are researching historical records, photographs, and archives of historic buildings in the Old City of Mosul, documenting buildings and spaces in 3D, and building a digital archive and database of cultural heritage located in the Old City (Silver et al. 2016).

9　LiDAR scanners are available on the iPadPro (4th generation) released in March 2020 and the iPhone 12/13 released thereafter.
10　The 3rd International Symposium on Digital Heritage https://www.youtube.com/watch?v=tPaoIBTOD1I

References

Arinta, R.R. & Andi, W.R.E. (2019). Natural Disaster Application on Big Data and Machine Learning: A Review, in 2019 4th International Conference on Information Technology, Information Systems and Electrical Engineering (ICITISEE). Presented at the 2019 4th International Conference on Information Technology, Information Systems and Electrical Engineering (ICITISEE), IEEE, Yogyakarta, Indonesia, pp. 249–254. https://doi.org/10.1109/ICITISEE48480.2019.9003984

Chatzistamatis, S., Kalaitzis, P., Chaidas, K., Chatzitheodorou. C., Papadopoulou, E. E., Tataris, G. & Soulakellis, N. (2018). Fusion of TLS and UAV photogrammetry data for post-earthquake 3D modeling of a cultural heritage church. The International Archives of the Photogrammetry, Remote Sensing and Spatial Information Sciences, XLII-3/W4, 143–150.

Chiabrando, F., Di Lolli, A., Patrucco, G., Spanò, A., Sammartano, G., and Teppati Losè, L.: Multitemporal 3d Modelling for Cultural Heritage Emergency during Seismic Events: Damage Assessment of S. Agostino Church in Amatrice (ri). Int. Arch. Photogramm. Remote Sens. Spatial Inf. Sci., XLII-5/W1, 69–76, https://doi.org/10.5194/isprs-archives-XLII-5-W1-69-2017, 2017

Gupta, R., Hosfelt, R., Sajeev, S., Patel, N., Goodman, B., Doshi, J., Heim, E., Choset, H. & Gaston, M. (2019). xBD: A Dataset for Assessing Building Damage from Satellite Imagery. arXiv:1911.09296 [cs.CV]. https://arxiv.org/abs/1911.09296

Kumar, P. et al. (2020). Detection of disaster-affected cultural heritage sites from social media images using deep learning Techniques. Journal on Computing and Cultural Heritage, 13(3), 1–31. doi:10.1145/3383314.

Li, J. et al. (2021). Risk management priority assessment of heritage sites in China based on entropy weight and TOPSIS. Journal of Cultural Heritage, 49, 10–18.

Nex, F. et al. (2019). Structural building damage detection with deep learning: Assessment of a state-of-the-art CNN in operational Conditions. Remote Sensing, 11(23), 2765.

Sarro Trigueros, R., Riquelme, A., Carlos García López-Davalillo, J., María Mateos Ruiz, R., Tomás, R., Luis Pastor Navarro, J., Cano, M. & Herrera García, G. (2018). Rockfall simulation based on UAV photogrammetry data obtained during an emergency declaration: Application at a cultural heritage site. Remote Sensing 10(12), 1923. https://doi.org/10.3390/rs10121923

Silver, M. et al. (2016). The CIPA database for saving the heritage of Syria. The International Archives of the Photogrammetry, Remote Sensing and Spatial Information Sciences, XLI-B5, 953–960. doi:10.5194/isprs-archives-XLI-B5-953-2016

Themistocleous, K. & Danezis, C. Editors: Diofantos G. Hadjimitsis, Kyriacos Themistocleous, Branka Cuca, Athos Agapiou, Vasiliki Lysandrou, Rosa Lasaponara, Nicola Masini, Gunter Schreier (2020). Monitoring cultural heritage sites affected by geo-hazards using in situ and SAR data: The Choirokoitia case study, in Remote Sensing for Archaeology and Cultural Landscapes. Springer Remote Sensing/Photogrammetry, Springer Nature Switzerland, Switzerland.

UNESCO and United Nations Institute for Training and Research (UNESCOUNITAR-UNOSAT) (2016). Satellite-Based Damage Assessment of Cultural Heritage Sites 2015 Summary Report of Iraq, Syria & Yemen, Nepal.

Valero, E. et al. (2019). Automated defect detection and classification in ashlar masonry walls using machine learning. Automation in Construction, 106, 102846.

Wang, N. et al. (2020). Autonomous damage segmentation and measurement of glazed tiles in historic buildings via deep learning. Computer-Aided Civil and Infrastructure Engineering, 35(3), 277–291.

Xintao, W., Liangbin, X., Chao, D. & Ying, S. (2021). 'Real-ESRGAN: Training Real-World Blind Super-Resolution with Pure Synthetic Data' arXiv:2107.10833 [eess.IV]

20 Integrating DRM Considerations into Heritage Management Systems

Barriers and Opportunities

Luisa De Marco

Background

The awareness of the need to provide specific responses to the impacts on the cultural heritage of disasters and other human-made catastrophes has developed as early as the 1970s in countries with a particularly deep-rooted culture of prevention.[1] This consciousness has progressively spread at the supranational level through international protection bodies between the 1980s and 1990s, leading to further evolution of thinking until today, when cultural heritage is considered an integral part and a driver of post-disaster recovery. Meanwhile, several decades have passed, and an untold number of catastrophic events occurred, with enormous damage to the natural and built environment and cultural heritage.

Significant progress has been made to understand how to cope with and reduce the effects of disasters on heritage, improve the post-disaster response, and reduce the probability of disaster occurrence. We better understand the damage mechanisms of certain exceptional events and the greater or lesser effectiveness of certain types of repair work, especially concerning seismic risks. We have learnt how much the post-disaster emergency management methods can affect, negatively or positively, the damaged cultural heritage, the communities, and their chances and potential for recovery. Increasingly reliable and sophisticated warning systems, methods of recording heritage damage, and cartographic systems to map tangible and intangible cultural heritage and cross-reference these data with territorial vulnerabilities to formulate risk maps have been developed. The interdependence between poor land management, or maintenance of the built heritage, and the increase in its susceptibility to hazards has been well understood, as well as the need to be prepared and 'trained' to cope with emergencies more effectively. In the wake of many frustrating experiences, it has been realised that disaster risk management for cultural heritage cannot be dealt with in isolation and separately from the day-to-day management of inheritance or the general disaster response and management system; otherwise, the measures will be ineffective.

A concise but compelling state of the art on knowledge and experience in the last two decades of the 20th century is represented by the ICCROM – UNESCO – ICOMOS handbook *Risk Preparedness: A Management Manual for World Cultural Heritage* (Stovel, 1998). We take this example because of its global impact, as it was one of the first instruments to support the management of World Heritage cultural properties: it already contained most recommendations that remain valid today and continue to be evoked as indispensable steps for adequate disaster risk preparation and management.

ICCROM issued a new resource manual in 2010 – Managing Disaster Risks for World Heritage[2] – which expanded consideration from preparedness to the whole risk management cycle and provided guidance on the need for disaster risk management plans for both cultural and natural World Heritage. It attests to an evolution of thinking and the effort to take stock of experiences in reducing the impacts of disasters on world heritage properties. These manuals provide essential guidance to address risks caused by exceptionally violent events to heritage.

Nevertheless, catastrophes and serious accidents continue to happen, and the toll paid by cultural heritage remains far too high. How come? The reasons why this happens are far less systematically explored and even less made public than studies and 'inventories' of research and good practices for the mitigation and prediction of, response to, and recovery from catastrophic events, be they singular events, prolonged or intermittent. Understanding where and why things did not work and disseminating knowledge of failures or 'bad practices' would help to build greater awareness of where 'mistakes' may lurk, what 'wrong' patterns recur, and where action is needed to reduce the likelihood of hazardous events and reduce their disastrous effects.

Research promoted in 2018 by the European Commission as part of the European Year of Cultural Heritage[3] highlighted that the lack of effectiveness of risk management and disaster response in the field of cultural heritage depends essentially on four factors:

- *The lack of coordination between and across the different (European, National, and Regional) strategies of risk management policies in most countries.*
- *The lack of alignment in the responsibility chain from policy making to practical application.*
- *The low current priority of cultural heritage in risk management planning.*
- *The lack of integration of cultural heritage protection measures into risk management strategies.*[4]

The study highlights a gap between policy making and implementation at an administration and management level, as well as in practical applications. Key recommendations include awareness-raising, strengthening the governance of disaster risk management through promoting cooperation among relevant administrations, supporting documentation of structural aspects of built heritage, investing in disaster risk reduction, and improving disaster risk preparedness. The study is interesting, at least concerning a large part of Europe comprised within the European Union, because it highlights how much remains to be done in terms of systemic and baseline action, such as streamlining approaches, making databases and geographic information systems interoperable, developing training programmes, even in a part of the world that has been making efforts over decades to build common policies in several spheres.

Also, within the World Heritage system, there was a growing realisation that guidance on disaster risk management needed to refocus on processes before instruments. Hence, after ten years since the issuance of the second ICCROM manual, in light of the lessons learned from several disasters which severely impacted heritage, the revision of the resource manual on disaster risk management has commenced as part of a more significant effort. It envisages updating in a coordinated manner all guidance documents and resources for World Heritage management produced by the Advisory Bodies to the World Heritage Convention within the Leadership Programme, jointly implemented by ICCROM and IUCN, with the participation of ICOMOS.

Recent Advancements and Challenges in Instruments and Approaches

The greatest advances at the operational level are represented by the more immediate preparation for catastrophic events, the development of more efficient warning systems, in methods and mechanisms to ensure a more rapid response in the event of a disaster, with significant developments both in terms of approach and in applicable technologies, especially in the field of documentation or virtual reality applications to reconstruction processes. On the other hand, more systematic prevention measures, i.e., aimed at reducing systemic and structural vulnerabilities of the cultural heritage and the frequency of occurrence of more typically human-induced disasters, appear much less developed and implemented. It is as if the focus is almost exclusively on improving the response to disastrous events, taking it for granted that disasters will occur rather than creating conditions that reduce the likelihood of their occurrence or the significance of their effects. The focus of applied research on the possibilities offered by technology in providing individual answers to specific and isolated problems is often functional to IT enterprises but also to applied research organisations and spin-offs of academic institutions, which have been progressively adopting a market approach, also due to lack of public funds for research. They try to position themselves in the market niche of disaster risk management and response (as well as in several other strategic sectors) and seek to develop 'disaster risk reduction' products or service packages. Complementarily, management entities tend to privilege these ad-hoc tools because they appear as prompt solutions to complex problems that would instead need an intersectoral approach, inter-agency dialogue, and coordination, in one word, a shift in management approach.

A misplaced trust tends to promote single, individually applied tools which will respond to disasters instead of what is needed: the interaction between different instruments, processes, behaviours, and decision-making mechanisms adapted to the specific governance context[5]. Experiences have shown that, when a disaster occurs, an effective response to the event, including the containment of its effects and impacts, often depends on a well-geared system of communication, coordination, and collaboration – be it formal or informal – between central and local administrations and between administrations of the same rank but belonging to different sectors, rather than on one specific tool or technology. At times, even where mitigation, preparedness, and response tools are available to the managing bodies of a heritage place, they are not known or sufficiently familiar to the staff to implement. When these tools have a high technological content incompatible with the conditions and capacities of the bodies in charge of their use, they may not be used even if they are in place. Similarly, when the turnover of personnel in technical–administrative entities is rapid and institutional memory is not built, it may happen that new staff does not even know that specific tools exist. In these situations, applied research and diffusion of tested good practices could assist in devising solutions adapted to conditions where sophisticated technologies cannot be used due to local circumstances.

The most recent approaches to risk management related to cultural heritage have focused on governance processes, rather than on tools, as a critical pillar to ensure effectiveness in responding to disasters and their impacts.[6] They recognise that integrating the specific needs of cultural heritage (or natural heritage, for that matter) among the fundamental factors in decision-making processes related to post-disaster emergencies is possible to ensure a more rapid response and a more effective recovery of cultural heritage well.[7] Awareness has been growing that documentation and damage assessment and disaster recovery conducted in parallel with general recovery processes did not help cultural heritage because, according to use, it could be treated as strategic infrastructure and therefore subjected to forms of recovery not respectful of its heritage characteristics, or it was not among the priorities and consequently left behind or even demolished for reasons of urgency or danger. Only if heritage becomes a driving factor in response to and recovery from disasters or, at least, is fully integrated into recovery processes, their effectiveness will also benefit heritage and communities alike. Adopting such a perspective also demands respect for the pace of a recovery which is inclusive, sustainable, equitable from a socio-economical point of view and compatible with the values and the characteristics of the impacted heritage. It is a major paradigm shift that demands profound changes in the governance of disaster preparedness and disaster response, but, most importantly, the approach to the management of heritage resources, which requires, to be effective, constant communication, coordination, inter-sectoral, and inter-institutional collaboration on all sides and well in advance of disaster occurrences. The requirements to make this paradigm operational remain challenging to apply and sometimes even to understand, particularly in countries where administrations still work 'in sylos' and a culture of inter-institutional cooperation has not developed yet. Often, these are also countries where disaster preparedness is low.

Weaknesses in Implementation and Effectiveness

The chronicle of disasters and the recurrence of the patterns of events show that the achievements of knowledge and experience struggle to spread and become a common asset and set of practices. In many cases, failures in preparing for or responding to disasters depend more on process hiccups, or lack of cooperation, than on a lack of the right tools. There seems to be a limited understanding that they need to be adapted to contexts and that their integration into a broader, coordinated system of actions, communication, and behaviours makes instruments effective.

Why is it that, despite the enormous economic and social costs of disasters, including on cultural heritage, measures that proved effective in reducing disaster occurrences remain unsystematic? The reasons are multiple, and here just a few of them are explored, suggesting some steps to improve the effectiveness of the response to disasters and possibly, contribute to reducing their frequency and impacts.

Prevention is regarded as the best practice *par excellence* in avoiding or reducing the likelihood of a disaster. It has been amply demonstrated that it is possible to reduce the vulnerability of the built heritage and the cultural heritage, and the territory if this is maintained by respecting its structural characteristics and components kept in efficient condition. However, maintenance is a systematic and widespread practice that struggles to establish itself, except in some contexts where this culture is already well-established. Yet, in most cases, maintenance remains much preached and little implemented. Everyday care of cultural heritage does not offer much visibility or trigger large-scale, appetible bids for enterprises, and it does not give the opportunities to 'cut ribbons' or make press conferences or high-level statements as large-scale restoration works do.

Furthermore, maintenance is often perceived as a permanent cost rather than an investment, whose benefit in significantly reducing possible future costs due to its absence is not understood. Suppose prevention works and potentially disastrous events do not cause damage. It appears challenging to estimate the benefits because it is not usual practice to quantify the hypothetical potential loss or damage and the costs of its eventual recovery. However, methods to assess different scenarios, even from a financial point of view, exist and can offer reliable projections in conjunction with disaster experiences.

Many factors are complicit in the difficulty of ensuring systemic maintenance of heritage places: they include policies aimed at supporting extraordinary interventions, rather than ordinary ones, through tax benefits or VAT reductions limited only to certain types of intervention (e.g., favouring replacements of building elements rather than repairs of existing ones. Moreover, the rationale of these policies is often not oriented to reducing disaster risks but developed only to sustain the construction sector.

In addition, prevention, beyond actions limited to the asset under attention, touches other sectors than that of cultural heritage management, such as land planning, infrastructures, the production of regulations, economic and agricultural policies, well-designed fiscal measures, and financial incentives. This change demands an intersectoral approach, the capacity of the cultural institutions to engage in constructive dialogue with other branches of national or regional administrations and show them the advantages of such an approach.

For instance, the problem of ensuring the maintenance of built cultural heritage can only be addressed and, at least partially, resolved through a considerable political and management effort at a higher level and a larger scale, for instance, by introducing legislation and policies regulating and supporting ordinary maintenance. We have been assisting at a flourishing of initiatives and support programmes to reduce energy consumption in buildings or cities; these could be coupled with programmes supporting risk assessment and maintenance. On the other hand, when larger territories and landscapes are concerned, ensuring their maintenance poses enormous challenges, which often relate to the reduction of the economic viability of certain traditional practices, causing the subsequent abandonment of rural areas, which then become vulnerable due to lack of use and lack of monitoring, and the loss of widespread agro-pastoral practices linked to subsistence economies that have disappeared in many parts of the world.[8] These practices appear difficult to recreate on a large scale, and it is not easy to identify viable adaptations that can remain sustainable in the long term. Where these practices still exist and have guaranteed or could still guarantee continuous monitoring and intervention in the territory, ensuring its 'resilience'; globalised socio-economic processes threaten them. They need extra efforts to be sustained and to prevent their disappearance. Ideally, international heritage designations should act as instruments to improve their appreciation and protection; in most cases, however, they risk accelerating transformations by speeding up their inclusion in the globalised economic circuits of tourism.

Despite these difficulties, maintenance remains a formidable tool for reducing vulnerability and could be strengthened with appropriate policies, tools, and mechanisms. These include fiscal support, guidance, bureaucratic simplification, and awareness-raising campaigns, first and foremost among decision-makers, to create the conditions for devising and implementing well-thought-through policies.

Furthermore, one cannot forget that disasters often unlock the release of financial resources in a short time that would not be otherwise made available in ordinary times and under exceptional utilisation conditions, where special procedures apply. Regular requirements, often those related to cultural heritage protection or public works, may be softened, under special legislation or regulatory frameworks, with the justification that recovery needs to proceed fast. All these circumstances generate a specific form of the economy around post-disaster recovery that may be seen as an 'opportunity' by the business sector and sometimes administrations and decision-makers. Hence, depending on the specific socio-economic and cultural contexts, the idea that disasters may offer development opportunities can gain implicit currency and may indirectly contribute to slowing down the already tricky path towards structural implementation of policies and measures to reduce vulnerabilities of the built environment and cultural heritage to disasters.

Managing risks and responses to disasters begins well before the event occurs. It demands clear governance architecture, where responsibilities and decision-making powers among the various actors – from central, regional, and local institutions to non-governmental organisations and civil society – are well set out and correspond to the reality on the ground. Effective disaster risk management also needs tested protocols, but more importantly, teamwork, everyday practice in communication and exchanges, a management environment propitious to efficacy and a willingness to 'make things happen'. These are not features that can build quickly in emergencies, although in some cases, the human factor was determinant to reduce at least some impacts of disastrous events.

However, in general, where governance and the management system are not transparent or have gaps, weaknesses, or bottlenecks, these are dramatically exposed during and after disasters in the phases of the immediate response and the recovery process. In other cases, management systems may seem well-articulated and ideally adequate to carry out their tasks in ordinary and extraordinary situations but to the test of facts, they reveal structural problems in the way they are conceived or function in practice. For instance, there may be a lack of adequate dissemination of newly established response protocols, limited understanding by decision-makers of the importance of prompt communication and coordination, or a lack of information-sharing on reforms undertaken at the central level with impacts on the local one.[9] A systematic analysis of what functioned in apparently weak or confused management systems and what, vice versa, did not work in situations where the management systems were deemed robust and well-articulated would offer valuable insights into factors that may enable or hinder effective responses to disastrous occurrences.

Some Proposals to Improve Mitigation, Preparedness, and Response to Disasters

The considerations above suggest that, whilst research in new technologies and tools for disaster response should indeed continue, there is a need, on the one side, to ensure that practices and measures that have contributed positively to the reduction of the impacts of disasters are systematically implemented.

For instance, the documentation of heritage resources and of their state of health, which seemed an obvious given among good heritage protection and conservation practice, has not become a widespread practice and, where it existed, has not been updated or digitised; often, archives are obsolete, unsafe and vulnerable themselves to disasters. This problem affects developing countries and has become an issue in wealthier parts of the world due to public budget shrinking and changed agendas for heritage, now focused more on promotion rather than on documentation and conservation. Documentation of heritage is vital to act promptly in case of disasters. It should be given priority within the overall management strategy through funding and developing appropriate documentation campaigns designed to create, update, and systematise the necessary baseline information for disaster risk reduction and response. Within the World Heritage system, documentation should receive high priority upstream, i.e., in preparation for the inclusion to the Tentative Lists, to ensure that properties, when inscribed, are adequately documented; for those already inscribed, monitoring processes should ensure that decision-makers at the national and property levels address documentation gaps urgently.

Whilst research has significantly developed along certain strands, others remain less explored, such as systematic comparative studies of disasters with similar characteristics and the differences and similarities in responses and impacts, under multiple perspectives. For instance, an accurate analysis carried out systematically on several cases, of how, for example, fires on individual monuments or small ensembles developed and resulted in widespread destruction and damage on the causes that triggered the fire events, turning them into disasters, would be beneficial and could lead to systemic actions to improve prevention and response measures.[10] Unfortunately, at present, the only genuinely accurate studies are those conducted as part of judicial investigations to ascertain possible human responsibility: it would be helpful if at least part of this extensive and detailed documentation could be made available when the aspect of establishing human responsibility is definitively closed, for ex-post analyses of the 'mechanics' of accidents and their transformation into disastrous events and the factors that lead to the disasters.

This type of study would usefully complement the analysis of good practices, which, in turn, is a more developed strand of research. Inventories and documentation of good practices, however, tend to remain to the surface and, in many cases, do not delve sufficiently deep into the circumstances to understand fully the factors and the causes of the 'positive outcome'. Nowadays, the accent is posed on the need to provide 'solutions' that managers can adapt to their context. Whilst the idea of distilling the experiences of others and learning lessons from them is undoubtedly positive, often there is a lack of sufficiently comprehensive background research to extract the critical lesson to be learnt, and good practices are presented but remain not adequately analysed and documented.

Similarly, not sufficiently systematic comparative research has been done on the difference in impacts of a potentially disastrous event on cultural heritage assets or areas where maintenance programmes exist and are implemented vis-à-vis with places where the built heritage suffered from obsolescence and neglect and on the differences in costs borne by society to recover from the events in the different instances. Developing projections of how much maintenance, if implemented, can reduce the effects of potentially disastrous events, and therefore reduce the impacts and the recovery costs, not limited to financial or economic burden but also extended to social and cultural 'costs', would be helpful to activate policies and processes that put ordinary maintenance and rehabilitation at the centre of disaster risk preparedness.

A culture of disaster prevention needs to be developed, and this also passes through the acquisition of the awareness that it can be built only through medium- and long-term processes: management effectiveness demands strategic vision but also and possibly even more everyday attention, humble work of communication, dialogue, coordination, and collaboration, with continuous adjustments to maintain the route towards the achievement of the critical objectives of management.

A change in focus from the role of the individual manager, seen as the protagonist of the action, to the importance of team working and team-building, and everyday care to the 'management engine' is essential to understanding and promoting the key qualities for good management, as a baseline and preparation to deal with disaster risks and response[11].

Medium- and long-term capacity-building strategies combined with training and drill programmes will be fundamental to achieving a prevention culture and embedding it stably into heritage management. In this sense, the undertaking of ICCROM, IUCN, and ICOMOS through the Leadership Programme to jointly revise the whole set of management-related resources manuals and guidance documents in conjunction with capacity-building activities offers opportunities to improve the situation: the resource manuals can benefit from experiences and issues encountered on the ground, thus integrating directly lessons learnt, site managers, decision-makers and resource persons have the chance to contribute to the creation of guidance tools, to reflect on their management experiences within a community of peers and strengthen their approach to management.

In Conclusion

Disaster risk reduction for cultural heritage has travelled a long way forward and, today, culture and cultural heritage are recognised among strategic sectors to be addressed for sound and long-lasting recovery. Much advancement has been attained in several areas, particularly in development of technologies, methodologies, devices, and tools. Notwithstanding these important achievements, much remain to be fulfilled to reduce risks, improve emergency response and achieve effective recovery. This depends on many factors, some of which relate to the increased exposure to risks of our cultural heritage, i.e., budget shrinking, reduction of dedicated staff, a disconnection between the currency gained by disaster risk reduction and how this is actually embedded in governance and management system. A renovated effort to respond to the disaster risk reduction for cultural heritage specifically within the more general disaster risk management framework demands coordinated action along three main strands: governance and management, research, and capacity building.

Under governance and management, more attention to systemic factors, such as maintenance, and everyday care for heritage but also for the management system, e.g., in establishing interinstitutional communication and cooperation mechanisms, would be strongly needed. In terms of research, exploring aspects of disasters that have not been sufficiently and systematically investigated, such as reasons for failures and 'bad practices', or comparative studies on effects of exceptional events on heritage places in good conservation conditions versus heritage in vulnerable conditions, would provide useful insights and basis for more informed decision-making at the strategic level. With regard to capacity building on disaster risk preparedness and reduction, this will need to become part of the toolkit of the good citizen, hence combined with systemic awareness-raising as baseline activities to develop a culture of risk prevention.

Key Points

- Despite advancement in both research and practice on DRM management-related topics, heritage still pays a high toll in these events – systematic exploration of the reasons for failures in reducing impacts of disasters on cultural heritage would be essential to understand how to improve DRM.
- Gaps between policymaking and policy implementation, lack of coordination among sectors, lack of integration of cultural heritage into disaster risk planning remain key factors to be addressed and this call for integrated planning strategies.
- Several spheres of research remain to be explored to understand how to reduce occurrence of disasters and to reduce their impacts on cultural heritage, i.e., effects of maintenance to enhance resilience of heritage places vis-à-vis of disasters, compare the overall costs of post-disaster recovery with investment in maintenance and preventive/preparedness actions.
- Technology can indeed be useful in DRM, but it does not suffice; humble actions, such as daily maintenance, and human capacity to address complexity remain fundamental in disaster risk reduction.
- Need for developing a culture of disaster prevention, an attitude to address potential risks which is widespread at all levels of society through educational programmes.

Notes

1. For a brief recapitulation of the main milestones of disaster risk reduction, the following websites can be consulted:
   https://www.undrr.org/about-undrr/history
   https://www.preventionweb.net/walk-though-history-disaster-risk-reduction (accessed on 28 March 2023).
2. UNESCO, ICCROM, ICOMOS, IUCN, *Managing Disaster Risk for World Heritage, Paris, 2010.*
3. The research outcomes are included in the report prepared for the European Commission by Alessandra Bonazza et al., *Safeguarding Cultural Heritage from Natural and Man-Made Disasters. A Comparative Analysis of Risk Management in the EU*, 2018.
4. Ibid, p. 9.
5. An interesting reflection on key challenges to management effectiveness and what shift is needed in current paradigms about management is developed by Jon Kohl and Steve McCool in the book *The Future Has Other Plans: Planning Holistically to Conserve Natural and Cultural Heritage.*
6. UNISDR, Sendai Framework for Disaster Risk Reduction 2015–2030, Geneva 2015.

7   Key guiding documents on the integration of culture into the overall recovery process include the Global Facility for Disaster Reduction and Recovery (GFDRR), European Union, World Bank, United Nations. Post-Disaster Needs Assessments Guidelines – Volume A, 2013, accessed at: https://www.gfdrr.org/en/publication/post-disaster-needs-assessments-guidelines-volume-2013 and the Global Facility for Disaster Reduction and Recovery (GFDRR), European Union, World Bank, United Nations. Post-Disaster Needs Assessments Guidelines – Volume B: Culture, 2017, accessed at: https://www.gfdrr.org/en/publication/post-disaster-needs-assessments-guidelines-volume-b-2
8   This is the case of several mountainous landscapes, which are today prone to landslides because they are no longer maintained and monitored by farmers as they were in the past. The lack of maintenance of the steep terraced territory and some local inappropriate earth movements were identified as major contributing causes to the landslides that occurred in the World Heritage property 'Porto Venere, Cinque Terre and the Islands Palmaria, Tino and Tinetto' following exceptional flash floods on 25 October 2011.
9   For instance, in Italy, the Ministry of Culture made a significant effort since 2012 to establish a functional structure operating at the central level, with specular branches operating at the territorial level to ensure prompt response for the safeguard of cultural heritage in case of disasters. The central structure works in coordination with Civil Protection and other responsible bodies at the national level. At the territorial level, operational branches of the Ministry of Culture ensure communication and coordination with agencies responsible for managing emergency situations. In some instances, where locally based emergency agencies were not fully aware of these new ad-hoc emergency structures activated by the Ministry of Culture at the territorial level, some tensions and bottlenecks have emerged in addressing initial phases of damage assessment on cultural heritage.
10 A number of destructive fires have occurred in monuments during the presence of building sites, often when no workers were on site, e.g., after the end of the working schedule or during weekends. Understanding better what causes triggered, in each case, the fire, and why such fire turned into a disaster of larger or lesser scale would be highly beneficial to understand what systemic weaknesses and increased vulnerabilities at monuments could be when they are the object of intervention works.
11 An interesting reflection on how efficacy/effectiveness is understood in western and Chinese world and how this impacts management models is developed by philosopher and sinologue François Jullien. See Jullien, F. (2005) Conférences sur l'efficacité, Paris 2005, PUF.

References

Alessandra Bonazza et al. (2018) *Safeguarding Cultural Heritage from Natural and Man-Made Disasters. A Comparative Analysis of Risk Management in the EU*, European Commission. https://data.europa.eu/doi/10.2766/224310

GFDRR, UNDP, EU, WB. (2015) Post Disaster Needs Assessment Guidelines: Volume A. https://www.gfdrr.org/en/publication/post-disaster-needs-assessments-guidelines-volume-2013

GFDRR, UNDP, EU, WB (2017) Post Disaster Needs Assessment Guidelines: Volume B – Culture, 2017. https://www.gfdrr.org/en/publication/post-disaster-needs-assessments-guidelines-volume-b-2

Jullien, F. (2005) *Conférences sur l'efficacité*, Paris 2005.

Kohl, J.M. and McCool, S.F. (2016) *The Future Has Other Plans: Planning Holistically to Conserve Natural and Cultural Heritage*.

Stovel H. (1998) *Risk Preparedness: A Management Manual for World Cultural Heritage*, Rome: ICCROM.

UNDRR (2015) *Sendai Framework for Disaster Risk Reduction 2015–2030*. https://www.undrr.org/publication/sendai-framework-disaster-risk-reduction-2015-2030

UNESCO (2010) *Managing Disaster Risk for World Heritage*.

21 Building Synergies for Cultural Heritage

Insights from Theory and Practice

Monia Del Pinto and Clinton Dean Jackson

Introduction. Synergies as Inherent to Heritage Protection

The notion of heritage is inherently associated with the act of protection and safeguarding since the idea of Cultural Heritage (CH) was first codified in the 20th century, with the emanation of the 1931 *Athens Charter for the Restoration of Historic Monuments* and the 1964 *International Charter for the Conservation and Restoration of Monuments and Sites* (Smith 2006, 2011). Throughout the following decades, principles regarding heritage safeguarding were consistently restated in the official documentation produced by UNESCO and related advisory bodies (Harrison 2013; Smith 2006).

Before WWII, the perceived threats to cultural property were associated with the 'impact of modernity', intended as physical, social, and economic transformations endangering established values and material artefacts of cultural significance. Conflict was included in the list of threats after WWII and prompted the UN's efforts towards the 1954 Hague Convention (UNESCO 1954). Over the following decades, the codified inventory of threats insisting on cultural heritage broadened, including illicit trafficking – added after the 1972 World Heritage Convention – the impact of disasters and the intertwined effects of urbanisation and climate change (UNESCO 1971, 2010, 1972).

Towards the end of 20th century, the established discourse on the influence and importance of heritage broadened, and so did the network of players involved in its definition and safeguard. If the 1954 Hague Convention gathered institutional representatives of the UN state members – setting forth principles and resolutions against the threats posed by a mutated society to the *grand* and *good* to be preserved for mankind (Smith 2011) – the most recent resolutions are increasingly informed by the experience from the ground, actively (re)shaping the definition of, and actions on, heritage.

Disaster Risk Management for Cultural Heritage (DRM for CH) is one of the areas where the presence of voices from the ground is being progressively incentivised with multi-stakeholders initiatives that bring together local communities and end users, academics, cultural heritage NGOs, humanitarian organisations, and a broad range of professionals from the built environment and cultural heritage sectors. Although these initiatives offer a virtuous example of building a multidisciplinary, inclusive, and radical approach to heritage protection, the required synergies between different actors are complex to establish and navigate, with values and priorities to be negotiated, and trade-offs considered, in each context.

The following sections will touch upon the growth of an integrated approach in the established heritage discourse, present a brief critical overview of synergies in disaster risk management for cultural heritage, and discuss the results of a survey on barriers and catalysts to synergies as recorded from the ground. A case study box describing a significant experience

DOI: 10.4324/9781003293019-26

from South Africa prompts further reflections converging into the final section, where possible ways forward are discussed.

Integrating Disaster Risk Management and Cultural Heritage Safeguard

Synergies for an integrated approach to tackle the threat of disaster on cultural heritage were first mentioned in the 1996 Declaration of Quebec, highlighting the importance of collaboration to build local capacity (Stovel 1998, pp. 119–122). Since then, their importance was reasserted in all the significant statements addressing the impact of Disaster Risk on Cultural Heritage. The 2005 Kobe/Tokyo declaration, compiled after the UNESCO World Conference on Disaster Reduction, marked the start of synergic action integrating CH protection and DRR – a step that was made possible as the conference gathered, and responded to, the concern expressed by practitioners from the field. Contributions from this diversified audience were recorded in the Kobe report:

> A question was posed to the audience for the session to find out how many came from the cultural heritage field and those who came from other areas of disaster management. <u>The majority attending the session were from other areas of disaster management</u>, which was seen as a positive first step in <u>opening a dialogue to better integrate heritage concerns into the field</u>.
>
> (UNESCO 2005, emphasis added)

While representing the start of a dialogue between multiple fields, this statement also prompts a reflection on a significant implication of its expression. Specifically, the statement articulates the intention of operating in DRM with attention to cultural heritage matters. However, it is important to notice that the integration of heritage concerns in DRM action is not equivalent to integrating DRR principles into heritage conservation action: this difference is key, as it determines what disciplinary and epistemological lenses are adopted, establishing a methodological hierarchy with one discipline being inevitably put before the other – an aspect that can impact the success of DRR or conservation of CH.[1]

UNESCO and the Protocols in Favour of Synergies

The codified practices to tackle disaster risk in heritage contexts were first compiled in the 1998 *Risk Preparedness: A Management Manual for the World Cultural Heritage*, a comprehensive text and part of the series of ICCROM and ICOMOS management guidelines for use by site officials (Stovel 1998). The text is meant to improve risk preparedness for cultural heritage by regulating the cross-disciplinary cooperation from the global to local scale and between a range of players (from NGOs to citizens), building upon protocols released by UNESCO and its advisory bodies over the previous decades (Stovel 1998, pp. 119–145). The idea of a shift from mono-disciplinary to synergic action aiming at mitigating disaster risk for cultural heritage initially met the resistance of both disaster risk reduction and heritage professionals, as well as the general public (Stovel 1998, p. 13). The recorded attitudinal barriers included: (i) a reactive, rather than proactive, attitude of professionals from the heritage conservation field, centred on response rather than risk mitigation, (ii) the perceived exceptional character of disaster preparedness measures that would not justify a systematic adoption of DRR measures, and (iii) reluctance from the general public, in that safeguard of heritage and of human life were perceived as mutually exclusive and competing items in allocation of resources (ibid). On the other hand,

Table 21.1 Summary of relevant documents regulating the protection and safeguard of cultural heritage in relation to disaster risk

Year	Context	Summary of relevant documents
1954	The Hague (Netherlands)	1954 Convention for the Protection of Cultural Property in the Event of Armed Conflict
1964	Venice (Italy)	1964 International Charter for the Conservation and Restoration of Monuments and Sites (Venice Charter)
1970	Paris (France)	1970 Convention against illicit trafficking
1972	Paris (France)	1972 Convention concerning the Protection of the World Cultural and Natural Heritage, UNESCO (1972)
1996	Quebec (Canada)	1996 Declaration of Quebec, 1st National Summit on Heritage and Risk Preparedness, Quebec City, Canada, 1996
1997	Kobe/Kyoto (Japan)	1997 The Kobe/Tokyo Declaration on Risk Preparedness for Cultural Heritage
1998	Assisi (Italy)	1998 Declaration of Assisi
2005	Kyoto (Japan)	2005 Kyoto Declaration on protection of cultural properties, historic areas, and their settings from loss in disasters
2006	Vilnius (Lithuania)	2006 Strategy Document for Reducing Risks from Disasters at World Heritage Properties. World Heritage Committee, 30th Session, Vilnius, Lithuania, 8–16 July 2006.
2015	Sendai (Japan)	2015–2030 Sendai Framework for DRR

Source: Author.

the counterarguments to these barriers looked at the call for inclusivity and cross-disciplinarity to better negotiate and achieve the preservation goals, ensuring the survival of values and practices (Stovel 1998).

In the following decades, cross-disciplinary action was further codified in official resolutions, such as the 2005 Kyoto Declaration, the 2006 Vilnius strategy document, and the 2015 Sendai framework (Table 21.1). The shared efforts informing these documents can be considered a manifestation of synergies among experts whose efforts converged in defining a shared approach. As a result of these efforts, the notion of *disaster risk* was first featured in the heritage discourse within the 1996 Quebec Declaration (Stovel 1998). After that, specific terminology and concepts become increasingly part of a shared, cross-disciplinary vocabulary, and a shift was recorded from a reactive to a proactive attitude of heritage professionals. Epitomising this change is the adoption, during the 2006 Vilnius convention, of the concept of '*Disaster Risk Reduction* for cultural heritage' (UNESCO and WHC, 2006) in replacement of the '*Risk Preparedness* for cultural heritage'. The latter was first introduced in the 1996 Declaration of Quebec and maintained in both the 1997 Kobe – Tokyo declaration and in the 1998 Assisi Declaration (Stovel 1998, pp. 119–143).

Insiders' Voices in Heritage Conservation and Disaster Risk Reduction

Integrated protection of cultural heritage is rooted in the practice of heritage safeguarding, which implies prioritising the preservation and conservation of the past. In recent decades, critical approaches to heritage studies highlighted the controversial aspects inherent in heritage conservation. From the perspective of critical heritage studies, preserving heritage can be looked at as the attempt to maintain the status quo by preventing the alteration of the grand symbols of a particular system – or at least to oppose the societal transformations threatening the power structures that certain specimens represent (Smith 2006). This conservative position underpins the 'heritage at risk rhetoric' (Rico 2014) that, in disaster contexts, promotes identical reconstruction and restoration of buildings and sites, hindering the emergence of new forms of heritage. In these cases, conservation appears to interfere with the constant heritage-making and unmaking process resulting from destruction and (re)construction dynamics within a disaster context. Yet, heritage conservation and safeguarding can be a way to ensure the survival of minor artefacts, buildings, sites, and cultural practices connected to endangered places and people that would otherwise disappear (ICOMOS 1999). It is the problematic case of non-Western cultures categorised as 'indigenous' (Ashley & Frank 2016) that, in the last two decades, entered the UNESCO agenda as forms of intangible and living heritage (UNESCO, 2003). Such an inclusive approach is still arbitrary, driven mainly by political agendas that still generate selective conservation.

With the latter point comes the issue of 'cultural significance' of what is labelled as heritage – particularly, what system and what voices are *authorised* to define what is heritage, and how to allow inclusivity in the current and future heritage discourse (Smith 2006, pp. 103–104). From an operative perspective, adopting a critical standpoint can be vital in rethinking action on heritage by ascribing authority to 'insiders' voices – overcoming the dualism of local, traditional knowledge versus expert academic knowledge. Acknowledging the validity of local instances would reverberate on heritage management strategies, with decisions on heritage not superimposed or codified by external authorities but defined and pursued from within.

A similar approach, underpinned by a global reflection on positionality and ontologies, is increasingly being adopted in disaster studies and DRR practice. This ongoing paradigm shift, rooted in de-colonising practice, values the presence of different ontologies in multi-cultural and multidisciplinary disaster research (Goodall et al. 2022) and questions the established hegemonic knowledge in favour of the long-neglected non-western epistemologies to give back power (i.e., voice and agency) to local actors (Gaillard 2019; Gaillard & Peek 2019).

The importance of these changes for synergies in CH becomes apparent when looking at the rate of success in DRM for CH projects involving local communities or citizens' groups: negotiating values and approaches, with external experts being respectfully supportive of, but not overpowering, the insiders, can be determinant for a project's success (ref section 5).

The Importance of Establishing Synergies: Horizontal and Vertical Knowledge Exchange

Interdisciplinarity allows holistically approaching a problem and proposing systemic, innovative solutions based on horizontal and vertical knowledge exchange (Danermark et al. 2002; Lélé & Norgaard 2005). Integrating DRR practice and CH safeguard implies drawing upon a multitude of expertise, and inherently requires cross-disciplinary understanding to unfold complex dynamics, including, but not limited to, disaster risk creation, vulnerabilisation processes, capacity and resilience building in heritage contexts (Del Pinto et al. 2021; Gaillard et al. 2019; Garcia 2021; Jigyasu et al. 2013).

From a practical standpoint, joined heritage safeguard and disaster risk reduction measures can be mutually enhancing, as shown by the mitigation of the physical vulnerability of heritage buildings following conservative structural retrofit, or by the capacity-building potential inherent in cultural practices (Gaillard 2010; ICCROM 2021). On the other hand, the absence of a cross-disciplinary approach might penalise safety in the name of preservation or vice versa: this is shown by disaster-blind heritage conservation measures that, in the name of authenticity, fail to address inherent weaknesses of heritage buildings in hazard-prone areas (Borri et al. 2017; Del Pinto 2021). Further examples are the adoption of invasive structural DRR measures threatening the integrity and authenticity of heritage artefacts.

From a theoretical standpoint, while establishing synergies between the heritage conservation and disaster risk reduction field, we are operating in the frame that Waterton and Watson (2017) define 'theories *in* heritage'.[2] In this process, we are sourcing theories outside the field and applying them within it to solve specific problems, as in the case of disaster risk management for cultural heritage sites. The operation of reframing and recontextualisation of knowledge inherent in this practice can sometimes evolve, gaining depth and breadth, and generate novel 'theories *of* heritage'. An example is provided in the paradigm shift turning heritage from *the object* of DRR measures into an *active agent* for DRR (UNESCO 2005, p. 8), after DRM action was incorporated in CH protection following the Quebec declaration.

The practice of outsourcing theory, which is inherent to shared heritage protection effort, has, however, inevitable side effects: when we bring together professionals that share little to no theoretical and/or methodological background, we must expect that the understanding of, and approaches to heritage safeguard – and consequently the idea of acceptable and necessary interventions on it – can significantly vary within a team of specialists.

The call for interdisciplinarity retains a twofold potential. If only based on horizontal knowledge exchange, cross-disciplinary action could reinforce the over-reliance on elites of experts governing decision-making, hence broadening the gap between decision-makers and end users. If underpinned by a vertical knowledge exchange, instead, synergies could offer a way to dialogically *rethink* the role of experts. This would neutralise the consolidated praxis that sees specialised knowledge in antagonism to local forms of knowledge (Harrison 2013) promoting, instead, constructive interaction between specialists and the general public. The latter approach characterises the most recent initiatives brought forward to maximise collective access to the heritage discourse and participatory practices, which open towards communities for consultation and bottom-up knowledge production and integrate professional and context-based know-how. Examples are offered by international training programmes led by the established cultural heritage NGOs and by the consequent projects carried out at different scales as well as in different contexts to implement the codified good practices (D-MUCH 2020; Hashem & Ambani 2021).

Exploring Synergies in Practice

The adoption of a synergic, multidisciplinary, and participatory approach is increasingly common in DRM for CH projects worldwide. However, although driven by globally acknowledged and shared principles, distinct projects show different rates of success, depending on specific and often unique conditions varying with the context.

To inform this chapter with an insight into what barriers and catalysts manifest on the ground, a survey was circulated in the network of CH and DRR professionals linked to ICCROM and Ritsumeikan University D-MUCH, operating in the field of disaster risk reduction and/or management for cultural heritage sites. The survey consisted of open-ended and close-ended questions where responders were asked to reflect on one or more DRM for CH projects they took part in. In the answers, participants had to specify (i) their professional background and role in the project, (ii) the type of additional stakeholders in the team and their role in the project, (iii) the team's relation to the context, (iv) perceived catalysts and barriers, and their impact on the successful design and implementation of the project, and (v) follow up strategies and learning points and how the experience informed future action.

A total of 33 respondents shared information on projects undertaken in 35 regions across Europe, Asia, Africa, South and North America, offering their perspectives as DRR or CH specialists, whether from within or from outside the context and reflecting on the outcomes. The results were analysed by operating cross-tabulation to compare responses and filtering to narrow down specific aspects surfacing from general results. Cross-tabulation allowed looking at the variation of challenges and catalysts along with the different relations of stakeholders' groups with the context. The following sections present the significant results of the survey, providing insights on barriers and catalysts recorded based on the stakeholders' background and relation to the context.

Barriers: What Hinders Synergies

Barriers indicate the elements that had an impact on hindering both design and implementation of a DRM for CH project. The following subsections discuss the results in relation to barriers in the design and the implementation stages and the outcomes of cross-tabulation used to assess variations in barriers along with the stakeholders' changing relation to the context.

Barriers in the Design Stage

The most common barrier recorded in the design stage was the lack of resources (budget and/or staff) indistinctly affecting 71% of the projects from Europe, Asia, North America, and Africa. In the multiple-choice answers, this aspect featured in combination with unclear regulatory frameworks and a lack of shared knowledge within the team (Figure 21.1). The second set of barriers featuring in 29% of answers was lack of shared knowledge within the team, associated with lack of teamwork (25.8%) – suggesting that the projects were designed by teams external to the context, maintaining an approach based on disciplinary segregation while narrowing problems and ideating solutions. Lack of teamwork paired with insufficient resources was otherwise mentioned in association with a lack of participatory methods – a combination suggesting insufficient engagement in interdisciplinary and context-based action.

Barriers in the Implementation Stage

In the implementation stage, the most common barrier was the lack of resources that, however, never features in isolation. Unclear regulatory frameworks and procedures – from the local to the national scale – are also pervasively mentioned, in combination with unclear implementation rules or lack of engagement with the context. The team's disconnection from the context and the local stakeholders' disengagement from the project can be looked at as a consequence of projects that are neither ideated nor perceived as 'needed' from within – but rather 'imposed' based on assumed values, generally from academic stance. This aspect, combined with insufficient consultation – can cause a misinterpretation of the relevant meanings to the public, reducing the effectiveness of the project.

What Barriers Emerge as the Stakeholders' Relation to the Context Changes

Cross-tabulating the disaggregated results in the pivot table (Figure 21.A1 in the Appendix) allowed looking at variations in recurring challenges during implementation based on the leading team's relation to the context. The most relevant result is that unclear regulatory framework and excessive bureaucracy are still the predominant barriers recorded by both in-house and

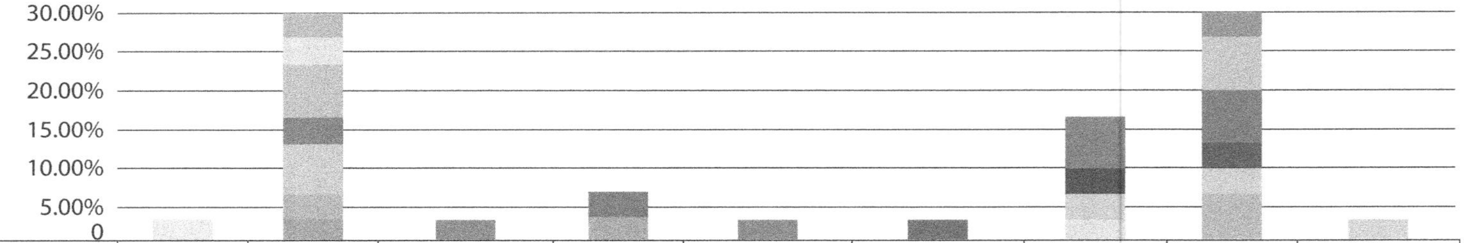

Stakeholders' relation to context / Catalysis	Project led by community members	Project led by local/in-house team	Project led by local/in-house team; project led by team of external professionals involving local professionals	Project led by a team external to the context	Project led by a team external to the context; project led by team of external professionals involving local professionals	Project led by a team external to the context; project led by team of local professionals involving external professionals	Project led by team of external professionals involving local professionals	Project led by team of local professionals involving external professionals	Project led by team of local professionals involving external professionals; project led by team of external professionals involving local professionals
Blank									
1; 3; 4; 5	3.33%	3.33%							
1; 3; 4		3.33%							
1; 3; 5; 6								3.33%	
1; 3		6.67%						6.67%	
1; 2; 3; 4; 5						3.33%			
1; 2; 3		3.33%	3.33%						
1; 2; 5				3.33%			6.67%	6.67%	
1; 2							3.33%	3.33%	
1;									
3									
8		6.67%							
2; 4; 5							3.33%	3.33%	
2; 4							3.33%		
2; 3		3.33%							
2;		3.33%		3.33%			6.67%	6.67%	3.33%
5									

List of catalysts from survey responses:
1- Trust and communication between the team and the recipients of the project
2- Employ of participatory methods involving all the stakeholder groups
3- Recipients' enthusiasm, interest and/or commitment (interest in participating)
4- Reduced bureaucracy
5- Clear regulations (local, regional, national, supranational)
6- Direct participation of the local community
7- Financial opportunities
8- not applicable

Figure 21.1 Overview of the barriers recorded in the design stage.

external teams – yet, some challenges emerged, specific to the stakeholders' relation to the context, that could prompt questions and indicate entry points to review current practices.

- *Projects led by an in-house team (local).* In addition to the general trends, one significant barrier emerging in projects led by an in-house team is a lack of participation and/or interest from local stakeholders. This is often associated lack of cooperation within the team and with the team's lack of engagement with the context – apparently a contradiction for an in-house team. The explanation could be a project conceived without adequate investigation in the design stage, not targeting the actual needs of the local community. Also, an in-house team disengaged from the context might be indicative of non-local organisations with their own (external) staff relocated in the context.
- *Projects led by a local in-house team and involving external professionals.* The main barriers emerging, in this case, include a lack of cooperation between the in-house and external team, combined with a lack of participation and/or interest from local stakeholders. Excessive bureaucracy is still a problem.
- *Projects led by team external to the context.* In this case, the emerging barriers were lack of participation and lack of engagement from local stakeholders. A combination of the two was also recorded along with the team's lack of engagement with the context, and lack of cooperation within the team. The disconnect with the context suggests a top-down approach, where the cultural values that are being safeguarded descend from an external stance only (i.e., academic), with little or no consideration for the meanings and values relevant at the local scale.
- *Projects Led by community members.* The cross-tabulation shows that bottom-up projects led by community members were mostly hindered by a lack of resources (both staff and budget) – a limitation which could be harder to overcome for community-led projects if not externally founded.

If compared to externally-led projects, the lack of resources for bottom-up initiatives suggests that projects led by external stakeholders might benefit from a flow of resources – yet this happens at the expense of adherence to contextual needs and represents a trade-off to be addressed to maximise the work of local communities. On the other hand, the disengagement of the public registered in top-down projects might be indicative of the consolidated habit among 'experts' to treat the public as a passive agent both from the heritage management professionals (Chirikure & Pwiti 2008) and disaster risk management experts (Del Pinto 2021, pp. 203–204).

Catalysts: What Facilitates Synergies

Catalysts indicate elements facilitating both the design and implementation of a DRM for CH project. The following subsections discuss the results of catalysts in the design and the implementation stage (4.2.1 and 4.2.2) and present the variations in catalysts along with the stakeholders' changing relation to the context (4.2.3).

Catalysts in the Design Stage

The most common catalyst facilitating the design stage was multidisciplinary knowledge, present in 73% of the answers, followed by pre-existing shared knowledge among the involved professionals and teamwork (46.7%). The acquisition of knowledge and the employ of participatory methods, as well as a shared language among the professionals, were mentioned respectively in 26% and 23% of the answers. Familiarity with the context appeared in 40% of the answers (Figure 21.2).

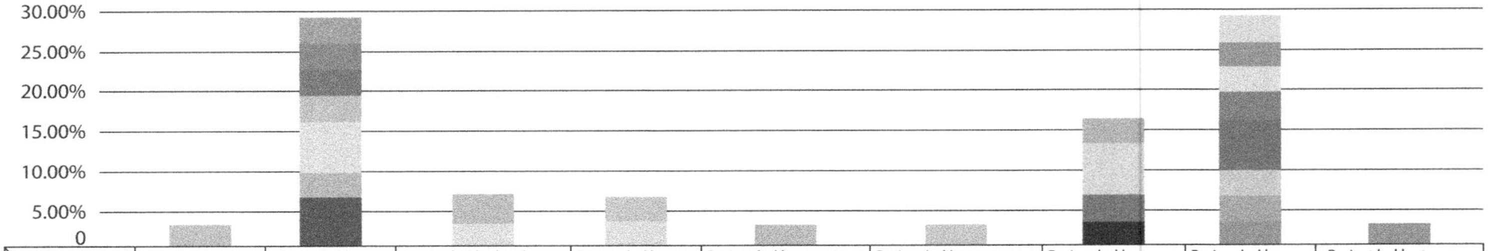

Figure 21.2 Overview of the catalysts recorded in the design stage.

Based on this general overview, the success of the design stage appears linked to the teams' knowledge of the mobilised disciplines rather than to knowledge – as well as involvement – of the context.

Catalysts in the Implementation Stage

The survey shows that it is not a single factor to facilitate the implementation – instead, it is a combination of multiple aspects that contribute to success (see also box 21.1). Trust and communication

---

Box 21.1 National Audit of Monuments & Memorials (NAMM) – A Prerequisite to Heritage Landscape Transformation Programme in South Africa

Background

The act of memorialising people and events through the construction of physical markers such as monuments, memorials, and, more specifically, statues is a long tradition, one which has left a lasting impact on the streetscape of modern South Africa. Considering the colonial and apartheid history of South Africa, these physical markers continue to serve as reminders of a painful past and the role that the leaders of that era played in shaping the zeitgeist of their time in power. The subjects of these are often associated with the colonial and apartheid power structures prior to the emergence of a democratic state of South Africa.

Post-apartheid the need for a new paradigm became more apparent to ensure that past inequalities were redressed and to ensure that the heritage landscape is representative of all inhabitants of South Africa. The impact these markers continue to have on the post-apartheid South African heritage landscape has been a concern for the nation's legislative bodies that are responsible for the management of heritage resources in the county, namely, the South African Heritage Resources Agency (SAHRA) and National Department of Sport, Arts and Culture (DSAC).

Towards the Project: Ideation and Drivers

To undertake this process of redress, a clear and comprehensive inventory of these resources is required to better understand the status quo. Whilst South Africa's legislation for the protection of the country's heritage specifically provides for the formulation and centralisation of data on heritage resources known as the inventory of the national estate, full implementation of this largely exists as an aspirational goal due to the non-functional nature of the legislated three-tier system of heritage management which places the management of heritage resources within a level of the South African governance structure appropriate to the significance a particular heritage resource holds (Jackson et al. 2019). Consequently, SAHRA lacked a comprehensive inventory of monuments and memorials across the country, a gap that further resulted in a compromised ability to manage and protect these resources.

Early attempts to generate this inventory of monuments and memorials this were made by both SAHRA and provincial/local authorities, however, these surveys were localised and did not address the need to hold a holistic understanding of the nature and spread across the country. The imperative for a national scale intervention was expressed in 2015 after the start of the Rhodes Must Fall campaign which saw the defacement of statues in public spaces. In response, a National Consultative Meeting on Transformation in the Heritage Landscape was hosted by the Minster of Arts and Culture (now Sport, Arts and Culture), which was attended by academics, students, heritage practitioners, and interest groups. Amongst the resolutions adopted at this forum was the need for a national audit of monuments and memorials to identify and better understand the nature of these resources across the country (Kubhecka 2016). Despite an attempt by SAHRA to start this process through exiting legislative triggers, it became apparent that many of the provincial authorities were not equipped to establish their own processes and protocols to manage such an emotive and pressing issue.

As the national body responsible for the coordination of heritage management in the country, SAHRA was tasked by the DSAC with undertaking the *National Audit of Monuments & Memorials (NAMM) – A Prerequisite to Heritage Landscape Transformation Programme* as an intervention managed at the national level to provide the necessary inventory of Monuments & Memorials across the country and empower decision-makers with the knowledge of where these resources are located, their densities, and condition. With renewed social consciousness stemming from the continuation of the Rhodes Must Fall movement and widespread discussions on decolonisation, the need to address the impact of these markers on the South African landscape became more important.

This project was eventually made possible through funding established by the Presidency of the Republic of South Africa under the broader Presidential Public Employment Stimulus Programme (PESP), a programme designed to provide short term employment in response to the economic impact of the ongoing Covid-19 pandemic (Republic of South Africa 2020).

Aims and Objectives of the Project

Considering the context in which the project was ideated and funded, the NAMM sought to achieve the following aims:

1 Identification and documentation of monuments and memorials across the country through the objectives of

　a Photographic recording
　b Capture of geolocation
　c Digitisation of interpretive data present on site
　d Inclusion of identified resources into the inventory of the national estate

2 Provision of employment opportunities to unemployed youth as part of the Covid-19 Presidential Public Employment Stimulus Programme.
3 Provide a baseline database for decision-making within the broader Heritage Landscape Transformation Programme.

Stakeholders

To achieve such an undertaking within this limited timeframe, SAHRA relied upon an extensive series of internal role players and external stakeholders to facilitate project guidance, planning, and implementation. The stakeholders support included political and authoritative guidance provided by the National Department of Sport, Arts and Culture, provincial level facilitation of local coordination, and day-to-day oversight of fieldwork teams.

While this project would not have been possible without the extensive stakeholder network, the complex governance framework, variable capacity of the stakeholders, and multi-layered reporting structures resulted in uncertainties within areas of project design.

Design Stage: Initial Barriers and Catalysts

Based upon the first project brief received from the national structures, initial conceptualisation, and planning for the NAMM included further project targets including audits of National flags flying at designated localities, and the recording of intangible heritage. This scope, however, changed during the planning process, favouring the focus on monuments and memorials. During this time, there was a great deal of uncertainty concerning the number of employment opportunities that would be available for the project, budget allocations, and the restrictions on the use of the budget, as initial directives indicated that any allocated budget could only be used for payment of salaries and not for any equipment or travel costs. A limitation that was thankfully removed as the project would not have been possible without equipment or Covid-19-related personal protective equipment.

Furthermore, the ability of more operational stakeholders to provide support varied depending on the resources and capacity available within each provincial level, as such a one-size-fits-all approach to engagement and expectation could not be adopted across similar stakeholder groups and necessitated a level of fluidity within project design and implementation that could accommodate variation on a case-by-case basis. For example, each of the provincial departments of Sport, Arts and Culture was requested to provide logistical support to assist field surveyors with their transportation needs. However, the capacity to provide this support varied between each province. Whilst some provinces were able to provide full transportation for survey teams across the province, others were unable to provide the same level of support and required that SAHRA provide remote intervention by facilitating vehicle rentals or shuttles on an ad-hoc basis as capacity constraints were encountered.

Project Implementation: Barriers and Catalysts

Consequently, project implementation included a constant feedback loop to trigger adjustment and refinements to the role out programme within each province. This was especially important when considering the timeframe constraints which precluded protracted negotiation with stakeholders.

Due to these variations, project implementation was not standard across the country with each of the nine provinces starting fieldwork at different times within the project lifespan (Table 21.2).

This staggered implementation resulted in a need to repeat many of the implementation steps on an individual basis. An example of this was the need to repeat training interventions to accommodate late start project teams.

Within the internal project structure itself the vast scope of such a project required that an internal team be appointed to guide functional areas of project implementation, including Project Management, Stakeholder Engagement, Administration, Finance, Human

Table 21.2 Start of fieldwork recordings per province

Province	Start of fieldwork recordings
KwaZulu Natal	18/01/2021
Free State	25/01/2021
Gauteng	25/01/2021
Limpopo	01/02/2021
Northern Cape	11/02/2021
Eastern Cape	22/02/2021
North West	22/02/2021
Western Cape	25/02/2021
Mpumalanga	02/03/2021

Source: Author.

Resources, Training, Research, and Data Collection. Importantly, SAHRA being a legislative entity needed to ensure the continuation of normal operations during this time. This resulted in split focus for the project team, often requiring that team members undertake project activities outside of specified areas of responsibility. The implication of this being that, at times, the activities outside of the specified area of responsibility received less intensive/rapid scrutiny resulting in delayed corrective measures being implemented or uncertainty in reporting structures to address time bound issues as they arose.

A factor to be considered within the implementation of this project was the global context of the Covid-19 pandemic. Whilst the funding provided under the PESP allowed for the targeted employment of some 260 beneficiaries to undertake the required survey work, it also introduced several challenges that needed to be addressed, including a requirement for all projects funded through this mechanism to be planned and implemented within a six-month timeframe. This was further compressed due to delays in project approval and funding allocation, which resulted in an effective period of four months to physically implement the nationwide survey.

These timeframes combined with prevailing restrictions imposed on gatherings resulted in an inability to provide comprehensive training with participants and stakeholders. This was, however, partially mitigated by taking advantage of readily available technology to enforce greater control over data collection.

Despite these barriers, the project would not have been possible without the support of the stakeholders regardless of their role or capacity to provide operational support. For example, the success of the Free State provincial survey, in terms of geographic coverage, would not have been possible without the direct involvement and support of the officials from Heritage Free State. In general, where provincial structures were able to be more directly involved, the surveys were able to roll out in a more structured manner than those which required higher levels of remote intervention from SAHRA to address barriers.

One factor that played a large facilitation role within this project was the adoption of technological interventions that provided the ability to perform remote oversight of projects, the data they collected, and rapid reporting to project sponsors. By employing the KoBoToolbox ecosystem, we were able to streamline data collection and post processing by eliminating the need for an extensive period of post survey data capture. Each survey team of two members was supplied with an Android smartphone with the KoboCollect mobile application preinstalled. This provided an effective platform which allowed surveyors to collect photographic, GPS, and textual data through a singular platform. Additionally, the ability to operate offline allowed surveyors to operate within areas of the country with limited internet connectivity and later sync the collected data once back in range of an internet connection. This allowed SAHRA to remotely monitor data collection in near time and consequently plan interventions such as additional training or redirect survey teams to address areas with insufficient geographic coverage (for further information on the tools used in this project, see Jackson et al. 2021).

Outcomes

Ultimately, the NAMM achieved its principal goals. At the close of the project on 30 April 2021, a total of 268 individuals had benefitted from short term employment opportunities, and 1 366 recordings were approved by the provincial teams for import into the inventory of the national estate. After a period of review, 1 149 of which were accepted by SAHRA to form the final population of monuments & memorials (inclusive of those already within the inventory). From the perspective of national coverage, 73% of the country's Metropolitan and Local Municipalities were subject to survey. A project of this scope had never been previously attempted by SAHRA, as such several lessons were learnt through the design and implementation process. Firstly, the project relied heavily on the inter-governmental framework. Differences in capacity and policy environments resulted varying delays in project implementation throughout the country, this necessitated a fluid approach to project implementation to accommodate the ever-changing environment. Secondly, internal project roles required refinement during project implementation due to capacitation constraints within SAHRA itself, resulting in changes to project area responsibility mid-stream. Finally, limitations in the available training opportunities resulted in inconsistencies within the data collected by the fieldwork teams. Whilst every effort was made to mitigate this through data collection mechanisms, further ratification and verification will be required.

between the team and the recipients were among the most common determinant (60%) combined either with recipients' enthusiasm or with the employ of participatory methods. The presence of trust and communication within the team was combined in 23.33% of cases with enthusiasm in the recipients and in 16.67% of the cases with the employ of participatory methods. The adoption of participatory approaches in 10% of cases was combined with other determinants – i.e., reduced bureaucracy, clear regulation, and recipients' enthusiasm. These results show that success in the implementation stage was rarely ascribed to one single reason: trust and communication alone were only mentioned in 10% of the responses; same for participatory methods alone, whereas clear regulation alone was considered the only catalyst in 6.67% of cases.

What Catalysts Emerge as the Stakeholders' Relation to the Context Changes

In the cross-tabulation of results (Figure 21.A2), while some of the answers refer to isolated catalysts – enthusiasm alone (3.33%), trust alone (10%), participatory methods (10%), clear regulation (6.77%) – the majority of answers reports combination of catalysts, with variations depending on the stakeholders' relation to the context. For example, trust and communication prevail in projects led by insiders (in-house team or local professionals), whereas the employ of participatory methods consistently features in projects led by outsiders, with or without local participants involved. In particular:

- *Projects entirely led by an in-house team from the leading institution.* Trust and communication between team and recipients were referred to as the primary catalyst (in 26.7%of cases). These were consistently paired with recipients' enthusiasm/interest/commitment and would also include either clear regulation and reduced bureaucracy or the employ of participatory methods. Even in the case of teams led by local stakeholders with the support of external participants, the elements of trust and communication feature among catalysts.
- *Projects led by team external to the context.* The success of these projects shows a component of trust and communication paired with the employ of participatory methods.

Recipients' enthusiasm and use of participatory methods involving all the stakeholders' groups were referred to as facilitators both in projects led by locals and in those led by external teams. While confirming the key role of information exchange evidenced in built environment rehabilitation projects (Fantazi et al. 2019) these results also imply for successful projects the neutralisation of possible language barriers.

Critical Reflections

The previous sections have shown that various factors can determine the effectiveness of interdisciplinary synergies and influence the success of DRM for CH projects.

Most of the elements affecting the rate of success reflect the rate of communication and engagement among stakeholders. These can manifest in varying teamwork skills within the group leading the project, or in different rates of engagement of the involved stakeholders throughout the project – from consultation to implementation. A reduced interaction, in turn, affects the enthusiasm of both experts and the general public. The outcomes of the survey restate some core themes present in literature, such as the experts' attitude towards the context, characterised by biases and assumptions, and the illusion of neutrality of technical knowledge.

Professionals' unpreparedness to operate in disaster and conflict areas features in the literature as a key issue for humanitarian action – a field requiring a holistic and human-centred approach. Charlesworth and Calame (2012), focusing on interventions in conflict areas, suggest that blindness towards the context is rooted in training built environment professionals and cultural heritage conservators. Operating in critical areas brings to light that professionals' action is prevalently *'aiming at neutrality and rooted in technical knowledge'* (ibidem, p. 172) as eloquently expressed in the following passage:

> [...] Architects, urban planners, and cultural-heritage conservators are similarly constrained by the conventional attitudes toward individual sites and structures cultivated in them in school. This educational system has generated a class of highly competent professionals that is generally unprepared to assist the citizens [...]
> (Charlesworth & Calame 2012, p. 172)

This observation can be extended to the broader built environment and heritage-related sub-disciplines and opens the question of inherent gaps in the corresponding sectors of higher education (Willems et al. 2018). The question of conventional training that lacks critical analysis of risk, conflict, and their impact and root causes is at the very core of the recent calls for pedagogical review of the BE curricula, in the name of interdisciplinarity, inclusivity, and social engagement (Chmutina & Von Meding 2022). A similar need surfaces in heritage sub-disciplines, such as archaeology, where higher education programmes neglect community engagement skills and prioritise the production of academic outputs and knowledge hardly useable within the dynamic action on the ground (Willems et al. 2018).

Professionals' neutrality and reliance on technical knowledge – underpinned by the false belief that academic knowledge is superior, and technical solutions are above political implications[3] – can be considered as possible predictors of (in)success in synergic action, in that they facilitate the operators' detachment from the context. We have seen from the survey that detachment fosters disengagement, culminating in projects that struggle to be successfully implemented due to a lack of communication of experts with local stakeholders – one of the barriers recorded in the survey and restated in literature (Fantazi et al. 2019).

Ways Forward

The virtuous examples from the survey and the case study suggest that ways forward are possible – on a practical level, building upon good practices of participation and vertical knowledge exchange, while on a theoretical level acknowledging the need for rethinking the established paradigms that drive knowledge production and its applications. For DRM and CH operators on the ground, possible action points include (i) proactively engaging with the context – from the project ideation to its implementation and follow up, and (ii) refraining from blindly adopting the heritage at risk rhetoric embedded in the academic stance. These changes are, in turn, linked to core action points in the spheres of training and positionality of professionals.

Training

Both in the built environment and cultural heritage training, the higher education curricula still largely reproduce and consolidate established biases – such as the assumed neutrality ascribed to technical knowledge, or the disproportion in agency between experts and non-experts, at the expenses of the latter. This suggests that a fundamental step to building effective synergies is the review of HE curricula in both content and methods by (i) questioning the hegemony of technical knowledge and tools as primary – and sometimes only – solution, (ii) integrating DRR principles in both BE and CH related sub-disciplines (Chmutina & Von Meding 2022), and (iii) critically re-evaluating the *endangerment sensibility* (Vidal & Diaz 2015) that underpins the narratives around built heritage.

Positionality

When preserving and managing heritage, we are not only operating on material assets but also on what Smith (2006, p. 88) defines as *cultural and social value and meaning*. Within this perspective, conservators and disaster experts carry the responsibility of defining meanings, values, and threats within a project. When cultural meanings are ascribed from an external stance (i.e., academic), a large portion of knowledge, values, and significance from the ground is at risk of being lost – failing the *'do no harm'* motto (Tandon 2018) ideally guiding the DRM for CH action. Literature shows how cultural heritage meanings change along with society, in the constant heritage-making and unmaking process that differs from the static idea of values and the past inherited from the Enlightenment (Ashley & Frank 2016; Harrison 2013). Such a static idea is still pervasively embedded in the academic stance and is shown when the action of specialists operating from a purely theoretical perspective ends up disconnected from the ground. In this regard, the need to adjust approaches once a project is implemented highlighted in the case study box is exemplary.

Being aware of the dynamic evolution and transformation of meanings, factoring the *perceived* threats in heritage risk assessment, and acknowledging the negotiation of social and cultural values driving the current heritage discourse, should be a guiding principle for effective disaster risk management action on cultural heritage. Besides informing on 'what to safeguard against what', in fact, knowing where cultural significance lies for communities also indicates what heritage elements retain potential for increasing local capacities from within.

Conclusions

This chapter proposed a reflection on interdisciplinary action in DRM for CH projects. Synergies among academic, practitioners, policymakers, and the general public, along with horizontal and vertical knowledge exchange, are increasingly incentivised and implemented from both the DRM and CH fields. Yet, against such a global, coral call to participatory and inclusive practices, the way to successfully implement multi-stakeholder collaboration in DRM for CH has shown recurring obstacles in practice.

The main barriers seem mostly linked to biases and gaps in both attitudes and knowledge of the involved stakeholders. This aspect appears associated with a more significant presence in the projects of outsider experts, whose action is rooted in training that fosters detachment from the context and disciplinary segregation. Projects that promote engagement by favouring collective participatory action show higher chances of being successfully implemented. This is more apparent when end users and local stakeholders actively partake in the project at all stages – consultation, ideation, implementation, and follow-up. In the presence of major constraints, as in the case of projects sponsored by the central governments, the need to operate within broader governance programmes and schedules, and the chain of bureaucracy, could restrict the breadth or depth of synergic action. Within this perspective, it is also not infrequent that projects involving disaster risk management for cultural heritage, although marketed as a priority, are treated as secondary in that perceived as a soft issue or not pressing on a government agenda.

Besides fostering good practices, possible ways to overcome the recorded obstacles are to address the attitudinal and epistemological changes highlighted from both the cultural heritage and disaster risk management disciplinary areas. These can be summarised as gaps and biases in higher education training, affecting the production and application of knowledge, and as a static and outdated conceptualisation of heritage that prioritises conservative positions (irrespective of the changes in meaning and values).

Key Points

- Multi-stakeholder and multidisciplinary cooperation are key in Disaster Risk Management for Cultural Heritage, both to safeguard cultural artefacts and practices from disaster risk and to enhance their potential to foster resilience and capacity building.
- Originally intended as cooperation among experts and based on horizontal knowledge exchange, cross-disciplinary practices are progressively opening towards local knowledge to inform action through ground up approaches.
- The experience from the ground shows that success of integrated action is favoured when local stakeholders lead, co-lead, or consistently inform, DRM for CH projects, complementing professional expertise with context-specific knowledge.

- The experts' attitudinal barriers to vertical knowledge exchange, along with lack of resources and the secondary role of DRM for CH action in policy agendas, still represents a limitation to effective multidisciplinary practice.
- The influence of training in creating and consolidating specialists' attitudinal barriers – whether conscious or unconscious – is proven in literature as a cross-cutting problem and reiterates the need to revise disciplinary epistemologies and Higher Education curricula in favour of inclusivity.
- In heritage studies, a paradigm shift is essential to enable a more dynamic definition and practice of heritage and conservation.

Notes

1 A significant example of DRR subordinated to heritage conservation is epitomised in the case of Norcia (Italy). The city is renowned for its religious heritage, which was heavily damaged after the 2016/17 earthquake. The buildings' vulnerability was not fully addressed over decades, even after past earthquakes, due to enforced preservation measures that limited the structural retrofit (Borri et al. 2017). With the religious heritage capitalised as a tourist destination, the lack of effective structural DRR measures in the name of authenticity put at risk not only the integrity of the built fabric but also the occupants of buildings and surrounding spaces (Del Pinto 2021).
2 Theories *in* heritage as posed by Waterton and Watson (2017) refers to the use in heritage field of external disciplines used to regulate the objects of heritage – encompassing archaeology, art history, architecture, management. Based on the classification provided by Waterton and Watson, '*theories in heritage are of short- to medium-range significance, used to explain and elucidate facets of heritage as found and experienced. They are usually sourced from outside the field and applied within it*' *(ibidem)*.
3 Professionals […] *are 'trained to be above partisanship […] but neutral space is typically not available for occupancy when you arrive in the divided city'* (Charlesworth & Calame 2012, p. 172).

Reading Suggestions

On the sources of theory in heritage studies, and to unpack the relation between heritage discourse and external disciplines:
Waterton, E. and Watson, S. (2017), "Framing theory: Towards a critical imagination in heritage studies", *International Journal of Heritage Studies*, Vol. 7258 October, pp. 546–561.
On the assumptions underpinning established preservation practices:
Vidal, F. and Diaz, N. (2015), "The endangerment sensibility", *Endangerment, Biodiversity and Culture (Routledge Environmental Humanities)*, pp. 1–40.
On knowledge creation and sharing in multi-cultural and multidisciplinary contexts:
Goodall, S., Li, Y., Chmutina, K., Dijkstra, T., Meng, X. and Jordan, C. (2022), "Exploring disaster ontologies from Chinese and Western perspectives: Commonalities and nuances", *Disaster Prevention and Management: An International Journal*, Vol. 31 No. 3, pp. 260–272.

References

Ashley, S.L.T. and Frank, S. (2016), "Introduction: Heritage-outside-in", *International Journal of Heritage Studies*, Routledge, Vol. 7258, pp. 1–13.
Borri, A., Sisti, R., Prota, A., Di Ludovico, M., Costantini, S., Berluzzi, M., De Maria, A., *et al.* (2017), "Analisi del danno degli edifici ordinari nel centro storico di Norcia a seguito dei sismi del 2016", pp. 17–21.
Charlesworth, R. and Calame, J. (2012), *Divided Cities: Belfast, Beirut, Jerusalem, Mostar, and Nicosia*.
Chirikure, S. and Pwiti, G. (2008), "Community involvement in archaeology and cultural heritage management. An assessment from case studies in Southern Africa and elsewhere", *Current Anthropology*, Vol. 49 No. 3, pp. 467–486.

Chmutina, K. and Von Meding, J. (2022), "Towards a liberatory pedagogy of disaster risk reduction among built environment educators", available at: https://doi.org/10.1108/DPM-02-2022-0041.

Danermark, B., Ekstrom, M., Jakobsen, L. and Karlsson, J.C. (2002), *Explaining Society. Critical Realism in the Social Sciences*, available at: https://doi.org/10.1177/089484530403100102.

Del Pinto, M. (2021), *Urban Form and Disaster Risk: The Role of Urban Public Open Spaces in Vulnerability of Earthquake-Prone Settlements Loughborough University.*, Loughborough University.

Del Pinto, M., Palaiologou, G., Chmutina, K. and Bosher, L. (2021), "Urban morphology in support of disaster risk reduction: towards theory and methods for a spatial approach to tackling urban vulnerability to earthquakes", *XXVIII International Seminar on Urban Form ISUF2021 : Urban Form and the Sustainable and Prosperous Cities 29th June–3rd July 2021, Glasgow*, pp. 354–362.

D-MUCH. (2020), "International Training Course (ITC) on Disaster Risk Management of Cultural Heritage", *Proceedings of XIV UNESCO Chair Programme on Cultural Heritage and Risk Management*.

Fantazi, I., Hecham, B.Z. and Petrisor, A.-I. (2019), "The impact of the absence of communication on the success of rehabilitation projects of the built heritage: The case of the old City of Constantine", *Present Environment and Sustainable Development*, Vol. 13 No. 1, pp. 225–239.

Gaillard, J.C. (2010), "Vulnerability, capacity and resilience: perspectives for climate and development policy", *Journal of International Development*, Vol. 22 No. 2, pp. 218–232.

Gaillard, J.C. (2019), "Disaster studies inside out", *Disasters*, Vol. 43 No. S1, pp. S7–S17.

Gaillard, J.C., Cadag, J.R.D. and Rampengan, M.M.F. (2019), "People's capacities in facing hazards and disasters: An overview", *Natural Hazards, Springer Netherlands*, Vol. 95 No. 3, pp. 863–876.

Gaillard, J. and Peek, L. (2019), "Disaster-zone research needs a code of conduct", *Nature*, Vol. 575 No. 7783, pp. 440–442.

Garcia, B.M. (2021), "Integrating culture in post-crisis urban recovery: Reflections on the power of cultural heritage to deal with crisis", *International Journal of Disaster Risk Reduction*, Elsevier Ltd, Vol. 60 March, available at: https://doi.org/10.1016/j.ijdrr.2021.102277.

Goodall, S., Li, Y., Chmutina, K., Dijkstra, T., Meng, X. and Jordan, C. (2022), "Exploring disaster ontologies from Chinese and western perspectives: Commonalities and nuances", *Disaster Prevention and Management: An International Journal*, Vol. 31 No. 3, pp. 260–272.

Harrison, R. (2013), *Heritage. Critical Approaches*. Routledge.

Hashem, Y. and Ambani, J. (2021), *A Story of Change. Success Stories and Lessons Learnt of the Culture Cannot Wait: Heritage for Peace and Resilience Project*. Rome: ICCROM, 2021

ICCROM. (2021), *World Heritage Capacity Building Strategy Internal Review Report*. Paris, 2021.

ICOMOS. (1999), "Charter on the built vernacular heritage", Ratified by the *ICOMOS 12th General Assembly, Mexico, October 1999*.

Jackson, C., Mofutsanyana, L. and Mlungwana, N. (2019), "A risk based approach to heritage management in South Africa", *The International Archives of the Photogrammetry, Remote Sensing and Spatial Information Sciences*, Vol. XLII September, pp. 591–597.

Jackson, C., Nkhasi-Lesaoana, M. and Mofutsanyana, L. (2021), "Rapid data collection for the audit of monuments and memorials in South Africa", *The International Archives of the Photogrammetry, Remote Sensing and Spatial Information Sciences*, Vol. XLVI September, pp. 313–320.

Jigyasu, R., Murthi, M., Boccardi, G., Marrion, C. and Douglas, D. (2013), *Heritage and Resilience. Issues and Opportunities for Reducing Disaster Risks*. Mumbai: Global Platform for Disaster Risk Reduction, 2013.

Kubhecka, T. (2016), *Activities Relating to the Working Group on Geographical Names as Cultural Heritage Task Team on Transformation of Heritage Landscape. Task Team on Transformation of Heritage Landscape*. United Nations Group of Experts on Geographical Names, twenty-ninth session, Working paper n. 86/16, Bangkok 25-29 April 2016.

Lélé, S. and Norgaard, R.B. (2005), "Practicing interdisciplinarity", *BioScience*, Vol. 55 No. 11, pp. 967–975.

Republic of South Africa. (2020), *Building a Society that Works. Public Investments in a Mass Employment Strategy to Build a New Economy*. Presidential Employment Stimulus, South Africa Presidency, 2020

Rico, T. (2014), "The limits of a 'heritage at risk' framework: The construction of post-disaster cultural heritage in Banda Aceh, Indonesia", *Journal of Social Archaeology*, Vol. 14 No. 2, pp. 157–176.

Smith, L. (2006), *Uses of Heritage*. Routledge: London, available at: https://doi.org/10.4324/9780203602263.

Smith, L. (2011), "El 'espajo patrimonial'. ¿Ilusión narcisista o reflexiones múltiples?", *Antípoda*, Vol. 12 June, pp. 39–63.

Stovel, H. (1998), *Risk Preparedness: A Management Manual for World Cultural Heritage*. Rome, ICCROM, 1998

Tandon, A. (2018), *First Aid to Cultural Heritage in Times of Crisis*. ICCROM.

UNESCO. (1954), *Convention for The Protection of Cultural Property in the Event of Armed Conflict with Regulations for the Execution*. Director-General of the United Nations Educational, Scientific and Cultural Organization, The Hague, 1954.

UNESCO. (1972), *Convention Concerning the Protection of the World Cultural and Natural Heritage* Paris, 1972.

UNESCO. (2003), *Convention for the Safeguarding of the Intangible Cultural Heritage* Convention for the Safeguarding of the Intangible Cultural Heritage, adopted 17th October 2003. Paris: France. The General Conference of the United Nations Educational, Scientific and Cultural Organization (UNESCO) meeting, Paris, 29 September to 17th October 2003, 32nd session.

UNESCO, ICCROM, and Agency for Cultural Affairs of Japan.. (2005), *KOBE REPORT Draft, Report of Session 3.3, Thematic Cluster 3 Cultural Heritage Risk Management*. World Conference on Disaster Reduction, Hyogo, 2005

UNESCO. (2010), *Managing Disaster Risks for World Heritage, World Heritage Resource Manual Series*. UNESCO, 2010.

UNESCO. (1971), *Assistance in Case of Natural Disasters. Resolutions Adopted on the Reports of the Third Committee*. Adopted at the 2018th plenary meeting, 14 Dec. 1971. In: Resolutions adopted by the General Assembly during its 26th session, 21 September-22 December 1971. - A/8429. - 1972. - p. 85–87. - (GAOR, 26th sess., Suppl. no. 29)

UNESCO and WHC. (2006), "Examination of the state of conservation of World Heritage Properties. Issues related to the state of conservation of World Heritage properties: Strategy for reducing Risk from Disasters at World Heritage Properties", *Convention Concerning the Protection of the World Cultural and Natural Heritage, Vilnius, Lithuania, 8–16 July 2006*.

Vidal, F. and Dias, N. (2015), "The endangerment sensibility", in Vidal, F, Dias, N, *Endangerment, Biodiversity and Culture (Routledge Environmental Humanities)*, New York, Routledge, 2016, pp. 1–40.

Waterton, E. and Watson, S. (2017), "Framing theory: Towards a critical imagination in heritage studies", *International Journal of Heritage Studies*, Vol. 7258 October, pp. 546–561.

Willems, A., Thomas, S., Castillo Mena, A., Čeginskas, V., Immonen, V., Kalakoski, I., Lähdesmäki, T., et al. (2018), "Teaching archaeological heritage management. Towards a change in paradigms", *Conservation and Management of Archaeological Sites*, Routledge, Vol. 20 No. 5–6, pp. 297–318.

Appendix

Graphs showing cross-tabulation of results

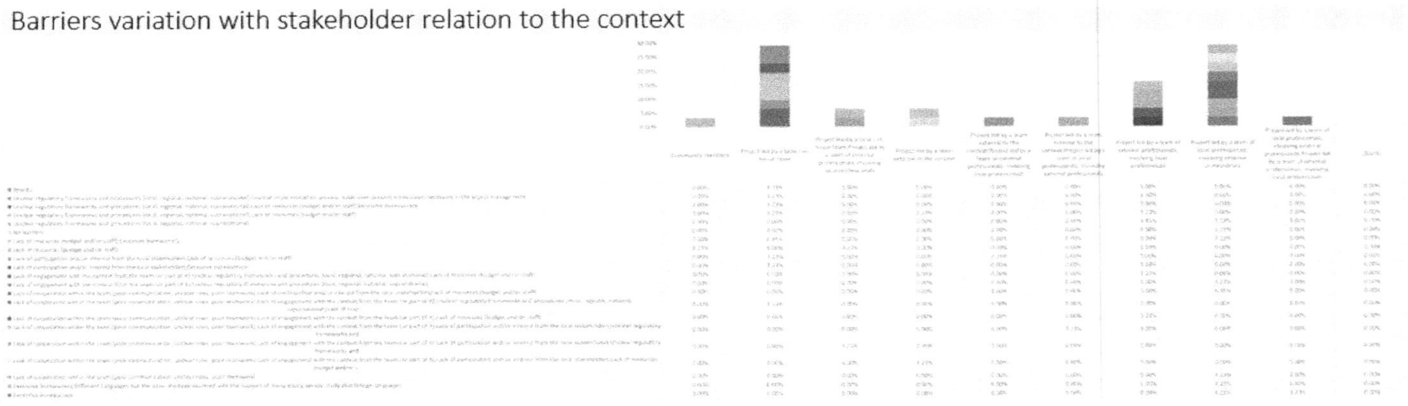

Barriers variation with stakeholder relation to the context

Figure 21.A1 Graph showing the distribution and type of barriers recorded with the stakeholder relation to the context.

Figure 21.A2 Graph showing the distribution and type of catalysts recorded with the stakeholder relation to the context.

Conclusions

Challenges and Opportunities for Disaster Risk Management of Cultural Heritage

Rohit Jigyasu and Ksenia Chmutina

The chapters in the handbook point towards several challenges and opportunities for disaster risk management of cultural heritage. First and foremost, we need to expand the predominant notion of cultural heritage beyond select monuments and sites, and include tangible and intangible, movable and immovable, and natural components, while recognizing continuity and change as an important characteristic of heritage, defining the significance and values. This broader understanding would remove misperception of heritage sector being elitist and enable better cooperation with the disaster management sector.

Cultural heritage is increasingly impacted by disasters caused by natural and human induced hazards, particularly those that are directly or indirectly related to climate change. In fact, climate change is indirectly continuing a colonial legacy of cultural oppression and destruction, most acutely felt by poorer and marginal communities, who are often key custodians of cultural heritage. Moreover, the relationship between cultural heritage, disasters and development in the era of disaster capitalism is a significant challenge. The SDG agenda prioritizes economic growth and avoids challenging the status quo, leading to cultural and territorial destruction in the name of development and progress. In fact, disaster capitalism drives cultural heritage destruction through the dismantling of institutions created to protect the territory, and cultural heritage, colonialism, and the financialization of space and life cannot be considered as separate domains. Therefore, understanding and addressing the underlying root causes of vulnerability and drivers of disaster risks that are linked to larger development paradigm are fundamental to reducing disaster risks to cultural heritage.

Cultural heritage is also increasingly at risk due to budget shrinking, reduction of dedicated staff, and a disconnection between the currency gained by disaster risk reduction and how this is actually embedded in governance and management systems that are often siloed with very little coordination among heritage, disaster risk management, environment, planning, and development sectors.

One of the most common issues is the lack of communication and engagement among stakeholders. This can manifest in varying teamwork skills within the group leading the project, or in different rates of engagement of the involved stakeholders throughout the project – from consultation to implementation. This issue is particularly relevant in disaster and conflict areas, where professionals' unpreparedness to operate is a key issue for humanitarian action – a field requiring a holistic and human-centred approach. Another common issue from the institutional point of view is the lack of coordination between decision-making bodies and lack of implementation of heritage sensitive guidelines and regulations, especially in historic urban areas, thereby risking the lives of local inhabitants including craftspeople and traders. Moreover, no formal institution recognizes these crafts or trades for their contribution to the development or building character and maintaining social cohesion of the city.

DOI: 10.4324/9781003293019-27

Other challenges include the institutional vulnerability of the disaster risk management sector particularly at the municipal level, due to lack of funding mechanisms and plans for disaster risk reduction and climate change adaptation. In fact, lack of resources is a pervasive issue that affects the success of DRM for cultural heritage projects. This issue is particularly relevant in the implementation stage, where the most common barrier is the lack of resources that however never features in isolation. Unclear regulatory frameworks and procedures – from the local to the national scale, in combination with unclear implementation rules or lack of engagement with the context also pose significant challenge. The disconnection from the context and the local stakeholders' disengagement from the project can be looked as a consequence of projects that are neither ideated nor perceived as 'needed' from within but rather 'imposed' based on assumed values given by external actors, generally from academic stance.

The use of new technologies such as 3D documentation technology, machine learning, satellite imagery, photogrammetry, and LIDAR can be effective in disaster risk management and cultural heritage conservation, but there are challenges in implementing these technologies, including data collection, damage severity assessment, economic affordability, and community relations. However, it would much more cost effective and sustainable in the long term if more attention is paid to systemic factors such as maintenance and every day care of heritage, establishing interinstitutional communication and cooperation mechanisms, and exploring aspects of disasters that have not been sufficiently and systematically investigated.

The chapters have pointed towards several important needs and considerations for successful planning and implementation of projects on disaster risk management of cultural heritage. Firstly, well established frameworks for disaster risk management should be adopted or adapted to the needs of those responsible for cultural heritage sites/assets. More importantly, there is need for developing a culture of disaster prevention and risk reduction through educational programmes and capacity building strategies.

Mainstreaming disaster risk management considerations into heritage management systems will also require integrated planning strategies and a shift in management approach towards governance processes rather than single, individually applied tools. The documentation of heritage resources and their state of health is vital to act promptly in case of disasters. Capacity building strategies combined with training and drill programmes, will be fundamental to achieving a prevention culture and embedding it firmly into heritage management. In fact, good management system can serve as a baseline and preparation to deal with disaster risks and response and necessitate coordinated action along three main strands: governance and management, research and capacity building. On the other hand, traditional conservation policies and practices may not always align with effective disaster mitigation strategies, and there needs to be efforts for achieving synergy between the two important goals of heritage conservation and disaster risk management.

Secondly, the importance of stakeholder engagement beyond traditional disaster risk management and cultural heritage disciplines and sectors and disciplines is required. Therefore, effective multi-disciplinary practice in DRM for cultural heritage projects requires synergies among academics, practitioners, policymakers, and the general public, along with horizontal and vertical knowledge exchange. Multi-stakeholder and multi-disciplinary cooperation is the key in DRM of cultural heritage, both to safeguard the cultural artifacts and practices from disaster risk and to enhance their potential to foster resilience and capacity building.

We also need to change the way we (re)develop our cities, infrastructure and buildings to integrate disaster risk management into cultural heritage through liking heritage with various urban sectors. It is also important to explore aspects of disasters that have not been sufficiently and systematically investigated, such as the reasons for failures and 'bad practices', or comparative studies on the effects of exceptional events on heritage places in good conservation versus vulnerable conditions.

Perhaps one of the most important considerations should be given to the role of traditional and indigenous knowledges for disaster risk reduction as well as response and recovery. This knowledge is embedded in vernacular building systems, intangible heritage, local community organization and communication methods, and need to actively considered for their potential role in building disaster resilience. Therefore, it is crucial to consider what cultural heritage can do for disaster risk management, and not just what disaster risk management can do for cultural heritage. This would require empowerment of communities and their participation in disaster risk management of cultural heritage of which they are the true bearers.

Index

Page numbers in *italics* indicate a figure and page numbers in **bold** indicate a table on the corresponding page.

3D documentation: and disaster risk management 279–282, 321; smartphone app *282*
3D laser scanning 281
1964 Venice Charter 92
2003 bushfires, Australia 129
2011 Sikkim Earthquake 106
2015 earthquake in Nepal *see* Gorkha Earthquake (Nepal, 2015), post-disaster reconstruction
2030 Agenda for Sustainable Development, 2015 31, 33, 42, 137, 180

Aboriginal Heritage Act, 1995 133
accidental tragedies 226
Action Plan 37
Act Related to the Reconstruction of Earthquake Affected Structures, 2015 212
Adamson, G.C.D. 82
adaptation **76**
Addendum 36–37
adopting monuments 52
aerial photography 281
Agamchchen, repairs and reconstruction 214
Agency for Cultural Affairs (ACA) 43; budget 49; prefecture-municipality 50
Agenda for Peace, 1992 65
Age of Enlightenment 132
agro-pastoral practices 293
Ahmed, B. 157
Air Quality Index (AQI) 140
Aldunce, P. 151
Alexander, B. 258
Alexander, D. E. 13
All Human Rights and Sustainable Development policy 32
anthropogenic climate change 235
Arab Spring 36
ARCH 50
art and cultural expressions, as tools for disaster risk management 247–248; art therapy, experiences and applications 255–257; *Kai-Kai and Treng-Treng* 257–258; *La Marca del Agua and Volver a Habitar* 253; mourning and recovery 250–253; Proyecta Memoria Foundation 248–249; Songs and Murals after 1939 Earthquake 249–250
Arxan-Chaihe Volcano area 232
Ashley, C. 121
Assam Style House 105, *106*
Australian Government Royal Commission into National Natural Disaster Arrangements Report 138
Australia's National Heritage List 129
authoritarianism 182
Authorized Heritage Discourse (AHD) 63, 265

Ba, O. 68
Bajo los escombros de Chillán 251
Balkan wars 92
Banda Aceh, Indonesia, tsunami 237, 272
barbarians 66
Basantpur Palace (Nau Tale) reconstruction 214, *215*
Bauddhanath 211
Bauman, Z. 151
Beck, U. 151
benefit transfer method 47
Bertram, C. 206
Bhaktapur, heritage reconstruction 211, 216–220
Bhaktapur Development Project (BDP) 217
Bhunga 110
Binns, T. 247
biodiversity assessment and management 149
Black heritage 268–269
Blue Mountain World Heritage area, NSW 129
Bosher, L.S. 12–14
Bourdieu, P. 184, 203
Boutros-Ghali, B. 64–65
Brink, E. 155
Buddhist Philosophy Promotion and Monastery Development Committee 212

324 Index

building back better (BBB) 47
bushfire smoke 140; *see also* 2003 bushfires, Australia

Capacity for Disaster Reduction Initiative (CADRI) 44
Carney, D. 121
cenotes in the Yucatán Peninsula 228
Chakraborty, R. 77
Chang, K. M. 157
Changu Narayan 211
Chilcote, R. 255
Chillán earthquake, 1939 249
Chinese Academy of Cultural Heritage 214
Chmutina, K. 12
CIGIDEN's arts and disaster unit 248
circular financing 52
Circular models Leveraging Investments in Cultural heritage adaptive reuse (CLIC) 52
Clarke, D. 42
climate change 75, 149, 179; climate stories 81–82; hazards, risks, and impacts 78, *79*; impacts on global heritage 79–80; and justice 77; knowledge systems 82–83; overcoming existing barriers 77–78; responses 81–85; stressors, impacts, and vulnerability 80–81; traditional ecological knowledge, role of 82–83; values-based climate change risk assessment 83–85
Climate Change 2021: The Physical Science Basis report 76
Climate Change Performance Index 139
climate justice and equity 76–77
Climate Vulnerability Index (CVI), Africa Project 83–84, *86*, 87
colonial heritage 270–271
Colonialism 133
community-driven planning of risk reduction 272
Convention Concerning the Protection of the World Natural and Cultural Heritage, 1972 35
Convention for the Safeguarding of the Intangible Cultural Heritage, 2003 36
Convention on the Means of Prohibiting and Preventing the Illicit Import, Export and Transfer of Ownership of Cultural Property, 1970 35
Convention on the Safeguarding of Intangible Cultural Heritage, 2003 35
convolutional neural networks (CNNs) 283, 287
Coronel, H. 256
Covid-19 pandemic 25–26, 308–310
Covid-19 syndemic 179
crépissage 69
cryptocurrencies 52

CSIRO (Commonwealth Scientific and Industrial Research Organisation) 140
cultural burning 135, 138
cultural cleansing 66
Cultural Emergency Response (CER) 46
cultural fires 137
cultural heritage (CH) 239, 298, 320; defined 165, 184; and development *see* disaster capitalism; for DRM 1–2; and DRR 30; intangible assets 186, **187**; in Japan and type of damage *285*; national heritage institute (Iphan) 187, *188*; natural and human induced hazards 1; policies 35–38; and post-disaster recovery *see* post-disaster recovery; pressures on 2; protection and conservation of 42; vulnerability to hazards/threats 15; world heritage properties **185**, *185*; *see also* heritage
Cultural Heritage Finance Alliance (CHiFA) 42
cultural properties 279; 3D documentation and disaster risk management 279–282; cultural heritage by combining machine learning and social media 283; deterioration monitoring of cultural heritage through machine learning 283; future developments 281–282; issues in implementing machine learning 284; machine learning 282; machine learning and disaster risk management 282–285; remote sensing 282–283; risk prediction of cultural heritage through machine learning 283
cultural rights 186
Culture/2030 Indicators 33
Culture in City Reconstruction and Recovery (CURE) Framework 47, 220; combination of public and private funds 47; use of 48
culture of prevention 247
Culture PDNA Guidelines 47
cyclic floods of Nile 233
cyclicity 227

Daly, C. 80
Damage and Loss Assessment (DaLA) 47
dark geocultural heritage' of disasters 226
dark heritage 225–227, 237; defined 225; destruction as process 235–237; heritage and natural hazards 232–235; landscape and heritage 227–230; Montserrat's Volcanic 'Exclusion Zone' 230–231
dark tourism 225
Daskon, C. 247
Dayton Peace Agreement 67
Deccan Plateau 101, *101*
Declaration of Quebec, 1996 298
Delgado, L. B. 247
deliberate death 226

Delica-Willison, Z. 153
Department of Archaeology (DoA) 219
DESARTES 248
Desbordes. Colectivo de Colectivos 253
Deutsche Gesellschaft für Internationale Zusammenarbeit (GIZ) 42
'dhaap' or 'baira' 119
Dhajji Diwaari 104
Dharahara 221n1
Dimitrovski, D. 226
dinosaur trees 139
Directorate-General for European Civil Protection and Humanitarian Aid Operations (DG-ECHO) 50
disaster capitalism 320; alternative development 178; Brazilian cities' profile 189, **189**; CH sector in Brazil 185–189, *188*; critical scholarship 179; cultural heritage 184–189; and development 181–183; economic growth or development 179–181; territory, disasters, and climate change 190–192
disaster colonialism 267
disaster preparedness 149
disaster resilience 112–113
disaster risk 10
disaster risk financing 41; for humanitarian action 48
disaster risk management (DRM) 257–258, 290, 321–322; instruments and approaches 290–292; milestones of disaster risk reduction 296n1; mitigation, preparedness, and response to disasters 294–295; weaknesses in implementation and effectiveness 292–294
disaster risk management for cultural heritage (DRM-CH) 2–4, 12, 298; challenges of 2–4; current status of 2; financing *see* financing DRM-CH; phases of 13–17; qualitative risk analysis *16*; risk management elements 15–16; role and power of culture and CH 42
Disaster Risk Mitigation Plan for Cultural Assets 96
Disaster Risk on Cultural Heritage 299
disaster risk reduction (DRR) 9; and cultural heritage 30; frameworks 213; guidance 270; human rights and sustainable development 31
disaster vulnerability, factors for increasing 108
distant justice model of ICC 69
documentation of heritage 294
Dongs 120
Douglass, Fredrick 207
draught, Australian Capital Territory 129
drone photos *280*
DRR *see* disaster risk reduction
dualisms 132–133

Dube, T. 157
Durbar Squares, Kathmandu 211, 214, 216
dynamic pressures 182

earthquakes: Bhuj, India, 2001, 2003 26; Bungamati, Nepal, 1934, 2015 118; central Italy and Myanmar 1; Chile, 1939 249; Ecuador, 2010 255–256; Gorkha, Nepal, 2015 92, 211–220; Great East Japan Earthquake (GEJE), 2011 49; Kashmir, India, 2005 103–104, *104*, *107*; Kobe, 1995 2; Kutchh, India 109; Latur, India 101; Marmara, Turkey, 1999 95, 97–98; Mexico, 2017 1; rehabilitation 114; resistant house *112*; Sikkim, India, 2011 106
ecological resilience 152
Ecuador, earthquake, 2010 255–256
EnergyStar® houses *269*
Environmental Resilience program 51
Environment Protection and Biodiversity Conservation Act 1999 133
episodic 'natural' events 181
Erikson, K. 207–208
Escuela México 250
EU Civil Protection Mechanism (EUCPM) 48, 50
EU Copernicus (2018) program 52
European Commission (EC) 48
European Investment Bank (EIB Institute) 42
European Union (EU) 35, 48, 212, 297n7
European Year of Cultural Heritage 48, 290
Evans, B. 152
evidence-based knowledge 134
exposure **76**

Fanon, F. 178
Federal Emergency Management Agency (FEMA) 51, 268, 270–271
Fernandes, F. 182
financing DRM-CH: brainstorming on opportunities and alternatives 51–53; capacity building and community engagement 52; case of Japan 49–50; challenge of finding finance 41–43; emergency preparedness and response 45–46; fundraising campaigns and public-private alliances 51–52; good practices 48–51; international support 43–48; protection and management 50; resilient recovery 46–48; risk understanding and risk reduction 43–45; technical support and knowledge exchange 52; *see also* disaster risk management for cultural heritage (DRM-CH)
financing impact on regional development of cultural heritage valorisation (FINCH) 51

fire, implications for DRM 137–141; disaster preparedness 139; disaster recovery 140–141; disaster response 139–140; disaster risk reduction 137, 139
'fire-smart' garden 141
fire-stick farming *see* cultural burning
Firesticks Alliance 135
First Aid and Resilience for Cultural Heritage in Times of Crisis (FAR) 38
First Aid to Cultural Heritage 272
First World War 92, 97
flooding in the city of La Plata, 2013 251
flood pillar, Ribe, Denmark *234*
floods 232
Folke, C. 152
Forde, S. 68
fragility, conflict, and violence (FCV) 48
Future of our Pasts: Engaging Cultural Heritage in Climate Action report 78

Gaillard, J. 153
Galeano, E. 180–181
Gans, Herbert J. 206
Gaudi, Antoni 203
genocide 226
geocultural heritage 228, *229*, 230, 239
geographical information systems (GIS) 266–267
geohazard forecasting 281
geoheritage 239
geological hazar 228
Geotechnical and Earthquake Investigation Directorate 95
German Corporation for International Cooperation agency (GIZ) 53n1
German Development Bank (KfW) 216
GFDRR Resilient Cultural Heritage Program 44
Giusti C. 228
Glasgow Climate Pact 132
Global Challenges Research Fund Collective Programme 51
Global Facility for Disaster Reduction and Recovery (GFDRR) 297n7
Global Heritage Fund (GHF) 51; *Non-Fungible Token* (NFT) 52
Global South 220
global warming 139
Global War on Terror 66
Gorkha Earthquake (Nepal, 2015), post-disaster reconstruction 92, 211–213; Bhaktapur 216–220; heritage reconstruction 212–213; international aid and expertise 214–220; Kathmandu 214–216; local planning and funding mechanisms 217–219; Patan (Lalitpur) 216
Graeber, D. 208
Great Fire of London 203
greenhouse gas emitters, Australia 139

Grey, M. S. 157
Gross Domestic Product (GDP) 180
Guerrero, X. 250
Guide to Developing Disaster Recovery Frameworks (GFDRR) 47
guthi 218–219
Guthi Sansthan 219
Guzmán, N. 256
Guzmán, Navas 256
Guzman, P. 80

habitus 181, 183
Hadizatou, T. 68
Hadzimuhamedovic, A. 69
Hague Convention, 1954 298
Hague Convention for the Protection of Cultural Property in the Event of Armed Conflict, 1954 35
Hall, S. 207–208
Hanumandhoka 214
Hayden, D. 205
hazards/threats 11–12, **12**, 15, **76**
Head, L. 132
Heian Kyo 204
Hendra Virus 136
Herculaneum 226
heritage: brass moulder 21, *22–23*; in building resiliency, intangible aspects 26–28; Chitpur 21; conservation frameworks 213; creative adaption to reduce risk 24–25; cultural *see* cultural heritage (CH); defined 20; disaster resilience 112; improved vernacular building system 112; intangible 20–21; irritants removal and up-grade building system 113; Kolkata 21; mainstreaming of vernacular buildings 113; milk trade 21, *25*; musician instruments 21, *25*; and natural hazards 232–235; paper mache masks to fight Covid-19 25–26; and peacebuilding *see* peacebuilding; retrofitting of existing vulnerable vernacular buildings 112, *113*; sensitisation of future building professionals 113; stone carvers 21, *22*; tangible 20; tea trading units 21, *24*; threat to cultural heritage 21; vernacular building heritage for tourism 113; vulnerabilities 21–24
Heritage Emergency National Task Force (HENTF) 51
heritage reconstruction, Gorkha Earthquake (Nepal, 2015) 211–213; Bhaktapur 216–220; heritage reconstruction 212–213; international aid and expertise 214–220; Kathmandu 214–216; Patan (Lalitpur) 216
HEritage Resilience Against CLimate Events on-Site (HERACLES) 52

Hewitt, K. 181
high-altitude Tibet 232
Higher Income Countries (HICs) 77
Historical Peninsula Plan 92–94, 96
Historical Peninsula Preservation Development Master Plan 94
Historic Cultural Heritage Act 1995 133
Hiwasaki, L. 157
Hollesen, J. 79
Holling, C. S. 152
Hsu, M. 264
human-induced disasters 291
Hunga Tonga-Hunga Ha'apai volcano 131
Hurricane Katrina, 2005 199, 206, 268
Hurricane Maria in Puerto Rico 232
hurricanes 232
Huxtable, Ada Louise 206
hybrid vernacular system 109
Hyogo Framework for Action (HFA) 34, 36, 165, 247
HYPERION project 51
hypocentres of memory *249*

ICCROM *see International Centre for the Study of the Preservation and Restoration of Cultural Property*
ICOMOS *see International Centre for the Study of the Preservation and Restoration of Cultural Property*
Iconem 287n1–2
ICORP on the Road 51
impacts **76**
Indian Ocean tsunami, 26 December 2004 131
Indian Ocean tsunami, 2005 237
Indonesian Development Planning Agency 205
InSAR ground-motion data 287n5
inseparability of nature and culture, DRM: in Australia's Bushfire Risk Management 135–137; burning bird and bogans 141; country's 'black summer' 129; cultural heritage items 129; fire as natural and cultural 137–141; impacts of bushfire disasters 131; (un)natural disasters 131–132; nature and culture 132–133; threat of fire 131; western science and traditional knowledge 133–135
Insurance Council of Australia 129
intangible cultural heritage (ICH) 33, 36
intangible heritage values 165
integrated protection of cultural heritage 300
Intergovernmental Panel on Climate Change (IPCC) 75, 147
International Assistance 45
International Centre for the Study of the Preservation and Restoration of Cultural Property (ICCROM) 36, 43, 166, 211, 272, 289–290, 295; First Aid and Resilience for Cultural Heritage in Times of Crisis (FAR) 46; Flagship Programme 38
International Committee of Architectural Photogrammetry (CIPA) 287n7
International Committee on Monuments and Sites (ICOMOS) 33, 36, 43, 75, 92, 166, 211, 272, 295; Guidance on Post Trauma Recovery and Reconstruction in World Heritage Sites 220
International Council of Museums (ICOM) 36
International Covenant on Economic, Social and Cultural Rights (1966) 31
International Criminal Court (ICC) 69; 'distant justice' model 69
The International Decade for Natural Disaster Reduction 41
international DRM organisations 166
International Federation of the Red Cross Societies (IFRC) 35
International Organization for Migration (IOM) 70
IPCC Representative Concentration Pathways (RCPs) 84
Istanbul, CH and urbanization in DRM: Bazaar and Ministry of Tourism 96; during Byzantine period *93*; Galata neighbourhood 96; *Geotechnical and Earthquake Investigation Directorate* 95; heritage sites 94; Historical Peninsula Plan 92–94, 96; Historical Peninsula Preservation Development Master Plan 94; historic preservation 92; *Istanbul Earthquake Master Plan* (IEMP) 95; Marmara earthquakes 95; Metropolitan Istanbul Sub-Region Master Plan 94; multiple hazards and crises 93; Ottoman Empire 94; planning team 96; site preservation 92, 94; *Strategic Plan for Disaster Mitigation in Istanbul* (SPDMI) 95; Turkish Republic, establishment of 94; *Urban Redevelopment Ignition, Local Redevelopment, and Land Readjustment* 95
Istanbul Earthquake Master Plan (IEMP) 95
Istanbul Metropolitan Municipality's Seismic Risk Mitigation and Emergency Preparedness Project (ISMEP) 96
IUCN 295

Jacobs, F. 264
Japanese Agency for Cultural Affairs (ACA) 43
Japanese National Institutes for Cultural Heritage (NICH) 43
Japan International Cooperation Agency (JICA) 212, 214
Joseph, J. 151–152

Kagunyu, A., 153
Kai-Kai and Treng-Treng 257–258
Kashmir earthquake, 2005 103–104, *104*, *107*
Kathmandu, reconstruction in 211, 214–216, *215*
Kathmandu Valley Preservation Trust (KVPT) 216
Kelman, I. 183
Kessenuma 199–202
key property values (KPV) 80
Kiandra Courthouse, 1890s 129, *130*
Kīlauea on Hawaii 234
Kilwa Kisiwani WH property *85*
Kinnaur district in Himachal Pradesh 120
Klein, N. 179, 190
Kluckhohn, Clyde 208
knowledge systems 82–83
Kobe/Tokyo declaration, 2015 299
KoboCollect 310
Kosciuszko National Park, Australian Alps 129, *130*
Krenak, A. 180
Kutchh earthquake, aftermath of 109
Kyoto Declaration on the Protection of Cultural Properties, 2005 265, 300
Kyotoku-Maru 202

Ladakh (Himalayan State), Northern India 104–105, *105*
La marca del agua 253
language 76
laser scanning 279, *280*
La Seu 203
Latin America and Caribbean (LAC) 181
Learning From Megadisasters 208
liberal peacebuilding 64
LIDAR 267, *280*, 281, 288n9, 321
Lin, P.S.S. 157
Little, A. 67
Local Development Committees (LDC) 219
local materials 100
loss of vernacular architecture 106; attributes of 106–107; carbon footprint 111; identity 110–111; impact of disasters on *107*, 107–108; and local context 100; local employment and economy 111; moving away from 108–109, *109*; water footprint 111
Low and Middle-Income (LMIC) 77
Luz, Maria de Lourdes 206–207

Ma, X. 123
machine learning 282, 321; cultural heritage 283; deterioration monitoring of cultural heritage 283; and disaster risk management 282–285; high-risk areas for human factors 283; issues in implementing 284; monitoring and conservation system for cultural properties *286*; Mosul's ancient city, reconstruction of 286–287; risk prediction of cultural heritage through 283; and social media 283
Making Cities Resilient (MCR) 44
Makondo, C. C. 153
Malawi, local knowledge for flood risk management in 154–155; accessible 154–155; and applicability for extreme events 155; dimensions of *154*; disaster risks 155; by local politics, with external stakeholders 155
Managing Country Together framework 134
Managing Disaster Risks for World Heritage 36, 44
Manas National Park (MNP) 119
manmade 11
Mansfield, M. 206
Marathwada, Maharashtra State, Western India 101, *102*
Marmara earthquakes 95
Martinussen, J. 178
Maya caves 228
Mayan pyramids 206
Merapi 228
Meteorological hazards 232
Metropolitan Istanbul Sub-Region Master Plan 94
Miller, D. M. S. 206
mitigation 10, **76**
Molina, F. 153
Moltke, H. Van 94
Montserrat's volcanic exclusion zone 230–231
Mosul's ancient city, reconstruction of 286–287, 287n8
Mumford, L. 207
Munjeri, D. 116
Mural 'Mexico to Chile' by Xavier Guerrero *251*
music, hazards and disasters in 258

National Audit of Monuments & Memorials (NAMM) 307–311
National Department of Sport, Arts and Culture (DSAC) 307
National Disaster Risk Reduction Framework 137
National heritage trusts 263
National Institute of Historic and Artistic Heritage 186
National Parks and Reserves Management Act 2002 133
National Trust for Historic Preservation 264, 274
NATO-campaign, Libya 66
natural disasters 41
natural fires 136, *136*
natural hazards 181; and disasters 232–235, 239; geo-hazards 12; hydro-meteorological hazards 12; processes 236; religious significance 234

natural landscape 227–228
Nature Conservation Act 2002 133
nature-culture dualism 133
negative heritage nominations 71
Nepal Ministry of Culture, Tourism and Civil Aviation 212
Nepal Reconstruction Authority (NRA) 212
Neruda, P. 250
NE States of India 105–106, *106*
New Constitution of 1988 186
New Orleans 199
New South Wales (NSW) 129
Ngootyoong Gunditj Ngootyoong Mara South West Management Plan 134
Nubia campaign 64

Office for the Coordination of Humanitarian Affairs (OCHA) 38
Office of the High Commissioner for Human Rights (OHCHR) 31
O'Keefe P. 12, 181
Olsson, L. 121
Open Eucalypt Forest 131
Organisation for Economic Co-operation and Development (OECD) 41
Orr, S.A. 78
Ottoman Empire 94
Outstanding Universal Value (OUV) 80
Ozdes, Gunduz 94

Palmyra 281
Parades of New Orleans 201–202
Paradise Tax 234
Paris Agreement 33, 132, 137
Paris Climate Agreement 116
Parks Victoria 134
Pasaje Dardo Rocha Cultural Centre 253
Pashupati 211
Pashupati Area Development Trust 212
Patan (Lalitpur), heritage reconstruction 211, 216
peacebuilding: defined 64; history of 63–66; memorialisation 70–71; professional discourse 63; rebuilding heritage in Balkans 68–70; rebuilding Timbuktu's heritage for peace 68–69; and reconstruction 67–68
peace dividend 65
photogrammetry 279–280, *280*, 321
physical heritage 228
piam 258
Pieterse, J. N. 178–179
planning disaster mitigation and reduction 263–264; heritage at risk 264–267; limitations and opportunities 263–264; planning protection of heritage at risk 268–272; planning reduction of disaster colonialism throughout Black Atlantic 271–272; US planning for resilient colonial heritage 270–271
Policy Guidance for Heritage and Development Actors 33
Political Capabilities for Equitable Resilience (POLCAPS) 51
Pompeii eruption of Italy's Mount Vesuvius 226
Pontchartrain Park Historic District (PPHD) 268
Post-Disaster Needs Assessment (PDNA) 35, 47, 212; damage and losses 53n7; guidelines 297n7
post-disaster recovery: cultural heritage and disasters 203–208; Former Disaster Management Center, Minamisanriku, Japan *200*, 200–201; Fukushima Prefecture 202; Great Fire of London 203; international frameworks for CH and disasters 208; Kessenuma 199–202; Kyotoku-Maru 202; La Seu 203; New Orleans 199; Parades of New Orleans 201–202
Post Disaster Recovery Framework (PDRF) 212
Prajapati, S. 216
preparedness 11
Preservation Resource Center of New Orleans 268
Preserve National Monuments 67
Presidential Public Employment Stimulus Programme (PESP) 308
Pressure and Release (PAR) framework 179, 182–183
Prince Claus Fund 46
Prost, Henri 94, 97
Protecting Cultural Heritage from the Consequences of Disasters (PROCULTHER) 50
PROTection of European Cultural HEritage from GeOhazards (PROTHEGO) project 52
PROTHEGO (PROTection of European Cultural HEritage from GeO-hazards) 281, 287n4
Proyecta Memoria Foundation 248–249

Quebec Declaration, 1996 300

Rajopadhyaya, A. D. 221n1
Rautela, P. 153
Reason, J. 121
'reconstruction' program 171
recovery 11
Reid, J. 152
remote sensing 282–283
residual risk 9–10
resilience 232
Responsibility to Protect (R2P) 66
revised or supplementary budget 49
Reynard E. 228
Ribot, J. 183
Ricco, T. 265, 272

risk 9–10, 232
risk mitigation planning, Istanbul 96
Risk Preparedness: A Management Manual for World Cultural Heritage (Stovel) 289
risks **76**
Rivera, D. 206, 264
Rodney, R.M. 140
Rogue Foundation 255–256
romanticisation of local knowledge: causes and associated problems 149–150; decolonial approaches to 156–158; in disaster studies 150–153; for DRR research, policy, and practice 158–159; as dynamic body of knowledges 153, 155; local knowledge, vulnerability, and resilience 150–153; in Malawi *see* Malawi, local knowledge for flood risk management in; natural hazards 147; risk reduction policy and practice 148; and socio-demographic changes 156; tangible and intangible heritage 148; traditional knowledge 147; Western or scientific knowledge 148
Ross River Virus 136
Royal Commission into National Natural Disaster Arrangements 138

Said, E. 178
Šakic Trogrlic, R. 153–154
Sant'anna, M. 186
Santos, A.L.V.D. 206–207
satellite imagery 321
SDG 11 1, 32, 165
Secretariat of National Defence (SEDENA) 170
Secretariat of the International Search and Rescue Advisory Group (INSARAG) 38
securitisation 66
self-insurance 172
Sendai Framework 212–213
Serrat, O. 122
Sesana, E. 78
SFDRR 173
shared heritage 67
Sherpa, P. Y. 77
shikinen zotai 116–117
Shiva Temple, repairs and reconstruction 214
Shuri Castle 281
Shuri Castle Digital Reconstruction 287n6
Silk Road system 136
Simpson, N.P. 76
sinkholes 232
Siqueiros, D. A. 250
smartphones 283
Smith, K. 18n3
Smith, L. 63, 267
Smith, L. J. 20
Smith, L. T. 156
Smith, N. 264

Smithsonian Cultural Rescue Initiative (SCRI) 46
Smithsonian Institution 51
smoking ceremonies 140
social media 283
solitary deaths 205
South African Heritage Resources Agency (SAHRA) 307–308
South Gujarat State, Western India 102, *102*
special disaster recovery budget 49
stereo camera principle 280
STORM project 50–51
Strategic Plan for Disaster Mitigation in Istanbul (SPDMI) 95, 97
Strategy for Reducing Disaster Risks at World Heritage Properties 36
stressors 80–81
Sukur Cultural Landscape (SCL) in Nigeria 83
sustainable development 149
Sustainable Development Goals (SDGs) 32, 42, 51, 116, 132, 178
Swayambhu 211
synergies for cultural heritage 298–299; barriers 303–305, *318*; catalysts 305–312, *319*; design stage, barriers in 303, *304*; design stage, catalyst in 305–307, *306*; documents regulating protection and safeguard of cultural heritage **300**; heritage conservation and disaster risk reduction 300–301; implementation stage, barriers in 303; implementation stage, catalyst in 307; importance of 301–302; integrating disaster risk management and cultural heritage safeguard 299–303; in Practice 302–303; reflections, critical 312–314; stakeholders' relation to context changes 311–312; stakeholders' relation to the context changes 303–305; UNESCO and protocols in favour of synergies 299–300
Syrian Civil War 281

Taq 104
Tasmanian State legislation 133
Tasmanian Wilderness World Heritage Area Management Plan 133
TCA *see* techno-centric approaches
techno-centric approaches (TCA): approaches to DRM of CH in Mexico 168–169; climate and weather-related hazards 165; for cultural heritage conservation 171–173; within cultural heritage conservation systems 167–171; to disaster risk management 166–167; post-disaster recovery 170; towards more integrated approaches to DRM-CH 173
Ten Essentials 44
Tengö, M. 157
terrestrial laser scanning (TLS) 281

territory and climate change: Brazilian cities' profile **191**; deterritorialisation 190; DRM plans 191; National Civil Defense 193, *194*; Ouro Preto 191–192; Outstanding Value of World Heritage Properties 192, *193*; reterritorialisation 190
Thomas, D.S.G. 153
Thunberg, G. 87
Timbuktu's mausoleums (Mali) 67–69
traditional ecological knowledge (TEK) 82–83
traditional livelihoods: Apple Orchards in Kinnaur, Himachal Pradesh, India 120; case of Bungamati in Nepal *117*, 118; for climate action 119; climate induced hazards and disasters 115; community 116; community-led water harvesting system 119–120; in context of culture 116–118, 124; defined 121; DRM planning 118; interacting with environment 121; intertwined tangible and intangible heritage *117*, 118; livelihood security 115; role of communities in conservation 122–123; sustainable livelihood framework 121, **122**, 123
Training Guide 44
transformative justice 64
traumascape 229
traumatic heritage 237
Trust Fund for Victims (TFV) 69
tsunami 131
Tumarkin, M. 229
Turkish Republic, establishment of 94

UAV photogrammetry 287n3
UK Research and Innovation 51
UN *see* United Nations
UNDRR (United Nations Disaster Risk Reduction) 131
UNESCO *see* United Nations Educational, Scientific and Cultural Organization
United Nations 297n7; Institute for Training and Research (UNITAR) 52; International Decade for Natural Disaster Reduction (IDNDR) 265; international strategy for Disaster Risk Reduction (UNDRR) 18n1, 271; Sendai Framework for Disaster Risk Reduction (SFDRR) 2015-2030 33–34, 41, 43, 116, 131, 137, 148, 165, 183; United Nations Declaration on the Rights of Indigenous Peoples, 2007 31; United Nations Development Programme (UNDP) 212; United Nations Framework Convention on Climate Change 33; United Nations International Strategy for Disaster Reduction (UNISDR) 263; University Institute for Environment and Human Security (UNU-EHS) 42; World Conference on Disaster Risk Reduction (WCDRR) 41, 43
United Nations Educational, Scientific and Cultural Organization (UNESCO) 208, 272, 279; categories 182; Chair for Cultural Heritage and Risk Management 44; Culture Conventions 45; Heritage Emergency Fund (HEF) 45; Institute for Statistics (UIS) 42; and the Operational Satellite Applications Programme (UNOSAT) 52, 281; *Policy on Engaging with Indigenous Peoples* 38; *Strategy for Action on Climate Change* 38; Tasmanian Wilderness 133; World Heritage Convention 133, 228; World Heritage List 42, 46, 65–66, 68, 70, 133, 184–185, 189, 228, 281, 287n4, 294
Universal Declaration of Human Rights, 1948 31
Urban Agenda Habitat III 183
Urban Redevelopment Ignition, Local Redevelopment, and Land Readjustment 95
US Fair Housing Act of 1949 268
US programs for hazards mitigation 273
Uttarakhand State, in Foothills of Himalaya, Northern India 102–103, *103*

values-based climate change risk assessment 83–85
Vatsala Durga Temple in Bhaktapur *218*
Vecvagars, K. 47
Venugopal, S. 121
vernacular built heritage and disaster resilience: aftermath of Kutchh earthquake 109; architecture, loss of *see* loss of vernacular architecture; attributes of 106–107; disaster resilience 112–113; earthquake resistant house *112*; factors for increasing disaster vulnerability 108; impact of disasters *107*, 107–108; increasing dilution 109–110, *110*; Kashmir State, Northern India 103–104, *104*; Ladakh (Himalayan State), Northern India 104–105, *105*; loss of vernacular architecture 110–114; Marathwada, Maharashtra State, Western India 101, *102*; NE States of India 105–106, *106*; South Gujarat State, Western India 102, *102*; Uttarakhand State, in Foothills of Himalaya, Northern India 102–103, *103*; vernacular architecture and local context 100; vernacular building systems 108–109, *109*
Victor Ray's theory of racialized organizations 273
Viles, H. A. 80
Village Civil Protection Committee (VCPC) 155

volcanic eruption, Chaitén, Chile, 2008 237
volcanic eruption, La Soufriére, St. Vincent, 2021 237
volcanic 'Exclusion Zone' 230
Von Meding, J. 12
Vousdoukas, M.I. 79
vulnerabilities 10, **76**, 232; assessing 81; carelessness of people 23–24; community 80; cycle *81*; decision-making bodies, lack of coordination 23; financial returns, lack of 23; lack of maintenance of buildings and public services 21–22; landscape-based approach 80; six-step vulnerability assessment framework 80
vulnerability and capacity assessments (VCA) 3

Wamsler, C. 155
Warsaw Recommendation on Recovery and Reconstruction of Cultural Heritage 37–38
Waterton, E. 302, 315n2
Watson, S. 302, 315n2

Western World 92
Whyte, K. 78
wilderness 133
Wisner, B., 151
World Bank 35, 297n7; Asian Development Bank (ADB) 212; City Strength Diagnostic program 45; Disaster Risk Financing and Insurance (DRFI) 41; Global Facility for Disaster Reduction and Recovery (GFDRR) 263, 271
World Commission on Environment and Development (WCED) 178
World Disaster Report 34–35
World Heritage (WH) 78
World Heritage Committee 70
World Heritage Convention (UNESCO), 1972 36, 45, 65, 217, 298; aftermath of Tamil Tigers' suicide bombings 65
World Monument Fund's Crisis Response Program (WMF) 46
Wren, Cristopher 203

zoonotic pathogens 136